SYNTHETIC POLYMERIC MEMBRANES

SYNTHETIC POLYMERIC MEMBRANES

A Structural Perspective

Second Edition

ROBERT E. KESTING

IRVINE, CALIFORNIA

19 85

A Wiley-Interscience Publication

JOHN WILEY & SONS

New York   Chichester   Brisbane   Toronto   Singapore

7301-106X

CHEMISTRY

Library of Congress Cataloging in Publication Data:

Kesting, Robert E., 1933–
   Synthetic polymeric membranes.
   "A Wiley-Interscience publication."
   Includes bibliographies and index.
   1. Membranes (Technology) 2. Polymers and
polymerization. I. Title.

TP159.M4K4 1985   660'.28424   85-6162
ISBN 0-471-80717-6

Printed in the United States of America

10 9 8 7 6 5 4 3 2 1

This book is dedicated to Loeb and Sourirajan on the occasion of the twenty-fifth anniversary of their invention of the first integrally skinned, hyperfiltration membrane—an event which not only focused scientific attention upon the advantages of membranes with inhomogeneity in depth but also served as the instrument that heralded the advent of the golden age of membranology.

PREFACE

The subject of this book is *synthetic polymeric membranes*, the thin polymer films in either solid or liquid states which act as semipermeable barriers for gaseous, liquid, or solid permeants. In this edition, as in the first, the membranes themselves are emphasized rather than the processes and applications in which they are utilized. This approach is expected to appeal to the growing ranks of membranologists who are in some way involved in membrane development or utilization. Included in this group are the chemists, polymer scientists, physicists, and biotechnologists for whom membranology is a primary area of interest. Also included, however, are professors and students in scientific and engineering disciplines, as well as workers in industry who wish to keep abreast of this dynamic and important field. At various levels, therefore, this book is intended for use as a reference source, a beginning text, and an advanced text.

This edition has been reorganized and extensively rewritten to substantiate the principles and concepts that were only adumbrated in the earlier work. Chapter 1 is an introduction, which explains my perspective and contains brief historical and economic overviews together with a consideration of the scattered membrane literature. Chapter 2 covers the utilization of membranes in separations and Chapter 3 in miscellaneous applications. Chapter 4, "Membrane Polymers," is of key importance because it delves into the essential characteristics and structural levels of the polymers from which membranes are made. Because this is a forward-looking book, I stress those materials which I consider to be of maximum present and future value rather than those which are merely of historical interest. Chapter 5, "Polymer Solutions," is important because most membranes are solvent cast. Chapters 6–9, which are concerned with the structure and synthesis of various classes of synthetic polymeric membranes, constitute the core of this book. These are topics which, for economic reasons, are seldom discussed in print, but which are undeniably the key to continued progress in the field. Chapter 10, "Biological Membranes," covers a class of membranes which is beginning to contribute both general concepts and practical innovations of some potential significance.

By virtue of my personal involvement in membrane research and development for more than two decades, I have been privileged to confer and correspond with many of the workers in the field. I wish to express my gratitude to each of them, not only for their original contributions, but also, in many instances, for discussions which have contributed significantly to my understanding of the subject. I wish particularly to acknowledge the contributions of R. Schultz, S. Sourirajan, S. Loeb, T. Matsuura, C. Cannon, A. Castro, J. Cadotte, M. Morrison, D. Hoernsche-meyer, C. W. Saltonstall, J. Henis, M. Tripodi, I. Cabasso, H. Hoehn, and J.-G. Helmcke. Special thanks are also due to my wife, Edith, for her secretarial help, patience, and understanding and to the publisher, without whom the concept of this book could not have come to fruition.

R. E. KESTING

Irvine, California
April 1985

CONTENTS

SYNTHETIC POLYMERIC MEMBRANES

1 INTRODUCTION

There are many different classes of practitioner in the field of membranology and each tends to see things from a different point of view. Categorization of such groups is necessarily subjective since unanimity within any circle is lacking and much overlap does occur. Nevertheless, even with these difficulties in mind it is possible to distinguish between a number of alternative approaches: the *empirical,* the *phenomenological,* the *fluid dynamical,* and the *structural.*

The empirical outlook is above all pragmatic. Membrane performance and stability are optimized by execution and evaluation of matrix experiments which cover everything from feed-solution pretreatment to environmental factors and membrane formulation. The phenomenological approach was developed during the early stages of this emerging scientific discipline. It is based upon quantitative analyses of material transport using mathematical models. Its utility lies in *post facto* rationalization and in limited extrapolations. It does not by itself point the way to the development of improved membranes.[1] Fluid dynamics is primarily concerned not with the membranes themselves but with the events which take place at the interfaces between the membrane and the external environment.[2] Its teachings have shown, for example, that membrane performance can be significantly modified by boundary and gel layers and that, in some instances, transport across the membrane may not even be the rate-determining step in a separation. The focus of the structurist is emphatically upon the membrane itself. As indicated by the book's subtitle, it is from the structural perspective that this treatise has been written.

1.1 STRUCTURAL VIEW OF MEMBRANOLOGY

The structurist holds that a thorough knowledge of membrane morphology is essential to the establishment of a causal relationship between structure and function. He or she is therefore not content merely to recognize that transport of a permeant

1

across the membrane occurs through pores but seeks rather to determine their exact nature and method of formation. The pursuit of this objective is tempered by the realization that the steric and chemical properties of membrane pores are dynamic rather than static in nature and hence are susceptible to modification by their interactions with solute, solvent, and environmental conditions.

Structurist methodology can perhaps be most easily understood by reference to the various morphological levels encountered in the specific example of integrally-skinned cellulose acetate hyperfiltration membranes (Table 1.1). It is significant that the various structural levels are present even in the nascent membranes, that is, in the sol precursors to the ultimate membrane gel.[3,4] Thus sol-state structures must be conceded structural as well as temporal primacy over those in the gel state.[5] In other words, the task of designing a membrane with given permeability, selectivity, and stability resolves itself into the preparation of suitably swollen and ordered sol structures whose integrity must be maintained throughout the sol $\rightarrow$ gel transition and beyond.

Although the ramifications of the structural levels listed in Table 1.1 are very broad, even they do not completely exhaust the possibilities. For example, the solubility of homogeneous cellulose acetates of various degrees of substitution (DS) is listed in Table 1.2.[6]

Homogeneously substituted cellulose acetates are prepared by the acetylation of cellulose to the *primary*, or (nearly) completely substituted, cellulose triacetate, followed by homogeneous deacetylation to *secondary* cellulose acetate in solution. If *heterogeneous* acetylation of cellulose to a given DS is attempted, the resultant products are nonuniform and can be treated as physical mixtures of cellulose triacetate and unreacted cellulose. The solubilities of such materials are quite distinct from those of their homogeneous counterparts. Furthermore, Tables 1.1 and 1.2 deal exclusively with cellulosic polymers which contain only acetyl and hydroxyl groups. If substitution with other acyl or carbamate moieties is effected, entrance is gained into the expanding universe of the *cellulose acetate mixed esters*. Indeed, further derivitization of cellulose acetate with various aliphatic and aromatic acyl groups, paraffinic as well as olefinic, ionogenic as well as nonionogenic, all have significant effects upon processing and end-use characteristics. Profound modifications are also possible if side-chain branching is introduced via graft copolymerization.[7]

The preceding discussion only summarizes the concerns of the structurist regarding a single class of membrane polymer in a membrane of a single physical type for one specific application. If this field is expanded to include all suitable synthetic membrane polymers in the several possible physical structures for all possible applications, it can be seen that the scope of the subject is quite far reaching.

1.2 HISTORICAL OVERVIEW

Although the use of naturally occurring polymers, such as cellulose, as materials in macrofiltration dates from antiquity, the history of synthetic polymeric membranes necessarily began after the invention by Schönbein[8] in 1846 of cellulose nitrate, the first synthetic (actually, *semisynthetic*) polymer. Throughout the first century of their existence, the history of polymeric membranes was concerned largely, but not exclusively, with cellulosic types. In 1855 Fick[9] utilized cellulose nitrate membranes to perform his classic studies, *Über Diffusion*. In the same year Lhermite[10] first stated the *solution* aspect of membrane transport, namely, that permeation occurs as a result of the interaction of the permeant species with the membrane. He recognized that solution and pore (capillary) theories were not mutually exclusive but merged without any abrupt change into one another. In 1860 Schumacher[11] developed cellulose nitrate membranes in a tubular form—test tubes were simply dipped into collodion solutions—which were to remain popular into the present century. Baranetzky developed the first flat membranes in 1872.[12] By varying cellulose nitrate concentration, Bechhold[13] prepared the first series of microfiltration membranes of graded pore size in 1906. He was also the first to define the relationship between the bubble point, surface tension, and pore radius. The concept of pore-size distribution was developed by Karplus[14] who combined the bubble point and the Hagen–Poiseuille permeability techniques.

Structurist interpretation of synthetic polymers can be said to have begun with the determination by Staudinger[15] in the 1920s that polymers were high-molecular-weight compounds. Meyer and Mark[16] utilized X-ray diffraction to establish the existence of long-range order (crystallinity) in dense polymer structures. They popularized the fringed micelle concept which has since been augmented by the lamellar-folded single-chain crystallite and the extended-chain configurations. Molecular-weight distribution has been studied by a variety of techniques, most conveniently by gel permeation chromatography developed by Benoit et al.[17] in 1966. Control of macromolecular architecture is a recent phenomenon. Branched polyethylene was developed by Swallow in 1935 and linear stereospecific polyethylene by Ziegler in 1953.[18] Random, block, and graft copolymers and the effects of their primary, secondary, and tertiary structures upon film and membrane structure and performance have been studied since 1960.

Although Manegold[19] discussed the possible systems consisting of matter and voids, unequivocal characterization of membrane morphology by direct microscopic observations awaited the advent of the electron microscope. The early studies of von Ardenne[20] were followed by the investigations of Maier and Beutelspacher[21] and Helmcke,[22] which for the first time revealed the true vacuole-like nature of the microstructure of microfiltration membranes.

Early attempts to control and vary membrane porosity were largely empirical in nature. Bechhold[23] noted that permeability varied inversely with the concentration of "membrane substance" in the casting solution. Bigelow and Gemberling[24] noted the effects of varying the drying time before immersion into a nonsolvent, and Malfitano[25] varied the alcohol–ether ratio of the solvent system and was among the first to employ annealing as a technique for increasing selectivity. Schoep[26] utilized glycerol and castor oil, nonvolatile plasticizers, to control gel structure and improve physical properties.

As the culmination of early attempts to fabricate membranes reproducibly, Zsigismondy et al.[27] and Elford[28] developed two practical series of graded cellulose nitrate membranes. Brown[29] produced a graded series of cellulose nitrate mem-

TABLE 1.1 STRUCTURAL LEVELS AND THEIR EFFECTS UPON INTEGRALLY-SKINNED CELLULOSE ACETATE HYPERFILTRATION MEMBRANES

Structural Level			Effects		
Order	Genus	Species	Sol	Gel	Function
Primary	Chemical group	DS acetyl DS hydroxyl	Solubility (see Table 1.2) Coil tightness (tertiary level) increases with increasing acetyl DS	Hydrophilic/hydrophobic balance Order increases with [acetyl] Capillary forces (membrane–water interaction)	Permeability Selectivity Hydrolytic stability Swelling Compaction Wettability
Secondary	Monomer/Chain segment	Cyclic anhydroglucose β-1,4 linkage between units Uniformity of substitution Equatorial substitution	Solubility Order (polymer–polymer interaction)	Order (via influence at tertiary level)	Selectivity Thermal stability Compaction resistance
Tertiary	Macromolecular	Conformation Rod (low MW) Random coil (high MW) MW MWD [α cellulose]	Solubility Viscosity Order Turbidity (tightness of coil increases with increasing DS acetyl)	Small pore = average interchain displacement Integrity increases with MW Order increases with narrowness of MWD Order increases with [α cellulose]	Selectivity Mechanical properties

4

Hierarchical level	Structure	Micelle-level characterization	Macromolecular (skin/substructure) characterization	Membrane/system characterization
Quaternary (microscopic)		Micelle size (turbidity) Micelle uniformity (order) Viscosity Stability Thixotropy		
Macromolecular aggregate (supermacro-molecular)	Skin: dense coalesced and compacted micelles Substructure: open cells		Skin: thickness and integrity (presence of large pores (defects) depends on extent of micelle coalescence) Substructure: porosity Small cells (ultragel) Large cells (microgel)	
Membrane (filter medium) — Configuration	Flat sheet Hollow-fiber Tube	Viscosity: hollow fiber >> flat sheet or tube	Porosity: flat sheet or tube >> hollow fiber	Permeability Selectivity Wet-dry reversibility Permeability: flat sheet or tube >> hollow fiber
Final (end use) — Packaging	Plate and frame Spiral wound element Pleated cartridge Hollow-fiber element Tubular array	n.a.[a]	n.a.	Packing density: hollow fiber > flat sheet >> tube Ease of cleaning: tube >> flat sheet > hollow fiber

[a]n.a. = Not applicable.

TABLE 1.2 SOLUBILITY OF
HOMOGENEOUS CELLULOSE ACETATES
OF VARYING DS[a]

DS	Solvent
2.8–3.0	Chloroform
2.2–2.8	Acetone
1.2–1.8	Methyl cellosolve
0.6–0.8	Water
0.0–0.6	Cellulose solvents

[a]From Ward.[6]

branes by swelling initially dense films in alcohol–water solutions of varying concentrations. He was also the first to employ cellulose acetate as a membrane material and to note the inhomogeneity in depth which was later to be of such significance to the development of the hyperfiltration process of desalination.

Bartell and van Loo,[30] Elford,[31] and Grabar et al.[32] pioneered attempts to describe the events occurring in the polymer solutions prior to and during the gelation process which account for membrane structure. In 1960 the important paper by Maier and Scheuermann,[33] *Über die Bildungsweise teildurchlässiger Membranen,* provided an excellent hypothesis which has since been utilized by the present author to accommodate every class of phase-inversion membrane within its general framework.

The development of ion-exchange membranes was retarded because of the difficulty of making ion exchangers in the shape of films.[34] Although Teorell[35] in 1935 and Meyer and Sievers[36] in 1936 developed the now generally accepted model for ion-selective membranes, it was not until 1949 that practical ion-exchange membranes were developed by Juda and McRae.[37] Although the first commercial microfiltration membranes were produced in Germany in 1927, it was not until the 1950s that commercial production began in the United States.

The golden age of membranology (1960–1980) can be said to have begun in 1960 with the invention by Loeb and Sourirajan[38] of the first integrally-skinned cellulose acetate hyperfiltration membrane. This development stimulated both commercial and academic interest, first in desalination by hyperfiltration and then in other membrane processes and applications.

During this two-decade period, significant progress was made in virtually every phase of membranology: applications, research tools, membrane formation processes, chemical structures, physical structures, configurations, and packaging (Table 1.3). Today, with the possible exception of chemical structure, which can undergo a virtually infinite number of combinations and permutations, the field of membranology can be said to have matured. However, even though the basic principles and methodology have already been established, tailor-making and optimization of membranes for a growing number of specific applications has only just begun and should continue well into the next century.

TABLE 1.3 SOME SIGNIFICANT DEVELOPMENTS DURING THE GOLDEN AGE OF MEMBRANOLOGY (1960–1980)

Phase	Development	Date or Period	Significance
Application	HF (hyperfiltration) desalination and water purification	1967–1980	Low-cost alternative to distillation
	Hemodialysis	1965–1975	Highest-volume membrane application
	ED (electrodialysis) in the chlorine-caustic cell	1970–1980	Decreased energy costs in NaOH/Cl$_2$ product cost
	UF (ultrafiltration) electrolytic paint recovery	1972	Cost savings and pollution control
	MF (microfiltration) sterilization of drugs and LVPs (large volume parenterals)	1970–1980	Efficient sterilization of thermally sensitive solutions
	MF particle removal from solvents and corrosive fluids in electronics industry	1975–1980	Higher yield of electrocomponents
	Gas separations	1980	Potentially more efficient gas separations
	Controlled release	1980	Constant delivery of drugs, herbicides, etc.
	Genetic engineering	1980	Convenient framework for microorganism culture and product separations
Polymer membrane research tools	Gel permeation chromatography (GPC)	1960	Routine study of MWD
	Scanning electron microscopy (SEM)	1965	Access to colloidal morphology
Membrane formation process	Wet method for skinned membranes	1960	Hyperfiltration becomes practical. Key event which focused attention on membranes
	Radiation track membranes	1963	Only commercial membranes with cylindrical pores

TABLE 1.3 (*Continued*)

Phase	Development	Date or Period	Significance
	Thin-film composites	1971	Separate optimization of skin and substructure for high performance
	Gel-spinning methods for hollow fibers	1971	HF and dialysis hollow fibers
	Wet-spinning methods for hollow fibers	1971	HF and UF hollow fibers
	Dry method for skinned membranes	1972	Easy handling, wet dry reversibility
	Melt/hydrolysis method for perfluorinated ionomers	1972	Energy savings in caustic/chlorine production
	Microporous stretched polypropylene	1974	Solvent-resistant MF membrane
	Microporous stretched poly(tetrafluoroethylene)	1976	Most solvent-resistant MF membrane
	Dry-spinning method for hollow fibers	1978, 1980	Economical production for dialysis, MF, and UF fibers
	Thermal phase-inversion process	1980	First phase-inversion polypropylene membrane for MF and controlled release
Chemical structure	Noncellulosic homopolymer membranes	1963–1980	Superior mechanical, thermal, and environmental resistance
	Ionomer membranes	1970–1980	Superior performance characteristics
	Copolymer membranes	1970–1980	Tailor-made polymers
	Blend membranes	1965–1980	Economical alternative to copolymer membranes
	Cross-linkable thermoplastic polymers	1970–1980	Hybrid thermoplastic/thermoset membranes for increased stability
	Interfacial thin-film polycondensates	1980	*In situ* high-performance thin films from monomers

8

TABLE 1.3 *(Continued)*

Phase	Development	Date or Period	Significance
Physical structure	Integrally-skinned ultragel	1960	First practical HF membrane
	M–S (Maier–Scheuermann) hypothesis for phase inversion	1960	Working hypothesis for formation of phase-inversion membranes
	Liquid surfactant membranes	1962	Improvement in HF membrane performance with feed additives
	Emulsion-type liquid membranes	1967	Unsupported liquid membranes
	Determination of skin structure in integrally-skinned membranes	1970–1973	Explained skin and skin-defect formation
	Integrally-skinned microgel	1972	Wet-dry reversible skinned membranes
	Interfacially formed thin-film composite	1980	High-performance HF composite
	Highly anisotropic skinless membranes	1980	High-throughput MF membranes
	Resistance-model membranes	1980	First commercial gas-separation membranes
	Immobilized liquid membranes	1980	Internally supported liquid membranes
Configurations and packaging	Hollow fibers	1970–1980	Highest packing density
	Spiral-wound elements	1968	Efficient utilization of skinned flat membranes at intermediate to high pressure
	Reusable plate and frame	1970	Economical membrane replacement for HF, UF, and MF
	Disposable plate and frame	1975	Economical alternative to hollow fibers for dialysis
	Pleated cartridges	1970	Efficient flat-sheet membrane cartridge for microfiltration

1.3 CONFIGURATIONS AND PACKAGING

Membranes are available in a variety of configurations: the *tube*, the *hollow fiber*, and the *flat membrane*. In the case of the tube, the feed solution is most often on the inside but it can also be on the outside. A collection of tubes, known as an array, is particularly advantageous with feed solutions which contain high concentrations of particulate matter that would tend to foul membranes in the other configurations. Both fouling and concentration/polarization can be minimized in tubes by circulation of the feed solution at a high velocity to promote turbulence. Furthermore, both exterior and interior surfaces of tubes can be readily cleaned. Because flow through cylindrical tubes is easy to control and is amenable to analysis, it is the configuration most often favored by fluid dynamicists. However, tubes tend to be expensive because they have the lowest packing density, that is, the lowest membrane area for a given package volume (Table 1.4). They also require the greatest amount of energy to circulate the feed solution. For these reasons tubes are only employed where the other configurations cannot be.

Hollow fibers are available in a variety of diameters and wall thicknesses. *Hollow fine fibers* with ODs of ~ 80 μm and walls which are ~ 20 μm thick and skinned at their outer surfaces are sturdy enough to act as their own pressure vessels. Pressure and feed are at the outside of the fiber. The high compaction resistance of these fibers is partially due to their chemical and physical structures, which in turn affect their moduli. It also varies inversely with porosity. For this reason such fibers invariably have lower void volumes and permeabilities than their flat and tubular counterparts. However, they need not be so productive on an area basis as other configurations because their packing density is much higher. One disadvantage of hollow fine fibers is that they are prone to fouling and difficult to clean once fouled. This is a consequence of the poor circulation of feed solution at the average fiber surface. This fact necessitates particle-free feeds which in turn require the use of extensive pretreatment. To minimize back pressure within the fiber

TABLE 1.4 AREA/VOLUME UTILIZATION
FOR HOLLOW-FIBER AND FILM MEMBRANES[a]

Membrane Configuration	A/V (ft^{-1})
Hollow fiber[b]	
OD 50 μm	12,000
OD 100 μm	6,000
OD 200 μm	3,000
OD 300 μm	2,000
Flat membrane, spiral wound[c]	150–250
Tubular membrane,[d] 0.5 in OD	50

[a]From Orofino.[39]
[b]Calculated with OD at 50% volume utilization (packing factor).
[c]Estimated from data of Gulf Environmental Systems.
[d]Estimated.

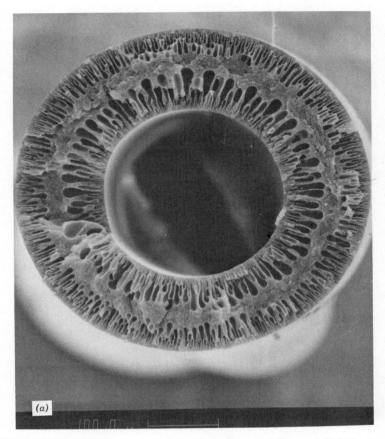

FIGURE 1.1.   Hollow fibers for gas separations: (*a*) developmental fibers and (*b*) high-pressure fibers (Monsanto).

lumen, somewhat larger fibers are required for gas separations (Fig. 1.1). Where, as in hemodialysis, the feed is located at the inside surface, still larger fibers known as *capillaries* are employed. A typical artificial kidney capillary has an outside diameter of 250 μm and walls which are 10–12 μm thick. The thinner walls are mandatory since resistance is proportional to total membrane thickness. They are feasible because only low-pressure differentials are encountered in dialysis. The largest capillaries utilized today have an OD of ~825 μm (Fig. 1.2). They are utilized in ultrafiltration of solutions which contain coarse particles.

Several types of flat membranes are produced: *unsupported* or *free standing* (consisting of the membrane matrix by itself), *reinforced* (containing a fabric mat within the membrane matrix), and *supported* (consisting of a composite structure in which the membrane matrix is bonded to a support fabric). Flat membranes can be incorporated into various *packages*. The oldest of these is the *reusable plate*

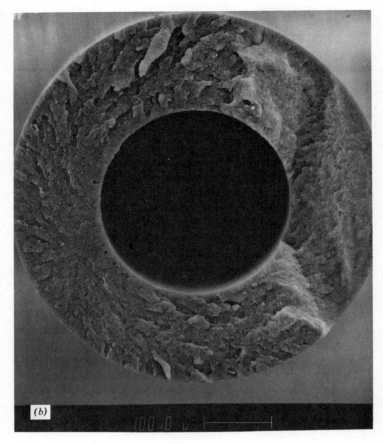

FIGURE 1.1. (*Continued*)

and frame. Plate and frame units are available which can be utilized in hyperfiltration, ultrafiltration, and microfiltration simply by insertion of the appropriate membrane. Higher capital and labor costs are at least partially offset by lower membrane costs and increased versatility. The *disposable plate and frame* is encountered today in the compact multiplate unit for kidney dialysis. It is slightly less efficient, but lower in cost than the capillary kidney. The *spiral wound element* is an efficient packaging device. It was initially, and is still most commonly, utilized for hyperfiltration, but is now also found in ultrafiltration applications (Fig. 1.3). It consists of a sandwich of two rectangular flat membranes with skin surfaces facing outward and sealed together at three sides of the rectangle. Within this two-membrane leaf is a material which serves as a product water channel. The open edge of the leaf is sealed to a product water tube into which holes have been drilled. A plastic net serves as a brine spacer to keep the membrane surfaces apart from one another. The membrane leaf is wound together with the brine spacer about the product water

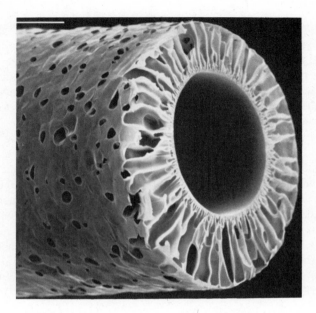

FIGURE 1.2. Capillary ultrafiltration membrane (Amicon).

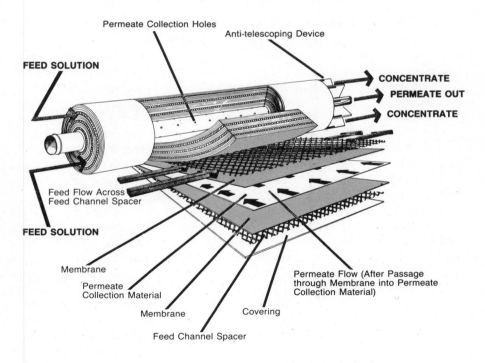

FIGURE 1.3. A spiral-wound element (Abcor).

13

tube to yield a cylindrical package which is then covered with a plastic tape or fiberglass, leaving the cylinder ends exposed. Several of these elements are placed in series within a pressure vessel to produce an assemblage known as a membrane *module*. The brine enters the open end of the element, traveling through and around the brine spacer which separates the rolled membrane leaves. A portion of the water permeates the membrane and follows the product water channel to the product water tube where it is collected.

The *pleated cartridge* is a convenient package which contains convoluted flat membranes. It was developed to increase the packing density so that disposable elements could become competitive with the cost-effective but cumbersome plate and frame units. It is utilized for microfiltration only, because it cannot withstand high pressures and because microfiltration is the only pressure-driven process whose membranes can accommodate the pore stretching which occurs at the apices of the convolutions. Because stretched pores will have a larger effective diameter, convoluted stock has a lower bubble point than the unconvoluted roll stock from which it is made. The result is that a 0.2-μm cartridge must contain a 0.17-μm membrane (bubble point 60–70 psi) if it is to exhibit the generally accepted 0.2-μm bubble point of 50 psi.

1.4 POLYMER MEMBRANE MARKETS

By comparison with the huge markets for polymers in the textile, paper, plastic, and elastomer industries, polymer membranes are a small business. Nevertheless, with the average 18% growth rate which has been forecast for the remainder of this decade, sales in the United States are expected to exceed $1 billion by 1990 (Table 1.5). Although 55–60% of membrane sales are currently for separation applications—primarily to produce purified fluids—it has been estimated that the growth of membranes as support materials, that is in nonseparation applications, will exceed that of membranes for use in separations by 1990.[40] This would appear to be a somewhat speculative assumption. However, the estimates for separation membranes in this forecast appear to be soundly based.

Porter[41] listed the annual worldwide market for pressure membrane filtration in 1982 as consisting of $300 million (microfiltration), $60 million (ultrafiltration), and $240 million (hyperfiltration). These processes are utilized in the following applications: potable water via desalination, ultrapure water and chemicals, and effluent treatment/concentration. By halving these values to obtain the approximate U.S. share of these markets, we arrive at the figure of $300 million. Because this figure also includes ancillary equipment such as pressure vessels and pretreatment, it must be further divided by a factor of 2 or 3 to obtain the actual membrane sales. For purposes of this discussion, a membrane includes all the value added up to and including the membrane element. This yields a value of between $100 and $150 million for the HF, UF, and MF membrane sales in the United States in 1982, a

TABLE 1.5 THE MARKETS FOR MEMBRANES[a]

Market	1980	1983	1988	1993	Average Annual Growth Rate 1983–1993
Separation membranes					
Potable water via desalination	24	33	57	97	11.4%
Ultrapure water and chemicals	10	15	33	72	17
Effluent treatment/concentration	33	47	112	267	19
Gas separation/enrichment	5	10	20	30	12
Electrochemistry	1	5	27	144	40
Dialysis/therapeutic	100	80	80	80	0
	173	190	329	690	13.7
Support Membranes					
Controlled release	80	132	354	964	22
Biotechnology/other specialty applications	3	5	22	100	35
	83	137	376	1064	22.7
Total	256	327	705	1754	18

[a]From Business Communications[40]; values in millions of dollars.

value which agrees fairly well with the 1983 figure of $95 million (33 + 15 + 47) obtained from Table 1.5

At present the largest single membrane separation process is hemodialysis, followed by microfiltration, hyperfiltration, and ultrafiltration. Gas separations appear at the threshhold of large-scale commercialization, awaiting only an upturn in the petrochemical industry. At the same time, electrochemical membrane processes can expect a substantial renascence as asbestos diaphragm cells inevitably yield to more-energy-efficient perfluorinated ionomer membranes in caustic/chlorine production.

1.5 MEMBRANE LITERATURE

Polymeric membranes can be viewed as a narrow subdivision or application of the broader field of polymer science. Therefore, an appreciation of the polymer literature should precede entry into the membrane field. The reader should also find the coverage of other related polymer subdivisions such as textiles, plastics, and elastomers useful.

The following works provide background information and in-depth coverage of many topics which are related to membranology:

1. *Encyclopedia of Polymer Science and Technology,* Wiley-Interscience, New York. Although most of this 1964 series is out of print, it is scheduled for replacement beginning in 1985. This series is an excellent source for in-depth coverage of

virtually every structural aspect of polymer science which is relevant to membranes.

2. F. Billmeyer, *Textbook of Polymer Science,* 3rd Ed. Wiley-Interscience, New York, 1984.

3. The High Polymer Series, Wiley-Interscience, New York. Of special interest is Vol. 5, 2nd Ed., *Cellulose and Cellulose Derivatives,* Pts. 1–3, E. Ott, H. Spurlin, and M. Graffin, Eds., 1955 and Pts. 4 and 5, N. Bikales and I. Segal, Eds., 1971.

4. D. van Krevelen, *Properties of Polymers,* 2nd Ed., Elsevier, Amsterdam and New York, 1980. This is a modern book written from a structural perspective.

5. J. Brydson, *Plastics Materials,* 2nd Ed., Van Nostrand Reinhold, New York, 1970. This is a highly readable and useful treatise of the principal classes of polymers.

6. P. Flory, *Principles of Polymer Chemistry,* Cornell University Press, Ithaca, NY, 1953. This classic presentation by a Nobel Laureate also contains an excellent historical overview.

7. P. Meares, *Polymer Structure and Bulk Properties,* D. Van Nostrand Co., Ltd., London, 1965.

8. F. Bovey and F. Winslow, Ed., *Macromolecules, An Introduction to Polymer Science,* Academic, New York, 1979.

9. H. Schnell, in *Chemistry and Physics of Polycarbonates* H. Mark and E. Immergut, Eds., Polymer Review Series No. 9, Interscience, New York, 1964.

10. P. Morgan, in *Condensation Polymers: By Interfacial and Solution Methods* H. Mark and E. Immergut, Eds., Polymer Review Series No. 10, Interscience, New York, 1965. This book was the inspiration for Cadotte's development of interfacially formed thin-film composites.

11. H. Mark, S. Atlas, and E. Cernia, Eds., *Man Made Fibers, Science and Technology,* Vols. 1–3, Interscience, New York, 1967. Because of the close relationship between the structure and synthesis of textile fibers and membranes, this trilogy is worthy of close scrutiny.

12. D. Paul and S. Newman, Eds., *Polymer Blends,* Vol. 1 and 2, Academic, New York, 1978.

13. K. Solec, Ed., *Polymer Compatibility and Incompatibility, Principles and Practices,* MMI Press, published under license by Harwood Academic Publishers, Chur, Switzerland, 1982.

14. A. Noshay and J. McGrath, *Block Copolymers: Overview and Critical Survey,* Academic, New York, 1977.

15. G. Molau, Ed., *Colloidal and Morphological Behavior of Block and Graft Copolymers,* Plenum, New York, 1971.

16. M. Kohan, *Nylon Plastics,* Wiley-Interscience, New York, 1973.

17. F. Helfferich, *Ion Exchange,* McGraw-Hill, New York, 1962. This is the classic treatise on the subject of ion exchange.

18. L. Holliday, Ed., *Ionic Polymers,* Wiley, New York, 1975.

19. A. Eisenberg and M. King, *Ion-Containing Polymers,* Academic, New York, 1977.

20. A. Eisenberg, Ed., *Ions in Polymers,* ACS Advances in Chemistry Series No. 187, American Chemical Society, Washington, D.C., 1980.

21. C. Carraher, Jr. and M. Tsuda, Eds., *Modification of Polymers,* ACS Symposium Series No. 121, American Chemical Society, Washington, D. C., 1980.

22. A. Barton, *Handbook of Solubility Parameters and other Cohesion Parameters,* CRC Press, Boca Raton,FL., 1983. This is the first exhaustive compilation of solvents which explains their behavior in addition to listing their physical characteristics.

23. J. Kavanau, *Water and Solute–Water Interactions,* Holden-Day, San Francisco, 1964.

24. H. Jellineck, *Water Structure at the Water–Polymer Interface,* Plenum, New York, 1972.

25. S. Rowland, Ed., *Water in Polymers,* ACS Symposium Series No. 127, American Chemical Society Washington, D.C., 1980.

Among the books which deal specifically with synthetic polymeric membranes are:

1. R. Kesting, *Synthetic Polymeric Membranes,* McGraw-Hill, New York, 1971. This first edition of the present work was the first and remains up until now, the only book which views the subject from a primarily structural perspective. Although the original is out of print, it has been recently reprinted and copies are available from the author.

2. H. Strathmann, *Trennung von Molekularen Mischungen mit Hilfe Synthetischer Membranen,* Steinkopf Verlag, Darmstadt, 1979. This is an excellent overview, written in German, of membrane separation processes with a strong structural component.

3. U. Merten, Ed., *Desalination by Reverse Osmosis,* MIT Press, Cambridge, MA, 1966.

4. S. T. Hwang and K. Kammermeyer, *Membranes in Separations,* Wiley-Interscience, New York, 1975. This is an excellent text which provides a unified treatise of many membrane separation processes from a phenomenological viewpoint.

5. R. Schlögl, *Stofftransport durch Membranen,* Steinkopf Verlag, Darmstadt, 1964. This highly theoretical book provides phenomenological models relevant to ionogenic membranes.

6. S. Sourirajan, *Reverse Osmosis,* Academic, New York, 1970. The importance of Sourirajan's invention, together with S. Loeb, of the first integrally-skinned cellulose acetate hyperfiltration membrane cannot be overestimated. His *ad hoc* studies qualify him as the most important empiricist in the field as well as the pathfinder who led the way into the golden age of membranology.

7. N. Lakshminarayanaiah, *Transport Phenomena in Membranes,* Academic, New York, 1969. This is a phenomenological treatment of transport through synthetic *and* certain biological membranes.

8. S. Sourirajan, Ed., *Reverse Osmosis and Synthetic Membranes,* NRCC Publ. No. 15627, Ottawa, Canada 1977. This is one of the few good multiauthor volumes in the field. It includes an excellent article on polyamide membranes by P. Blais.

9. A. Cooper, Ed., *Ultrafiltration Membranes and Applications,* Plenum, New York, 1980. This extensive book covers state-of-the-art ultrafiltration from an empirical point of view.

10. R. Lacey and S. Loeb, Eds., *Industrial Processing with Membranes,* Wiley-Interscience, New York, 1972. Electrically-driven and pressure-driven processes are covered in this book. Of particular interest are two chapters by Reid and Lonsdale on hyperfiltration from a phenomenological viewpoint.

11. H. Lonsdale and H. Podall, Eds., *Reverse Osmosis Membrane Research,* Plenum, New York, 1972. Although a symposium volume, this book is an interesting compilation taken at the midpoint of the golden age of membranology. It covers the subject from the phenomenological, fluid dynamical, empirical,and structural, viewpoints and has some key articles, namely by Klein and Smith on solubility parameters, by King et al. on cellulose acetate–triacetate blend membranes, and another on cross-linkable cellulose acetate methacrylate membranes. Integrally-skinned polyamide and some early composite membranes are also covered.

12. H. Hopfenberg, Ed., *Permeability of Plastic Films and Coatings to Gases, Vapors and Liquids,* Plenum, New York, 1974. This is another interesting symposium volume which has an extensive section on transport theory and sections on industrial membranes and membrane-moderated biomedical devices.

13. A. Turbak, Ed., *Synthetic Membranes,* Vols. 1 and 2, ACS Symposium Series Nos. 153 and 154, American Chemical Society, Washington, D.C., 1981. This two-volume set was compiled on the occasion of the symposium marking the 20th anniversary of Loeb and Sourirajan's invention of the first integrally-skinned cellulose acetate hyperfiltration membrane. Volume 1 covers desalination from virtually every point of view. Loeb discusses the origin of the Loeb–Sourirajan membrane and 25 other papers cover several aspects of hyperfiltration and ultrafiltration. The only unrelated paper is one by the present author and his colleagues which describes the first highly anisotropic microfiltration membrane. Volume 2 covers hyperfiltration and ultrafiltration applications.

14. T. Brock, *Membrane Filtration: A User's Guide and Reference Manual,* Science Technology, Industries, Madison, WI, 1983. This book is a good source for

the myriad analytical applications, primarily of microfiltration membranes, in such areas as viability counting of bacteria and viruses and biomedical research.

15. J. Flinn, Ed., *Membrane Science and Technology, Industrial, Biological, and Waste Treatment Processes,* Plenum, New York, 1970. This book contains several articles on concentration/polarization problems in ultrafiltration and hyperfiltration.

16. T. Fendler, *Membrane Mimetic Chemistry,* Wiley-Interscience, 1982. This book covers the emerging field of synthetic biological membranes. Micelles, monolayers, bilayers, vesicles, and others, which are found in biological membranes, are imitated and turned to other uses.

17. *Polymeric Delivery Systems,* MMI monograph Vol. 5, Gordon and Breach, New York, 1978. This monograph covers the use of membranes as support devices in drug delivery systems.

18. M. Jain and R. Wagner, *Introduction to Biological Membranes,* Wiley, New York, 1980. This is an excellent text dealing with the structures, functions, and origins of biological membranes.

Many membrane symposia are presented in book form. Symposia sponsored by the American Chemical Society were formerly published by Plenum Press and Marcel Dekker but are now usually made available in the ACS Symposium Series or in the ACS Advances in Chemistry Series.

Membrane and polymer U.S. patents are covered, albeit in an incomplete and haphazard fashion, by Noyes Data Corporation, Park Ridge, New Jersey:

19. *Industrial Membranes. Design and Applications,* 1972.

20. *Membrane Technology and Industrial Separations,* Chemical Technology Review No. 69, 1976.

21. *Membrane and Ultrafiltration Technology. Recent Advances,* Chemical Technology Review No. 147, 1980.

22. *Hollow Fibers Manufacture and Applications,* Chemical Technology Review No. 194, 1981.

A selective, but by no means all-inclusive, list of the journals and periodicals which cover significant polymer and/or membrane developments includes:

1. *Journal of Applied Polymer Science*

2. *Desalination*

3. *Journal of Membrane Science*

4. *Journal of Colloid and Interface Science* (formerly *Journal of Colloid Science*)

5. *Kolloid Zeitschrift und Zeitschrift für Polymere* (formerly *Kolloid Zeitschrift*)

6. *die Makromolekulare Chemie*

7. *Journal of the American Institute of Chemical Engineering*

8. *Macromolecules*

9. *Industrial and Engineering Chemistry* (and its various subdivisions).

Much of the hyperfiltration membrane research which was funded by the Office of Saline Water (U.S. Department of the Interior) is to be found in a series of OSW research and development reports (the "Green Reports"). More recent reports must be obtained from the National Technical Information Service in Bethesda, Maryland.

Important meetings and symposia dealing with membranology are sponsored by:

1. The American Chemical Society, semiannual national meetings.

2. The European Federation of Chemical Engineering (Working Party on Fresh Water from the Sea), sesquiannual meetings. This meeting is now co-sponsored by the International Desalination and Environmental Association (IDEA) and the Water Supply Improvement Association (WSIA).

3. The Gordon Research Conferences. There is one meeting dealing with separation science and a new one dealing with hyperfiltration and ultrafiltration membranes.

Traditionally, new developments in microfiltration membranes are found somewhat apart from the mainstream of membranology:

1. Parenteral Drug Association, annual meetings.

2. Filtration Society, national and international meetings on an irregular basis.

REFERENCES

1. W. Pusch, personal communication.
2. P. Brian, in *Desalination by Reverse Osmosis,* U. Merten, Ed., MIT Press, Cambridge, MA, 1966.
3. R. Kesting, *J. Appl. Polym. Sci.,* **17,** 1771 (1973).
4. M. Panar, H. Hoehn, and R. Hebert, *Macromolecules,* **6,** 777 (1973).
5. R. Kesting, in *Cellulose and Cellulose Derivatives,* Vol. 5, Pt. 5, N. Bikales and L. Segal, Eds., Chap. XIX F.1, Wiley-Interscience, New York, 1971.
6. K. Ward, Jr., in *Modified Cellulosics,* R. Rowell and R. Young, Eds., Academic, New York, 1978.
7. J. Wellons, J. Williams, and V. Stannett, *J. Polym. Sci. Part A-1,* **5,** 1341 (1967).
8. C. Schönbein, British Patent 11,402 (1846).
9. A. Fick, *Ann. Phys. Chemie,* **94,** 59 (1855).
10. M. Lhermite, *Ann. Chim. Phys.,* **43**(3), 420 (1855).
11. W. Schumacher, *Ann. Phys. Chemie,* **110,** 337 (1860).
12. J. Baranetzky, *Pogg. Ann.,* **147,** 195 (1872).
13. H. Bechhold, *Biochem. Z.,* **6,** 379 (1907).
14. H. Karplus, cited by F. Erbe, *Zolloid Z.,* **63,** 277 (1933).
15. H. Staudinger, *Chem. Ber.,* **53,** 1073 (1920).
16. K. Meyer and H. Mark, *Chem. Ber.,* **61,** 593 (1928).
17. H. Benoit, Z. Grubisic, P. Rempp, D. Decker, and J. G. Zilliox (in French), *J. Chem. Phys.,* **63,** 1507 (1966).

18. F. McMillan, *The Chain Straighteners*, The McMillan Press Ltd., London, 1979.

19. E. Manegold, *Kolloid Z.*, **80**, 253 (1937).

20. M. von Ardenne, *Elektronenübermikroskopie*, Springer Verlag, Berlin, 1940, p. 350.

21. K. Maier and H. Beutelspacher, *Naturwissenschaften*, **40**, 605 (1953).

22. J.-G. Helmcke, *Kolloid Z.*, **135**, 29, 101, 106 (1954).

23. H. Bechhold, *Z. Phys. Chem.*, **60**, 257 (1907).

24. S. Bigelow and A. Gemberling, *J. Am. Chem. Soc.*, **29**, 1576 (1907).

25. C. Malfitano, *Z. Phys. Chem.*, **48**, 243 (1910).

26. A. Schoep, *Kolloid Z.*, **8**, 80 (1911).

27. R. Zsigismondy, E. Wilke-Doerenfurt, and A. von Galecky, *Chem. Ber.*, **45**, 570 (1912).

28. W. Elford, *Proc. Roy. Soc. London, Ser.*, *B* **106**, 216 (1930).

29. W. Brown, *Biochem. J.*, **9**, 591 (1915); **11**, 40 (1917).

30. F. Bartell and M. van Loo, *J. Phys. Chem.*, **28**, 161 (1924).

31. W. Elford, *Trans. Faraday Soc.*, **33**, 1094 (1935).

32. P. Grabar, S. Levenson, and S. Schneierson., *Ann. Inst. Pasteur*, **64**, 275 (1940).

33. K. Maier and E. Scheuermann, *Kolloid Z.*, **171**, 122 (1960).

34. P. Helfferich, *Ion Exchange*, McGraw-Hill, New York, 1962.

35. T. Teorell, *Proc. Soc. Exp. Biol. Med.*, **33**, 282 (1935).

36. K. Meyer and J. Sievers, *Helv. Chim. Acta*, **19**, 649, 665, 987 (1936).

37. W. Juda and W. McRae, U.S. Patent 2,636,851 (1953).

38. S. Loeb and S. Sourirajan, UCLA Report 60-60 (1960).

39. T. Orofino, in *Reverse Osmosis and Synthetic Membranes*, NRCC Publ. No. 15627, Ottawa, Canada 1977.

40. Business Communications, Stamford, CT cited in *Chem. Week*, September 28, 1983, p. 22.

41. M. Porter, personal communication.

2 MEMBRANE SEPARATION PROCESSES

irrespect in of direct in

To think of membranes is generally to think of separations. Indeed, as we have seen, approximately 60% of synthetic polymeric membranes are today employed as semipermeable barrier layers which permit certain components of solutions or suspensions to permeate more rapidly than others. The absolute rate at which a permeant traverses a membrane is known as *permeability*, and the rate at which two different species permeate relative to one another is *selectivity*. Permeability and selectivity are the primary, but by no means the sole, determinants of the practicality of any membrane separation. In this chapter each of the principal separation processes will be considered. This will be done in such a way as to support the central theme of this treatise, namely the relationship between structure and function.

There are only six or seven commercially significant membrane separation processes and these are driven by only three forces: gradients of concentration, electricity, and pressure (Table 2.1). The steric qualities of both membrane and solute are always involved, which in turn implies that sieving plays a significant role in every such separation. However, as the sizes of both solute and membrane pores decrease, other factors such as diffusion and solution come into play.

2.1 CONCENTRATION-DRIVEN PROCESSES

Diffusion refers to the migration of a substance across a concentration gradient. It is perhaps most easily understood in the case of gaseous diffusion, wherein concentration can be replaced with pressure. Bringing together two gases at opposite sides of a permeable interface will cause the gases to cross this interface even if pressures are initially equal. This is so because the partial pressure of a gas in a

TABLE 2.1 CHARACTERISTICS OF PRINCIPAL MEMBRANE SEPARATION PROCESSES

Process	Objective or Desired Product	Driving Force	Nature of Critical Solute and Membrane Parameters	Separation Mechanism(s)	Primary Species Transported
Gas, vapor, and organic liquid permeation	Product enriched in (or depleted of) various components	Concentration gradient (pressure + temperature assisted)	Steric/Solubility	Diffusion/Solution	All
Dialysis	Solutions of macrosolutes free of microsolutes	Concentration gradient	Steric/Solubility	Diffusion/Solution sieving	Microsolute
Electrodialysis	1. Solvent which is free of ionic solutes 2. Concentrated solution of ionic solutes 3. Ion replacement 4. Metathesis 5. Separation of electrolysis products 6. Fractionation of electrolytes	Electrical current	Ion mobility (including steric + valence factors) Ion-exchange capacity (membrane)	Counterion transport through macroionic membranes	Microions
Microfiltration	Sterile, particle-free solutions	Pressure	Steric	Sieving	Solution
Ultrafiltration	1. Solutions of macrosolutes free of microsolutes 2. Solutions of individual macrosolutes	Pressure	Steric	Sieving	1. Microsolute solution 2. Solutions of smaller macrosolutes
Hyperfiltration	1. Solvent which is free of all solutes or 2. Concentrated solution	Effective pressure	Steric/Solubility	Preferential sorption/Capillary flow	Solvent

23

mixture is independent of any other gases which may be present. For similar reasons diffusion occurs in condensed states as well. Fick's first law of diffusion[1] states that the flux J is proportional to the concentration gradient:

$$J = -D \, \partial c / \partial x$$

where the proportionality factor D is called the *diffusion coefficient*. A formal relationship exists between the van't Hoff equation[2] for osmotic pressure, $\pi V = nRT$, and the ideal gas law, $PV = nRT$. For the latter, pressure is the result of collisions between gas molecules with the walls of the container, whereas for the former, pressure is due to the greater frequency of collision of solvent molecules with the membrane on the side with the lower solute concentration. On the more concentrated side, a lesser fraction of the collisions is due to solvent molecules (the rest being due to solute). The net effect is a higher solvent pressure on the less concentrated side of the membrane, which forces a flow of solvent into the more concentrated solution.

Solution refers to the mixing of two components on the molecular level. Both steric and polar factors are involved (Chapter 5). Insofar as membrane permeation is concerned, solution refers to the role which physicochemical interaction between permeant and membrane plays in the transport of the permeant.

2.1.1 Separation of Gases, Vapors, and Organic Liquids

As in all membrane separations, gas and vapor permeability is a function of membrane properties, the nature of the permeant species, and the interaction between membrane and permeant species. The permeability coefficient P can therefore be expressed as the product of three separate factors A, B, and C:

$$P = ABC$$

where A is a function of the membrane's physical and chemical structure; B is a function of gas properties, such as size, shape, and polarity; and C is a function due to membrane–gas interaction. Since A and B determine the diffusional characteristics of a particular gas through a given membrane, they can be combined in a single diffusion coefficient D. It is customary to refer to C as the solubility coefficient S, whence

$$P = DS$$

The rate flux J_1 at which gas 1 permeates a membrane may be written as

$$J_1 = P_1 A \, \Delta p_1 / l$$

where J_1 is the steady-state flux of the gas in cubic centimeters at STP per second, P_1 is the permeability coefficient, A is the membrane area in square centimeters,

Δp_1 is the partial pressure differential for the gas across the membrane in centimeters of mercury, and l is the membrane thickness in centimeters. Therefore, P_1 has the units of cc STP-cm/cm^2-s-cm Hg and $P_1 = J_1$ for a 1-cm-thick membrane with a surface area of 1 cm^2 at a partial pressure differential of 1 cm Hg.

The permeability coefficient P for a given polymer–permeant system is obtained by increasing the flux of the gas or vapor under a pressure difference Δp_1 across the membrane. Concentration,[3-6] volume,[7-9] and pressure vacuum[10, 11] methods have been utilized for determining gas flux. The separation of P into diffusion and solubility coefficients is accomplished by the time-lag method of Daynes[11] and Barrer.[12] The straight line on the pressure versus time plot (Fig. 2.1) yields an intercept, on the x axis, known as the time lag. The time lag is related to the diffusion constant D as $D = l^2/d\theta$, where θ is expressed in seconds and l is the thickness in centimeters. P is obtained from the slope of the steady-state portion of the curve, and S can be calculated from the ratio P/D since $P = DS$.

$$P = J\, d\theta/A\, \Delta p_1$$

The selectivity or separation factor α_2^1 at a given temperature of gas 1 relative to gas 2 is equal to the ratio of their permeability coefficients at equal Δp. The selectivity of condensed gases is often expressed as the enrichment factor σ.

Because permeation through polymeric membranes may involve separation and motion of macromolecular chain segments, any factor which serves to restrict in-

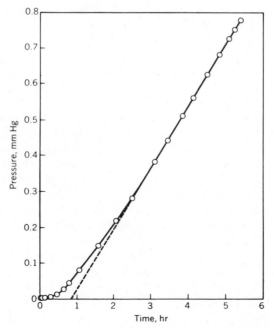

FIGURE 2.1. Time-lag method (from Barrer[12]).

terchain displacement will also reduce permeability. Therefore since the presence of polar groups on the polymer gives rise to strong cohesive forces between polymer chains, permeability of noncondensible gases through polar membranes will be lower, everything else being equal, than through nonpolar membranes. A factor of considerable importance is the degree and type of crystallinity. High cohesive forces, low chain flexibility, and a high degree of molecular symmetry favor the formation of crystallites. Although all these properties may be present simultaneously, as in the case of cellulose triacetate, one property may suffice, for example, molecular symmetry in the case of polyethylene. A crystalline or even a paracrystalline domain is a region of high molecular order and dense packing. Permeant molecules are insoluble in polymer crystallites and poorly soluble in paracrystalline domains, so they tend to permeate through disordered regions. The permeation sites may be either amorphous material or other interstices between crystallites. Crystallites and other ordered regions reduce permeability in two ways: (1) by reducing the volume fraction of the membrane available for solution of the permeant species; and (2) by forcing diffusion to occur in tortuous paths between and around crystallites. For noncondensible gases the amount of sorbed permeant is directly proportional to the fraction of the membrane in the amorphous phase[13-15]:

$$S = S_a X_a$$

S_a and X_a are the solubility coefficients and volume fractions of the amorphous polymer, respectively.

The effect of crystallinity on the diffusion coefficient is more complex than on the solubility coefficient[16-18]:

$$D = D_a/\tau$$

Here D_a is the diffusion constant for the amorphous portion of the polymer, and τ is the tortuosity factor which accounts for the necessity of bypassing crystallites.

The reduction in diffusivity is a function not only of the volume concentration of crystallites but also of their size and shape. In highly crystalline polymers, such as linear polyethylene ($\sim 70\%$ crystalline) with lamellar crystallites, the diffusivity of a gas in the crystalline polymer is an order of magnitude lower than in the amorphous polymer. When coupled with the fact that only about one-third of the crystalline polymer is capable of dissolving the permeant, its overall permeability is only about $\frac{1}{30}$ that of its amorphous counterpart. Furthermore, the presence of crystallites imposes certain restrictions upon the movement and displacement of intervening amorphous material. If the diffusing molecules are sufficiently large, they may encounter pathways—even in the amorphous phase—which are too narrow to permit their passage. For this reason the size of the permeant species can be a significant factor in both permeability and permselectivity. Thus whereas hydrogen is about $\frac{1}{30}$ as permeable in a crystalline as in an amorphous polyethylene, methane will be less than $\frac{1}{100}$ as permeable. Correlations have been developed to predict the influence of crystallinity upon the solubility and crystallinity of noncondensible gases.[19-21]

Considering now the restrictive effect of crystallites upon the motion of polymer chains in the amorphous phase, the diffusion constant for a semicrystalline polymer can be described as

$$D = D_a/\beta$$

where β is the chain-immobilization factor. Therefore, $1/\beta$ is the fractional reduction in diffusivity due to the decreased freedom of chain movement in the amorphous phase. The permeability constant P for a semicrystalline membrane is therefore

$$P = S_a X_a D_a / \tau \beta$$

Because crystallite size and shape are dependent upon crystallization conditions, such conditions can have an important bearing upon membrane permeability and permselectivity. Solution casting from thermodynamically good solvents leads to membranes of lower crystallinity and hence higher permeability than casting from poor solvents. The thermal history of both solution-cast and melt-extruded membranes is also of importance. A linear polyethylene membrane which has been slowly cooled from the melt exhibits lower gas permeability than the same polymer quenched from the melt and subsequently annealed at a high temperature in such a way that both membranes have the same degree of crystallinity.[22] The differences are related to the presence of more perfect thin lamellar crystallites in the former membrane and of defective thicker lamellae in the latter. High-temperature annealing causes the polymer to crystallize with less strain, thereby reducing the number of brittle intercrystalline linkages in the extended-chain configuration. As a result, a membrane annealed at a high temperature will subsequently swell to a greater extent and hence be more permeable than one annealed at a low temperature.[22, 23]

Where membrane–permeant interaction occurs, the effects of crystalline morphology on permeability are even more pronounced than they are in noncondensible gases, the reason being that crystallites act as virtual cross-links to impose a swelling limit upon the amorphous regions. Diffusivity, of course, increases greatly with increasing degree of swelling. Permselectivity is higher in crystalline than in amorphous membranes because of the more limited swelling. A well-known example is the greater permselectivity of crystalline over amorphous polyethylene to the isomeric o-, m-, and p-xylenes (Fig. 2.2).

Annealing a membrane while it is swollen with a permeant can effect up to an order-of-magnitude increase in permeability with little loss in permselectivity. This solvent-annealing technique offers the potential of combining the permselectivity characteristic of crystalline membranes with the permeability of amorphous types. Nor is thermal treatment the only means for controlling a membrane's permeability characteristics. Studies have shown that uniaxial stretching below the melting point can result in a significant reduction of swelling with a negligible increase in crystallinity.[24, 25] Although permeability decreases by at least two orders of magnitude, permselectivity increases dramatically.

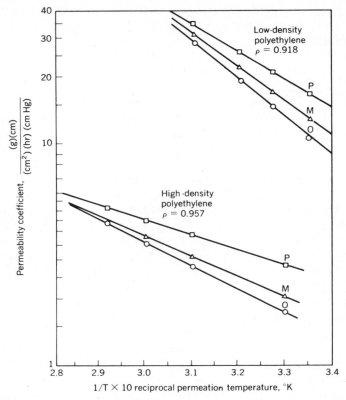

FIGURE 2.2. Permeability coefficients of liquid xylene isomers in high- and low-density polyethylene as a function of temperature (from Michaels et al.[22]).

Closely related to the influence of crystallinity is the question of glassy versus rubbery membranes, that is, the effect of the glass transition temperature T_g upon permeability and permselectivity. Below T_g the membrane is in the glassy state and may contain rigid voids which can trap permeant molecules, thereby contributing little to the diffusive process. Below this temperature also certain chains have such restricted motion that activated diffusion is possible. Above the T_g, on the other hand, chain mobility and diffusivity increase (Fig. 2.3). Plasticization increases permeability because by reducing cohesive forces between polymer chains it reduces T_g and leads to increased diffusivity.[26] Since copolymerization provides internal plasticization, it also results in increased permeability.

In noncondensible gases, membrane structure is not perturbed by the permeant species. The solubility of a permeant gas in the amorphous phase of a membrane depends on the critical temperature of the gas (since this value determines the cohesion of gas molecules) and upon the polymeric membrane. In general, the higher the Hildebrand solubility parameter δ, the lower the solubility of permeant gases. However at high pressures, gases such as CO_2 and NH_3 act to solvate polar polymer groups.

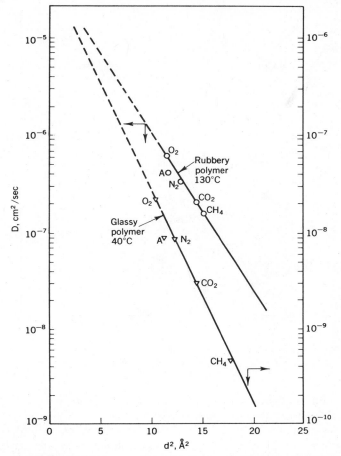

FIGURE 2.3. Correlation of diffusion constants of gases in poly(ethylene terephthalate) above and below T_g (from Michaels et al.[20]).

The low degree of membrane–solute and solute–solute interaction character-istic of permanent gases supports separation primarily on the basis of permeant size (Fig. 2.4). This is not to say that solubility does not play a significant role in some cases. Gases such as H_2S and CO_2 with high solubility coefficients and relatively large size, tend to permeate faster than CH_4, CO, N_2, and C_2H_6 whose sizes may be small, but whose solubility coefficients are also low.

The hydrogen permeabilaity coefficients and H_2/N_2 selectivities for several classes of dense isotropic polymer films are found in Table 2.2. Since the perme-ability coefficients for even the most permeable of gases such as H_2 and He are low, processes other than membrane processes have long dominated gas separa-tions. Recently, however, substantial gains have resulted by: (1) increasing the par-tial pressure of gases on the feed side; (2) decreasing the effective thickness of the membrane; and (3) increasing the membrane area. Although in the strictest sense

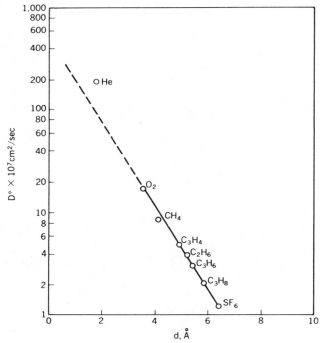

FIGURE 2.4. Correlation of diffusion constants (25°C) of gases in amorphous polyethylene (from Michaels and Bixler[21]).

gas permeation is a concentration-driven rather than a pressure-driven process, in practice, pressure is utilized to increase the gas concentration on the feed side of the membrane and thus indirectly to increase gas permeation. After the feasibility of utilizing integrally-skinned membranes with a thin dense barrier layer had been

TABLE 2.2 HYDROGEN PERMEABILITIES AND H₂/N₂ SELECTIVITIES FOR VARIOUS CLASSES OF POLYMERS[a]

Polymer	$P_{H_2} \times 10^{-9}$ (cm³ STP-cm/cm²-s-cm/Hg)	$\alpha_{N_2}^{H_2}$
Silicone rubbers	100–500	1.5–3.0
Hydrocarbon rubbers	50–300	2.0–4.0
Polyphenylene oxides	50–100	10–20
Substituted polysulfones	20–70	15–25
Polycarbonates, polysulfones	0.5–20	25–75
Polyesters, nylons	0.5–3.0	50–150
Acrylonitrile copolymers (high-concentration acrylonitrile)	0.1–1.0	100 to > 1000

[a]From Henis and Tripodi[27]; © 1983 by the AAAS.

demonstrated for desalination by hyperfiltration, analogous membranes were developed for use in gas separations. A 2000-fold increase in permeability results if a skinned microporous membrane with a 500-Å skin is utilized instead of a dense film of the same material 100 μm thick. However, because the skin layer of integrally-skinned membranes invariably contains a small number of defects (Chapter 7), its selectivity is less than that of the corresponding dense films. This problem has been solved by plugging defects with permeable elastomers in what has become known as resistance-model (RM) composites.[27] Finally, membrane area has been augmented by increasing packing density through the use of spiral, and especially, hollow-fiber elements (see Table 1.4). To minimize pressure drops, large-bore (~ 200 μm) fibers have been developed. Whereas earlier work centered on the investigation of a broad range of thick elastomeric and semicrystalline dense films, current research interest is centered on integrally-skinned membranes of high T_g polymers in the glassy state. Examples are the polysulfones[28] and the aromatic polyimides.[29] Since gas permeation occurs through the minute pores created by skewed interchain displacements in the dense glassy state, subtle techniques for increasing the displacement in fine increments are currently being sought[30] (Chapter 4).

Among the applications for membrane gas separations are the removal of hydrogen from ammonia purge, methanol purge, and naphtha hydrotreater purge gases, the adjustment of the H_2/CO ratio in synthesis gas, and the recovery of high-purity hydrogen. Other applications are the separation of H_2S and CO_2 from methane and the recovery of carbon dioxide for use as a miscible feed for enhanced secondary oil recovery. At present, interest in membranes for gas separations is closely tied to the fortunes of the petroleum industry.

The permselectivity of polymeric membranes to condensible gases and vapors is more complex owing to increasing membrane–solute and solute–solute interactions. For gases which tend to interact strongly with the membrane and swell it, sorption by the membranes is energetically favorable. The solubility factor in permeation becomes of much greater importance than for permanent gases. Depending on the closeness between the solubilty parameters of the permeant and the polymeric membrane:

1. An amorphous membrane will swell (or dissolve if the permeant concentration is high enough).
2. A covalently cross-linked polymer will swell to varying extents.
3. A semicrystalline polymer (whose crystallites function as virtual cross-links) will swell to varying extents.

The relationship between gas sorption and its partial pressure becomes more complicated since macromolecular displacement in the membrane will be affected by permeant concentration. Mixtures of swelling permeants are even more complex since solute a–solute b–membrane interactions can influence solute a–membrane interactions. When a membrane is appreciably solvated by a permeant, diffusivity increases rapidly with the amount of sorbed permeant The permeability of a given membrane to a swelling permeant will therefore be strongly concentration dependent. The overall permeabilities of membranes to swelling permeants are generally several orders of magnitude higher than to permanent gases. The permselectivity of swollen membranes decreases with increasing degree of swelling because the increasing interchain displacement lessens the possibility of separations on the basis of size.

Water vapor can permeate through hydrophilic polymeric membranes as a swelling permeant and through hydrophobic membranes as a clustering permeant. In the former it leads to increasing permeability with increasing concentration; in the latter to permeability constants which are independent of pressure (Fig. 2.5).

A special case of ordinary gas permeation by diffusion is the process known as *pervaporation* or *liquid permeation* in which a membrane separates an upstream solution in the liquid state from downstream permeants in the gaseous state. The downstream side is maintained at a lower pressure to ensure the absence of liquid. Although there is a very strong dependence of diffusion rate on solvent concentration in the polymer film, there does not seem to be any basic difference between the values obtained for pervaporation and those for vapor diffusion. Stannett and Yasuda[31] utilized membranes equilibrated with the permeant and found no difference in permeability for liquid versus vapor from solutions of benzene and cyclohexane through polyethylene and acetone and acetonitrile through rubber. Thus when vapor and liquid permeation rates do differ, diffusion cannot be the rate-limiting step.

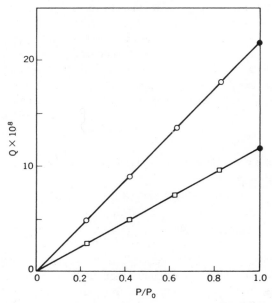

FIGURE 2.5. Water-vapor transmission as a function of relative vapor pressure: ○ = vapor polyethylene (0.922 density); ▢ = vapor polypropylene (0.907 density); and • = liquid water (from Stannett and Yasuda[31]; © 1965).

The details of the separation mechanism in the pervaporation process have been a matter of some dispute, with some workers[31-34] claiming that the principles of gas permeation do not adequately account for the high pervaporation rates which are observed; other workers[35] contend that difficulties attendant upon the measurement of pressures and temperatures of saturated vapor, of removing vapor from the product side of the cell, and of maintaining steady-state conditions account for the anomalous results sometimes reported. There is, however, universal agreement that the separation of organic liquids, including aceotropes, by pervaporation is an area of significant potential.

Schematic representations of the process and of a laboratory assembly which demonstrates its essential requirements are found in Figures 2.6 and 2.7. In the generalized example shown in Figure 2.6 the charge pressurized sufficiently to maintain it in the liquid state contains a 50:50 mixture of two different types of molecules. The permeate contains 80% of the enriched component. Both batch and continuous operation are possible. The type of membrane will vary with the nature of the liquid components. The selectivity of a given mixture can be reversed by changing the chemical nature of the polymeric membrane (Table 2.3). In the example cited, a polar membrane yields a permeate enriched in methanol, whereas a nonpolar membrane produces a permeate enriched in benzene. Permeability and permselectivity also vary with degree and type of crystallinity, plasticizer, and operation above or below T_g. As long as the charge mixture is in the liquid state, the charge pressure affects neither rate nor selectivity in liquid permeation (Table 2.4). The pressure differential between feed and product sides is likewise without effect on the rate of permeation granted only that the product side is maintained in the vapor phase (Table 2.5). The reason for this is the large concentration difference between feed and product sides, which is of such an order that it is negligibly influenced by changes in permeate pressure.

It has been discovered that the activation energy for the permeation of large species is approximately equal to that of the viscous flow of the polymeric species in the membrane.[36] Thus temperature furnishes the energy necessary to increase

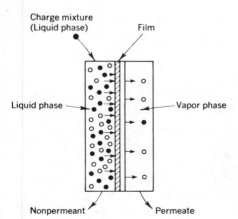

FIGURE 2.6. Schematic diagram of liquid permeation, showing unit divided into two compartments by the film. The more permeable molecules are indicated by open circles (from Binning et al.[32]; reprinted with permission from *Industrial Engineering Chemistry*, © 1961 American Chemical Society).

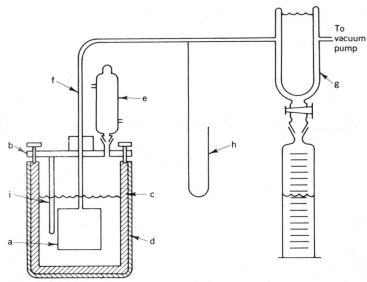

FIGURE 2.7. Laboratory assembly for liquid permeation (from Binning et al.[32]; reprinted with permission from *Industrial Engineering Chemistry*, © 1961 American Chemical Society).

interchain displacement and accelerate the passage of permeate (Table 2.6). In common with other membrane processes, permeability varies inversely with membrane thickness, whereas selectivity is unaffected thereby. However, the latter property refers only to membranes where structural integrity is unimpaired. As membrane thickness decreases, structural integrity becomes increasingly difficult both to maintain and to obtain.

The structure of the permeant species strongly influences membrane permse-

TABLE 2.3 REVERSAL OF SELECTIVITY[a,b]

| Film | Composition (vol %)[c] | | Permeation Rate (gal/ft²h) × 10³ |
	Charge	Permeate	
A	32 M	69.5 M	105
	68 B	30.5 B	
B	32 M	10 M	104
	68 B	90 B	

[a]Binning et al.[32]. Reprinted with permission from industrial engineering chemistry, © 1961 American Chemical Society.
[b]Benzene–methyl alcohol mixtures can be separated by selectively permeating either component; liquid charge: 60°C, 1 atm; film thickness: 1 mil; permeate zone pressure: 40 mm Hg.
[c]M = methyl alcohol; B = benzene.

TABLE 2.4 CHARGE PRESSURE DOES NOT AFFECT RATE OF SELECTIVITY IN LIQUID PERMEATION[a,b]

Charge Pressure (psig)	Permeation Composition (vol % n-heptane)	Permeation Rate for 1-mil Film (gal/ft^{2}h) $\times$ 10^3
15	75	140
115	75	140

[a]From Binning et al.[32] Reprinted with permission from *Industrial Engineering Chemistry*, © 1961 American Chemical Society.
[b]Charge composition: 50:50 n-heptane–isooctane; operating temperature: 100 °C.

lectivity. The inclusion of Ag within a membrane[37] greatly increases the solubility and hence permeability of olefins because of the formation of olefin–Ag complexes. Branching inhibits and unsaturation increases permeability in hydrocarbons containing a given number of carbon atoms. The permeation rate of n-hexane is 100 times that of 2,2-dimethylbutane and one-third that of 1-hexene (Table 2.7).

Although pervaporation has generally been the process explored for the membrane separation of organic liquids, this is not the only option available. The downstream side may also contain an acceptor liquid.[38] The latter must not permeate the membrane and must exhibit low viscosity, heat capacity, and vapor pressure. Furthermore, to minimize the energy required to separate the permeant from this liquid, there should be a minimal difference between the operating temperature and the separation temperature.

TABLE 2.5 PERMEATION RATE REMAINS CONSTANT OVER A WIDE RANGE OF PRESSURE DIFFERENTIALS ACROSS THE FILM[a,b]

Pressure on Permeate Side of Film (mm Hg)	Pressure Differential $p_2 - p_1$ Across Film (mm Hg)	Permeation Rate (gal/(ft^2)(h) $\times$ 10^3) 1-mil Film	0.55-mil Film	Air in Permeate (0.55-mil film) (mole %)
20	740	—	236	2.4
40	720	—	233	2.4
50	710	134		
100	660	119	230	2.5
200	560	125	230	2.9
300	460	121	236	2.5
400	360	125	226	2.3
500	260	116	230	2.7

[a]From Binning et al.[32] Reprinted with permission from *Industrial Engineering Chemistry*, © 1961 American Chemical Society.
[b]Charge: pure n-heptane; charge pressure: 760 mm Hg; temperature: 99°C.

TABLE 2.6 HIGHER TEMPERATURES INCREASE THE PERMEATION RATE AND MODERATELY REDUCE SELECTIVITY[a,b]

Film Thickness (mils)	Permeation Temp. (°C)	Permeate Composition (vol % n-heptane)	Permeation Rate (gal/ft^{2}h) $\times$ 10^3
0.8	70	79	78
	80	78	105
	90	76	144
	100	75	205
1.0	70	77	58
	80	77	80
	90	75	112
	100	75	156
1.4	70	76	33
	80	77	50
	90	75	69
	100	75	93
1.9	70	76	22
	80	76	33
	90	77	47
	100	75	66

[a]From Binning et al.[32] Reprinted with permission from *Industrial Engineering Chemistry*, © 1961 American Chemical Society.
[b]Charge composition: 50:50 vol % n-heptane–isooctane; charge: liquid phase, 1 atm; permeate zone pressure: 35 mm Hg.

TABLE 2.7 OLEFINS PERMEATE FASTER THAN PARAFFINS AND UNBRANCHED HYDROCARBONS FASTER THAN BRANCHED ISOMERS[a,b]

Hydrocarbon	Permeation Rate (mL/m^{2}h) per 0.001-cm film thickness
n-Hexane	479
n-Heptane	197
n-Octane	145
n-Nonane	31
2-Methylpentane	145
3-Methylpentane	115
2,2-Dimethylbutane	5
1-Hexene	1,437
2-Heptene	532

[a]From Binning et al.[32] Reprinted with permission from *Industrial Engineering Chemistry*, © 1961 American Chemical Society.
[b]Liquid charge: 52°C, 1 atm; film thickness: 0.0046 cm; permeate zone pressure: 360 mm Hg.

2.1.2 Dialysis

Dialysis is the diffusion-controlled, concentration-driven process in which solute from a more concentrated (feed) solution permeates a membrane to enter a less concentrated (dialysate) solution. Dialysis is related to ultrafiltration in that both are utilized to remove microsolutes from solutions of macromolecules (Table 2.1). However, in the former the net flow is of solute *per se,* whereas in the latter solute flux is coupled with that of solvent. To the extent that diffusion is free, that is, no specific steric or solubility influences are operative, the only effects of the membrane are to reduce the area through which diffusion occurs and (where the length of the pores exceeds the membrane thickness) to reduce the concentration gradient.

For a solute to pass from the bulk feed solution and permeate a membrane to enter the bulk dialysate, it must overcome the total resistance R_t which consists of the membrane resistance R_m and the resistances R_f and R_d from the two boundary layers consisting of feed-side and dialysate-side liquid films adjacent to the membrane surfaces:

$$R_t = R_f + R_m + R_d$$

If both boundary layers are dissipated by turbulent flow or laminar flow in a thin channel, then only the resistance due to the membrane need be considered. The permeability or flux J_s of solute through a membrane is inversely proportional to membrane resistance or thickness l and directly proportional to the concentration gradient Δc and membrane area A. Hence

$$J_s = - kA\,\Delta c/l$$

where k is the proportionality constant known as the permeability coefficient for the particular solute–membrane pair.

Hemodialysis, the removal of low-molecular-weight solutes such as urea and creatinine from the blood of patients with chronic uremia, is the most important use of dialysis today. Hemodialysis is also believed to represent the largest single application of membranes to separations. Because the driving force for dialysis is a concentration gradient, it is not a particularly rapid process. Nevertheless, it is still also used in the pharmaceutical industry to remove salts, in the rayon industry to recover caustic from rayon steep liquor, and in the metallurgical industry to remove spent acid.

The most commonly encountered dialysis membranes are finely porous (MW cutoffs ~ 1000) isotropic hydrogels of cellulose which are produced by the extrusion of cuprammonium hydroxide solutions of cotton linters cellulose in both capillary and flat-sheet form into aqueous salt solutions. Cellulose capillaries are also made by the deacetylation of cellulose acetate fibers. The recent introduction of *N*-methylmorpholine *N*-oxide (NMMO) as a cellulose solvent may prompt further activity in the now mature field of cellulose dialysis membranes.

2.2 ELECTROMEMBRANE PROCESSES

Salts or electrolytes are composed of positively charged (cations) and negatively charged (anions) ions which dissociate in water. If a direct electric current is passed through the solution, cations and anions will conduct the current and move in opposite directions. The speed and direction of flow of the ions will depend upon the current potential and density, as well as upon the resistance of both solutions and membranes and the characteristics of the individual ions, such as charge classification and valence. The only commercially significant electromembrane process is *electrodialysis*, which is utilized to deplete (or concentrate) aqueous solutions containing low-formula-weight ionic solutes.

Although there are many variations,[39] the most widely encountered form of electrodialysis is the transport of ions through selective cation and anion exchange membranes as a result of the passage of an electrical current. Electrodialysis can be used in the separation of electrolytes from nonelectrolytes,[40] in depletion[41] or concentration of electrolytes,[42] ion replacement,[43] metathesis reactions,[39] the fractionation of electrolytes,[44] and the separation of electrolysis products.[45]

Before considering the specific nature of electrodialysis, transport of matter and electricity across ionic-exchange membranes will be reviewed. Three separate groups of theories based upon different models can be distinguished. Although advocates of the various approaches often consider the other approaches to be deficient in one or more aspects, the truth is that they represent points of view rather than mutual exclusivity. Helfferich[46] has enumerated and categorized the various groups of theories. Those of the first group view the membrane as a two-dimensional surface separating two liquid phases and interposing varying resistances to the transport of the various permeant species. The driving forces for material transport across the membrane are the differences in chemical potential between the separated liquid phases.[47-50] The second group of theories views the membrane as if it were a homogeneous phase of finite thickness. The driving forces are the local gradients of the general chemical potentials in the layer.[51-71] These gradients can be influenced by convection. The third group of theories views the membrane as a series of potential-energy barriers so that the membrane is considered as an inhomogeneous layer.[72-74] Inhomogeneity results because of the greater probability of the occupation by a solute particle of a position between the activation thresholds. The driving forces are the differences between the transition probabilities in opposite directions normal to the membrane. Theories of the second and third groups yield substantially identical results when the density of fixed charges is relatively high.

A number of different phenomena are encountered in material transport across ion-exchange membranes. Of primary importance is the fact that the permeabilities of counterions, co-ions, and nonelectrolytes differ, particularly at high membrane capacities and dilute solution concentrations. As previously noted, these differences are related to diffusion, membrane–solute interaction, and steric factors. Counterion concentration in the membrane exceeds co-ion concentration. In dilute solutions, the membrane counterion may be essentially independent of solution concentration and exceed that of the co-ion by several orders of magnitude. The concentration of counterions in solution, on the other hand, is lower, and that of co-ions, higher than in the membrane.

The rate-determining step on the transport of solutes can be controlled by diffusion through the membrane itself or by diffusion through the film of liquid adjacent to the membrane. Film diffusion control tends to occur where membrane diffusion coefficients are high (or where the membrane is very thin), where poor agitation results in thick films, and where there is little difference between the concentration of species in solution and in the membrane. For the latter reason film diffusion control is quite common in the case of counterions.

The difference in electric potential between two electrolyte solutions separated by a permeable or a semipermeable membrane is known as the *membrane potential*. The membrane potential depends upon the properties of both fixed and mobile ions and is usually independent of membrane thickness or cross-sectional area. The membrane potential is the sum of the diffusion potential within the membrane, the Donnan (phase-boundary) potential, and (where partial or total film control obtains) the diffusion potentials in the films. Ideal membrane potentials E_i can be calculated considering the case of hypothetical membrane permeable only to cations in the cell

$$\text{Ag} \mid \text{AgCl} \mid \text{NaCl}(a_1) \parallel \text{membrane} \parallel \text{NaCl}(a_2) \mid \text{AgCl} \mid \text{Ag}$$

$$E_i = \frac{2RT}{\mathcal{F}} \ln \frac{a_1}{a_2}$$

Here a_1 and a_2 are mean activities for NaCl and $\mathcal{F}$ is the amount of electricity in faradays.

Measured potentials which are lower than ideal potenials show that the ideal cell reaction does not adequately describe the real situation. Deviations from ideality result because of water and/or coanion transfer.

Bi-ionic potentials result from cells which contain two electrolyte solutions AX and BX such that

$$\text{AX} \parallel \text{membrane} \parallel \text{BX}$$

Multi-ionic potentials result from cells in which both solutions contain various A, B, C, . . . and K, L, M, . . ., respectively:

$$\text{AX, BX, . . .} \parallel \text{membrane} \parallel \text{KX, LX}$$

Bi- and multi-ionic potentials are quite complex, consisting as they do of many interacting phenomena, such as interdiffusion and Donnan potentials. In general the electric potential for the cation-exchange membrane tends to be more positive in the more dilute solution, or the solution containing counterions of lower mobility, valence, or affinity for the membrane. The opposite situation obtains for anion-exchange membranes. Film diffusion control can, of course, complicate the situation considerably.

Diffusion of solvent (osmosis) across ion-exchange membranes can occur in a different manner than with uncharged membranes, in which the net flow (in the absence of applied pressure) is always from the more dilute to the more concentrated side of the membrane. In such cases the osmotic transfer of solvent is a strictly colligative function of the relative solute concentration on either side of the membrane. For ion-exchange membranes such normal osmosis may be replaced by *anomalous osmosis*, in which solvent transfer may occur in the normal direction but be quantitatively greater than that expected on the basis of concentration differences (anomalous positive osmosis) or in the opposite direction (anomalous negative osmosis).[68, 75, 76] Whether positive or negative, anomalous osmosis is the result of solute diffusion and cannot occur in its absence. The diffusion of solute creates the electric field and gives rise to strong diffusion potentials when the mobilities of the counterions and co-ions differ considerably. When the counterion diffuses more swiftly, the electrically charged pore liquid move toward the concentrated solution, giving rise to anomalous positive osmosis. When the co-ion is faster, the pore liquid travels in the opposite direction, resulting in negative osmosis where the electric field is stronger than the pressure.

The osmotic properties of ion-exchange membranes are of considerable importance because where resistance to solvent flow and ionic mobility are low, the electrolyte flux may be greatly affected. Strong positive osmosis may carry an electrolyte from the dilute to the concentrated solution (*incongruent salt flux*). Although early investigations ascribe anomalous osmosis to structural inhomogeneities in the membrane,[75, 76] Schlögl contends that anomalous osmosis is the rule rather than the exception for ionic solutes and charged membranes irrespective of structure.[68]

Three important phenomena occur when excess hydrostatic pressure is applied forcing pore liquid through the membrane: streaming potential, streaming current, and electrolyte filtration. Since the pore liquid bears an electric charge, its displacement results in an electric potential difference—the *streaming potential* (Fig. 2.8). The streaming potential partially balances the effect of pressure, thereby reducing the flow across the membrane. It also inhibits counterion mobility and increases co-ion mobility so that both species are carried along with the solvent. If the streaming potential is short-circuited by attaching reversible electrodes both to membrane surfaces and to one another, an electric current, the *streaming current*, will result from the charge transfer due to excess counterion flux. Highly perm-

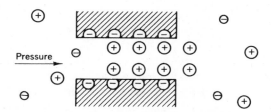

FIGURE 2.8. Origin of the streaming potential (from Helfferich[46]).

selective membranes strongly exclude the co-ion because of the Donnan potential. Since solvent transfer is not so strongly hindered, the electrolyte is partially retained (filtered) by the membrane. Filter action is more pronounced for electrolytes with counterions of low valence and co-ions of high valence, conditions which generally tend to inhibit the uptake of electrolyte by the membrane. Electrolyte concentration in the Nernst film at high pressure at the membrane surface and depletion in the film at the low-pressure surface causes the filter action and the streaming potential to decline with time.

The fact that in electrodialysis the electric current transfers many more counterions than co-ions permits this process to be utilized for removing electrolytes from solutions. Consider the cell

$$\text{cathode}^- \; \text{Ag} \,|\, \text{AgCl} \,|\, \text{NaCl} \; \left\|\; \begin{array}{c} \text{cation-exchange} \\ \text{membrane} \end{array} \;\right\| \; \text{NaCl} \,\|\, \text{AgCl} \,|\, \text{Ag anode}^+$$

In the ideal case only the counterion Na^+ will be transported. One faraday of electricity passed through the cell will produce 1 Cl^- at the cathode, transfer 1 Na^+ across the membrane, and consume 1 of Cl^- at the anode. The current efficiency, that is, the change in equivalent content per faraday, is unity. In the practical case, the current efficiency is somewhat less than unity because of solvent transfer and incomplete co-ion retention. Solvent transfer as a result of convection of the pore liquid during the passage of an electric current is known as *electroosmosis*. Since counterion transference numbers decrease with increasing solution concentration, current efficiency does also. For this reason demineralization by electrodialysis is most economical for the case of dilute solutions. However, if the solutions are too dilute (< 200–400 ppm NaCl), solution resistance will be too high for energy-efficient separation. The accumulation of electrolyte in one compartment and depletion in the other counteracts transference because of increasing opposition from diffusion. The transfer of electric current is proportional to the current density and independent of the membrane thickness. Since the opposing rate of diffusion is inversely proportional to the membrane thickness, the utilization of thick membranes and high current density should act to increase current efficiency. Unfortunately, these two conditions necessitate a higher operating voltage and result in higher resistive losses owing to the evolution of heat. Furthermore, if the current density exceeds a certain critical value, the current efficiency decreases abruptly. The critical current density is the value at which the ohmic resistance, and therefore the voltage drop, increases to the point where water is dissociated (Fig. 2.9). The H^+ or OH^- ions in the case of cation- or anion-exchange membranes, respectively, then compete with the counterion for transport across the membrane, while the OH^- or H^+ are transferred back into solution. Current efficiency is therefore decreased because the dissociation energy of water is added to that required for counterion transference. Although the corrosive effects of electrolytic products have long been a problem, reactionless electrodes have recently been described by Kedem et al.[78] in which circulated carbon particles eliminate the formation of acid, base, and chlorine.

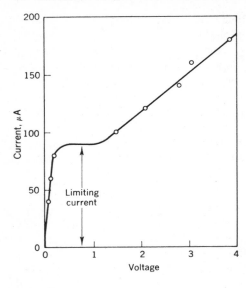

FIGURE 2.9. Experimental current–voltage characteristics of an ion-exchange membrane (from Peers[77]).

For demineralization on a large scale, multicompartment cells with large numbers of alternating cation- and anion-exchange membranes in series are commonly employed (Fig. 2.10). The electric current causes electrolyte depletion in every other cell and electrolyte accumulation in the intervening cells. The energy required depends on both the current efficiency and the operating voltage. For current to be passed through the cell to effect depletion of the more dilute solution, the applied voltage must be higher than the sum of the membrane potentials. This voltage will increase with increasing ohmic resistance so that intermembrane distance must be minimized. Cell resistance increases with progressive depletion of the dilute solutions and with increasing current density. The excess energy required for the actual electrodialysis situation is irreversibly consumed in co-ion transference, compensation for diffusion, and heat production.

If only anion-exchange or cation-exchange membranes are utilized in electrodialysis, continuous anion or cation replacement is possible. An example of the former is citrus juice sweetening in which citrate ions in the juice are replaced by

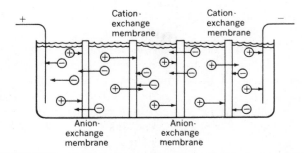

FIGURE 2.10. Multicomponent cell for electrodialytic demineralization.

hydroxyl ions from a caustic solution. A significant portion of the citrus crop is converted into juice for human consumption. Since all natural products are extremely complex systems which vary according to climate, local soil conditions, and so on, it is to be expected that a certain percentage of any given fruit will produce juice which is more sour than the norm. Although such products can be sweetened by blending or by the addition of sugar, the end result is not equivalent to the desired premium product, nor can it be marketed as such. Therefore, to upgrade this costly item, means for eliminating the excess citric acid responsible for the sour taste have been sought. A particularly promising approach is the electrodialytic removal of the citrate ion by the utilization of an all-anion-exchange-membrane electrodialysis stack (Fig. 2.11).

Alternate compartments contain citrus juice and potassium hydroxide solutions. The passage of current causes the citrate ions in the juice to permeate the anion-exchange membranes and enter the potassium hydroxide compartments. The hydrogen ions remain and are neutralized by hydroxyl ions, which enter the juice compartments from the potassium hydroxide compartments. The extent of acid removal can be controlled by regulating the residence time of juice in the system and/or the current density. This deacidification process was first proposed and investigated by Kilburn[79] and later developed through the pilot plant stage.[80, 81]

Anion-exchange membranes were specially designed for this application since some of the citrus juice components are highly reactive. Because of the large amount of pulp in citrus juice, cell width is increased to 0.25 in. and the membrane supports are removed from the juice compartments. The membrane is supported instead by operating the juice compartments at slightly higher pressures than the caustic compartments. Loss in current efficiency because of the buildup of juice

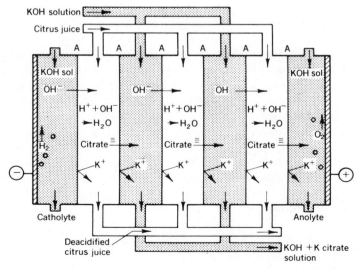

FIGURE 2.11. Electrodialysis of citrus juices (from Zang[43]).

TABLE 2.8 PILOT PLANT OPERATION FOR ELECTRODIALYTIC DEACIDIFICATION OF GRAPEFRUIT JUICE[a]

Feed temperature	92°F
Feed acidity	1.52%
Product acidity	0.90%
Production rate	95 gal/h
Cell velocity	
Product	0.3 ft/s
KOH	0.1 ft/s
Voltage	176 V
Current	122 A
Current density	13 A/ft^2
Current efficiency	70%
dc Energy consumption	0.22 kW h/gal

[a]From Zang.[42] Reprinted with permission.

constituents at membrane surfaces is overcome by periodically cleaning the surfaces by reversing the current flow at frequent intervals. Data for a pilot plant run in which grapefruit juice was sweetened give an idea of the magnitude of the task as well as approximate operations and energy requirements (Table 2.8).

Among the other applications of electrodialysis to food treatment which have been investigated to date are: (1) demineralization of dairy whey for use as baby food (a commercial process); and (2) ion exchange of Na^+ for Ca^+, K^+, or Mg^+ to produce low-sodium milk.

2.3   PRESSURE-DRIVEN PROCESSES

Pressure-driven membrane separations are a continuum of processes designed to separate suspended or dissolved particles of different sizes by utilization of membranes containing appropriately sized pores. In order of decreasing particle and pore size these are *microfiltration* (MF), *ultrafiltration* (UF), and *hyperfiltration* (HF). The last named is also known as *reverse osmosis* (RO). Each of these processes employs a porous membrane which inhibits the passage of dissolved or suspended particles based primarily (MF and UF), or partially (HF), on the size and shape of the permeant species and the membrane pores. A summary of the differences between the processes and the membranes which they utilize is found in Table 2.9.

2.3.1   Microfiltration

Microfiltration, the oldest of the pressure-driven processes, covers the pore-diameter range between 0.1 and 10 μm. Industrially, it is utilized to sterilize, that is, to

TABLE 2.9 SUMMARY OF PRESSURE-DRIVEN SEPARATION PROCESSES AND MEMBRANES

	Process Features			Membrane Characteristics			
Filtration Process	Maximum Feed Conc. (M)	MW Cutoff (daltons)	Operational Pressure Range (psi)	Structure	Porosity[a] (%)	Pore-Size Range (Å)	Pore Density (no./cm²)
Micro (MF)	10%	n.a.[b] (very high)	10–30	Several types, all skinless	~70	10^3–10^5	10^9 (0.2-μm phase inversion)
Ultra (UF)	10^{-3}	10^3–10^6	50–100	Integrally-skinned	~60	10–1000	10^{11}
Hyper (HF)	1	10–10^{2c}	100–800	Two types: Integrally-skinned	~50	Two ranges: small, 1–10	Small ($>10^{12}$)
				Thin-film composite		Large, variable 20–100	Large (variable no.)

[a]For phase-inversion membranes only.
[b]n.a. = Not applicable.
[c]Cutoff not determined solely by molecular (or formula) weight.

remove viable microorganisms such as bacteria and yeast cells from aqueous solutions. It is also utilized to remove inanimate particular matter from both aqueous and nonaqueous suspensions. Because MF membranes have relatively large pores, there is relatively little resistance to flow and low ($\sim$ 30 psi) pressures suffice as a driving force. The high porosity of MF membranes is still another reason why low pressures are utilized, since such membranes are subject to distension under pressure.

Several types of microfiltration membranes are commercially available:

1. Phase inversion (the oldest and most widely utilized) consists of skinless porous open-celled matrices whose structures in depth are either isotropic or anisotropic (Chapter 7).
2. Crystalline or semicrystalline films into which pores have been introduced and maintained via stretching and annealing, respectively (Chapter 8).
3. Dense films which have been irradiated by fission fragments from radioactive elements to produce tracks which are subsequently enlarged by caustic etching (Chapter 8).
4. Sintered membranes (Chapter 8).

The pores (surface openings) and cells (subsurface voids) are irregular apertures between neighboring enclosed spaces. Particles of a certain size permeate through irregularly shaped openings from the surface into the membrane interior (Fig. 2.12). The sieving effect exercised upon a particle in a solution or suspension therefore depends on the pore- and cell-size distributions, the number of openings in cell walls, and the number of superimposed cell layers. Pall and Kirnbauer[83] have determined that the log reduction value (LRV)

$$(\log_{10}) \ \text{LRV} = \frac{\text{(number of bacteria in feed)}}{\text{(number of bacteria in the product)}}$$

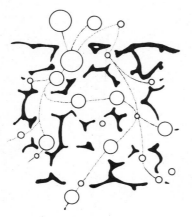

FIGURE 2.12. Schematic representation of solute permeation through a microfiltration membrane (from Helmcke[82]).

increases as the number of layers of membrane is increased. Individual particles must pass through a large number of openings in the cell walls before they permeate the membranes, and it is to be expected that sieving will depend on the narrowest dimensions of these openings. From the point of view both of the membrane and of the suspended particles, permselectivity is not absolute but depends on the laws of statistics. Furthermore, membrane separations depend not solely on pore size but also on the specific properties of the suspension being filtered. Interactions between the surface of the particles, the membranes, and the nature of the colloidal medium will influence separations as will concentration and pressure.

There are several independent methods for determining pore statistics:

1. The bubble-point technique.
2. The Hagen–Poiseuille, or solvent-permeability, method.
3. Direct observation with standard and scanning electron microscopes.

Bechhold[84] first evaluated pore size by measuring the pressure necessary to blow air through a water-filled membrane. He employed Cantor's[85] relationship

$$r = 2\sigma/P \tag{1}$$

where r is the radius of capillary, σ the surface tension (water/air), a nd P the pressure.

In practice a filter apparatus is placed upside down so that air can impinge on the membrane from beneath. Bubbles of air are then observed as they penetrate the membrane into an overlying layer of water. Representative values of pore radii versus pressure are shown in Table 2.10.

Because the larger pores open at lower pressures, this method tends to yield high values. This weakness, however, may be used to advantage in qualitatively estimating the pore-size distribution. If the number of pores which are permeable to air increases substantially with but a small increase in pressure, a narrow pore-size distribution is indicated. On the other hand, a gradual increase in the number of air-permeable pores is indicative of a broad pore-size distribution. The bubble-point method is strictly valid when the imbibed medium completely wets the membrane and when the ratio of the pore diameter to that of the permeant species is large. As this ratio diminishes the validity of Eq. (1) decreases, particularly when

TABLE 2.10 REPRESENTATIVE PORE RADII BY THE BUBBLE-POINT METHOD[a]

Pressure (atm)	Pore Radius (nm)
0.1	150
1	15
10	1.5
100	0.15

[a]From Bechhold et al.[84]

there is interaction between the permeant species and the membrane. Moreover, because of the high water–air surface tension of 73 dyn/cm, narrow pores require a relatively high pressure before air will permeate. High pressures in turn cause plastic flow in the membranes, which result in pore-size changes in a time-dependent manner. In recognition of these difficulties, Bechhold et al.[86] modified Bechhold's original bubble-pressure method by reducing the interfacial tensions between the imbibed and penetrating media. Whereas the earlier system had consisted of water as the imbibed and air as the penetrating medium, a new system, namely isobutyl alcohol–water, was chosen (for application to collodion membranes) in which σ varied from 1.6 dyn/cm at 3°C to 1.8 dyn/cm at 37°C. This system permits the measurement, at a given pressure, of pores which were smaller by a factor of 40 than those measured by the original bubble-pressure technique. Today, kerosene, rather than isobutanol–water, is utilized as the wetting fluid with pore diameter in micrometers and pressure P in psi; the equation pore diameter $(\mu m) = 5/P$ holds. Conveniently the kerosene bubble point is exactly half the water bubble point.

The concept is, of course, applicable to other than nitrocellulose membranes. In general, it is preferable to utilize as the imbibed medium that liquid which exhibits the lower contact angle with the membrane, that is, that which more readily wets the membrane. To render observation of the droplets of the penetrating liquid possible, there should be a substantial difference between the indices of refraction of the two liquids.

A weakness of the bubble-pressure method is that the values obtained vary with the rate of pressure increase.[87-89] The values of the largest pores, that is, those which become permeable first and at the lowest pressure, decrease with increasing rate of pressure increase. Furthermore, these values also decrease with increasing capillary length, increasing viscosity, and decreasing interfacial tension. To eliminate these possible sources of error Bechhold et al.[86] developed the relationship

$$r = \frac{2\sigma}{d}\left(1 + \frac{21}{\sigma}\sqrt{A\frac{\eta_1 + \eta_2}{2}}\right)$$

where l = length of the capillary,
$\quad A = dP/dT$,
$\quad \eta_1$ = viscosity of the imbibed phase, and
$\quad \eta_2$ = viscosity of the penetrating phase.

Therefore, for a given system, by varying the rate of pressure increase A, both pore radius and length can be determined. The pore radii which are determined by this method lie between the maximum and the average values.

Instruments are now commercially avialable which utilize a variation of the bubble-point technique known as the *mercury-intrusion method*.[90] Equation (1) is modified to take into account the contact angle θ (mercury/membrane):

$$r = \frac{-2\sigma \cos\theta}{P}$$

TABLE 2.11 POROMETER DATA FOR MILLIPORE MEMBRANES
(AVERAGE FOR SINGLE MEMBRANE)[a]

Criterion	Sample A	Sample B
Bulk volume at 1.32 psi (cm^3)	0.2655	0.2628
Bulk volume at 1015 psi (cm^3)	0.0564	0.0547
Total void volume (porosity), (%)	79	79
Weight of membrane (g)	0.0850	0.0826
Density of membrane at 1015 psi (g/cm^3)	1.51	1.51
Maximum in pore-radius distribution curve (μm)	0.48	0.58

[a]From Honold and Skau[90]; © 1954 by the AAAS.

A membrane is placed in a porometer chamber, and pressure is applied, thereby forcing mercury into the pores. It is assumed that all void space is full at the highest pressure (usually about 75 atm). This assumption, of course, is valid only for cases in which all the voids are of the open-cell variety. From the weights of the membranes at the lowest and highest pressures, bulk densities are obtained from which the void volume (porosity) can be obtained by difference (Table 2.11).

Pore radius r can also be obtained by measuring the volume of water permeating in a given time at constant pressure from the Hagen–Poiseuille relationship[91]:

$$J = \frac{n\pi r^4 APt}{8\eta d} \tag{2}$$

where J = effluent flux,
n = number of pores per square centimeter,
A = surface area of membrane (cm^2),
P = pressure,
t = time,
η = viscosity of flowing liquid, and
d = length of the capillary (= membrane thickness).

For laminar flow in which the Reynolds number Re < 2300 is assumed, solving Eq. (2) for r gives

$$r = \sqrt[4]{\frac{8J\eta d}{n\pi APt}} \tag{3}$$

Further, if the void volume $V = n\pi r^2$,

$$r = \sqrt[2]{\frac{8J\eta d}{VAPt}} \tag{4}$$

In terms of the permeability constant

$$P_c = \frac{J}{APt} = \frac{n\pi r^4}{8\eta d}$$

we have

$$r = \sqrt[4]{\frac{8P\eta d}{n\pi}}$$

Therefore, pore radii can be determined from permeability measurements in two ways: using the pore density n in Eq. (3) and the void volume V in Eq. (4).

The number of pores per square centimeter of membrane surface is known as the *pore density n*. The value of n established by direct microscopic investigation is referred to as n_{obs}. However, there remains the question of the density of pores n_{eff} actually utilized during filtration.

In those cases where some pores do not penetrate the membrane, low values of r will be obtained. This situation is certain to arise when a fraction of the surface openings lead to dead ends. This problem is minimal in the high-porosity membrane filters in commercial use today. It is also a factor where pore blockage occurs owing to the presence of impurities, such as dust in the distilled water.[92] The latter is one cause of the *filter effect*,[93] in which the permeability of a membrane to distilled water decreases at constant temperature and pressure.

The void volume (or porosity) V is that fraction of the membrane volume which is not occupied by the polymer substrate. Where all voids are interconnecting and the membrane is completely wet by an imbibed liquid such as water, the void volume can be calculated both from the density of the void-free polymer and from the difference between the wet and dry weights of the membrane. However, where closed voids occur and where the membrane framework itself (as distinguished from the larger voids within the membrane) is porous, a distinction should be made between the measured void volume V_{obs} and the effective void volume V_{eff}, with the latter being the parameter to be inserted into Eq. (4).

The bubble-pressure and solvent-permeability methods can be combined to yield a measure of the *pore-size distribution*. Erbe[94] first applied Karplus's technique, which makes use of the fact that at a certain minimum pressure the largest pores become permeable whereas smaller pores remain impermeable. Therefore, the volume of liquid which first permeates through the membrane is that which has permeated through the largest pores. As the pressure is increased, progressively smaller pores become permeable and the liquid permeates through both large and small pores. Eventually, a pressure is reached at which further increases effect only an increase in permeability proportional to the increase in pressure. At this point no additional pores are being opened.

Because of the importance of this concept, the details of its practical application will be discussed. The method requires two steps, the first of which is the deter-

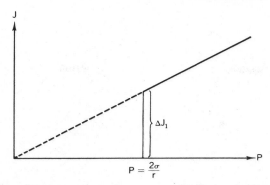

FIGURE 2.13. Effluent-flux–pressure curve for one or more capillaries of a given radius (from Erbe[94]).

mination of the effluent-flux–pressure curve. The second step is the construction of a number of subsidiary curves: (1) The pore-density–pore-radius curve; (2) the pore-area–pore-radius curve; and (3) the effluent-flux–pore-radius curve. The first curve is utilized to determine how many pores of the largest, second largest, and so on, sizes are found in the membrane; the second is used to determine the total surface area occupied by pores of the various sizes; and the third determines the volume of liquid which at constant pressure permeates the pores of different sizes.

The effluent-flux–pressure curve is based upon the simplified assumption that a membrane is composed of a film of constant thickness through which parallel circular capillaries penetrate perpendicular to the surface. The radius of the capillaries varies between the maximum value r_{max} and a minimum value $r_{min} > 0$.

Consider first a single capillary with radius r and length d which is filled with a liquid I and brought into contact with an immiscible fluid phase II (either liquid or gas) (Fig. 2.13). If liquid I thoroughly wets the capillary wall and II possesses a lesser affinity fr the capillary wall than I, Eq. (1) applies.

Between $P = 0$ and $P = P_1$ the effluent flux $J = 0$. At $P_1 = 2\,\sigma/r$ the effluent flux increases from $J_0 = 0$ to $J_1 = P_1\,(\pi r^4/8\eta d)$ (Hagen–Poiseuille), where $J_1 - J_0 = \Delta J$. With a further increase in pressure, J increases proportionally to the increase in pressure. For n parallel and identical capillaries, the curve has the same form but

$$\Delta J_1 = n P_1 \frac{\pi r}{8\eta d}$$

Consider next a system consisting of two capillaries of equal length but of different radii, r_1 and r_2 ($r_1 > r_2$) (Fig. 2.14). The effluent flux through each capillary is not influenced by the presence of the other.

Through the capillary with the larger radius r_1, $\Delta J_1 = P_1\,(\pi r^4/8\eta d)$. At pressures greater than P_1, J increases along curve I. Through the second capillary, with the smaller radius, an additional flow begins at $P_2 = 2\sigma/r_2$, so that

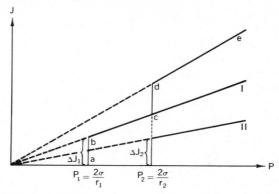

FIGURE 2.14. Effluent-flux–pressure curve for a system of two different capillaries (from Erbe[94]).

$$\Delta J_2 = P_2 \frac{\pi r_2^4}{8\eta d}$$

With further increases in pressure, effluent volume increases along curve II.

Since the volumes through both capillaries are independent of one another, the total effluent volume is the sum of those through the two individual capillaries. The curve abcde graphically portrays the total effluent volumes for both pressures. The same can be obtained by superimposing the effluent volume J_2 upon curve I at pressure P_2 and drawing a straight line from d through the origin.

Let us consider now the example of a porous membrane. Because of the large number of pores of various radii, a continuous curve replaces the step function (Fig. 2.15).

At pressures below $P_1 = 2\sigma/r_{max}$ the membrane is impermeable. At P_1 flow through the largest pores begins. At higher pressures, smaller and smaller pores

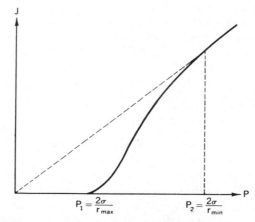

FIGURE 2.15. Effluent-flux–pressure curve for a microfiltration membrane (from Erbe[94]).

become permeable until finally, at $P_2 = 2\sigma/r_{min}$, the smallest pores become permeable. With further increases in pressure, J increases proportionally to P according to the Hagen–Poiseuille law. The effluent-flux–pressure function for membranes is generally an S-shaped curve, whose maximum slope corresponds to the range of maximum pore density.

After the effluent-flux–pressure curve has been determined, it is necessary to convert it into a pore-density–pore-radius curve. This is done by combining Eqs. (1) and (2) and solving for n; whence

$$n = \frac{\eta d}{2\pi\sigma^4} P_1^3 \, \Delta J_1 \quad \text{and} \quad r_1 = \frac{2\sigma}{P_1}$$

The desired values r and n can be obtained from the known values η, d, and σ and the graphically obtained values P_1 and J_1. In like manner

$$n_2 = \frac{\eta d}{2\pi\sigma^4} P_2^3 \, \Delta J_2 \quad \text{and} \quad r_2 = \frac{2\sigma}{P_2}$$

If the length of the capillaries is unknown, evaluation of the factor $\eta d/2\sigma^4$ may be omitted; then relative rather than absolute values of n will be obtained.

For graphical evaluation of the effluent-flux–pressure curve experimentally obtained for a porous membrane, the continuous curve (Fig. 2.15) is replaced by a step curve (Fig.2.16) in which the abscissa is divided into intervals, for example, AB, and the continuous curve A′B′ is replaced by a stepped curve B′A′. This means that the pores really present in the interval AB and those whose radii vary between $r_3 = 2\sigma/P_3$ and $r_4 = 2\sigma/P_4$ are replaced by a certain number $n_{3,4}$ of capillaries with the common radius $r_{3,4} = \frac{1}{2}(r_3 + r_4)$. The number $n_{3,4}$ therefore represents the number of pores of radius $r_{3,4}$ which permit the same volume of liquid to permeate as the sum of the actual pores in the interval AB having radii between r_3 and r_4.

So that the intervals on the abscissa can be compared to one another, they must all include equally broad ranges of pore radii; that is, $r_1 r_2$ must equal $r_2 r_3$, ... In addition, the points at which the perpendicular rises in the step curve occur must be placed so that their corresponding radii r_1, r_2, r_3, . . . , will be the arithmetic means of the two bounding radii. Therefore the values r_1, $r_{1,2}$, r_2, $r_{2,3}$, . . . , and likewise the related values $1/P_1$, $1/P_{1,2}$, $1/P_2$, $1/P_{2,3}$, $1/P_3$, . . . , will constitute an arithmetical series, in which a constant interval between the members can be chosen at will. Since P values are represented on the effluent-flux–pressure curve, the abscissal interval must be so chosen that the values of P_1, $P_{1,2}$, P_2, $P_{2,3}$, P_3, . . . , form a series of regularly increasing intervals. Then P_1, P_2, P_3, . . . , are the abscissal values for the initial and terminal points of the intervals, and $P_{1,2}$, $P_{2,3}$, $P_{3,4}$, . . . , are the abscissal values for the points of the perpendicular rise of the step. Hence the total number of pores between $r_{max} = r_1 = 2\sigma/P_1$ and $r_2 = 2\sigma/P_2$ can be approximated by $n_{1,2} = (d/2\pi\sigma^4)P_{1,2}^3 \, \Delta J_{1,2}$ pores of radius $r_{1,2} = 2\sigma/P_{1,2}$. The total number of pores between $r_2 = 2\sigma/P_2$ and $r_3 = 2\sigma/P_3$ can be

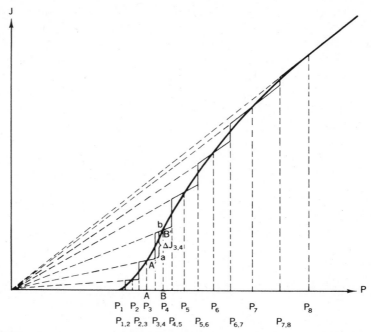

FIGURE 2.16. Graphical solution of the effluent-flux–pressure curve for a microfiltration membrane (from Erbe[94]).

similarly calculated, as can the intervals in the entire curve up to r_{min}. The number of pores n can be plotted versus the pore radii to yield the pore-density–pore-radii curve (Fig. 2.17).

The surface bounded by the curve and the abscissa is a measure of the total number of pores, and the portion of the surface between two abscissal values is a measure of the number of pores whose radii lie between these values.

The total surface area A occupied by pores of a given range of radii can be obtained by multiplying the surface area of an individual pore of a given radius by the total number of such pores. Hence

$$A_{1,2} = n_{1,2}\,\pi r_{1,2}^2 = \frac{nd}{2\pi\sigma^4}\,P_{1,2}^2\,\Delta J_{1,2}\,\pi\,\frac{4\sigma^2}{P_{1,2}^2} = \frac{2nd}{\sigma^2}\,P_{1,2}\,\Delta J_{1,2}$$

If the values are plotted versus pore-radius intervals, the pore-area–pore-radius curve results (Fig. 2.18). The total surface area occupied by all the pores, as well as that occupied by pores of a given range of radii, can be read from this curve. By plotting $J_{1,2}, J_{2,3}, \ldots$, versus the pore radii, the effluent-flux–pore-radius curve can be constructed (Fig. 2.19). If the pressure is taken as unity, the ordinates of this curve represent the absolute values of the effluent volumes. Knowledge of the effluent volumes which permeate pores of different radii is important to the understanding of the sieving effect in porous membranes.

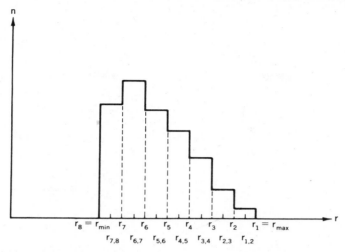

FIGURE 2.17. Pore-density–pore-radius curve for a microfiltration membrane (from Erbe[94]).

The mathematics of the combined bubble-pressure and permeability methods has been treated by Grabar and Niktine,[95] and form the basis for the ASTM method, D-2499. The average diameters derived from permeability measurements and the statistical averages determined by the bubble-pressure technique do not correspond exactly. Furthmore, both these values differ appreciably from those corresponding to the maximum pore density and from the values obtained from electron microscopy.

The porosity of an MF membrane is determined by weighing dry membranes of *uniform thickness* and comparing the weights of these membranes to the weights of equal volumes of dense films of the membrane polymer(s). The latter can be calculated from the specific gravities of the materials in question.

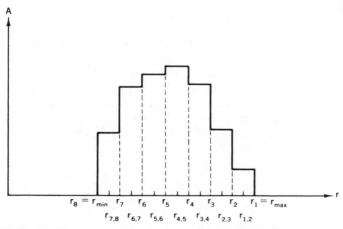

FIGURE 2.18. Pore-area–pore-radius curve for a microfiltration membrane (from Erbe[94]).

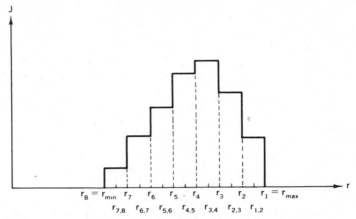

FIGURE 2.19. Effluent-flux–pore-radius curve for a microfiltration membrane (from Erbe[94]).

The pore-size distribution of MF membranes can either be determined manually as described above or with an apparatus[96] that (1) simultaneously measures the flow of air versus pressure through a wet membrane and through a dry membrane to yield wet–dry flow ratios which are then (2) differentiated with respect to time to yield a pore-flow-distribution curve.

Microfiltration membranes are generally utilized in a dead-end configuration with the suspension directed at right angles to the membrane surface. Under these circumstances, membrane pores become progressively blocked by filtered particles as filtration progresses. The narrower the pore-flow distribution and the closer the correspondence between pore size and the size of the filtered particle, the more pronounced will be the tendency of the membrane pores to become blocked. As a result of blocking, permeability gradually decreases with time. When, by convention, the permeability at a given pressure decreases to one-fifth of its original value, the filtration is terminated and the filters replaced.

The amount of filtrate which has been collected to this point is known as the *filtration capacity* or *throughput*. There are several methods for increasing throughput. One is to employ *cross-flow filtration* in which the suspension is circulated parallel to the membrane surface rather than at right angles to it. However, although this procedure tends to hinder the deposition of blocking substance at the membrane surface, it requires additional energy and more expensive equipment to effect the circulation. The alternative solution is to *prefilter* the suspension to remove those particles which block the final filter. Prefiltration utilizes filters with pore sizes larger than those of the final filter. Unfortunately, although prefilters increase the throughput of the final filter, it is most costly to utilize two (or more) filters in series than to utilize a single filter.

One answer to this dilemma is the highly anisotropic MF membrane[97] which is, in effect, a prefilter–filter combination in a single membrane (Fig. 2.20). This skinless membrane consists of an integral bilayer in which one layer—the fine-structured final filter—amounts to one-third the thickness of the membrane and the other

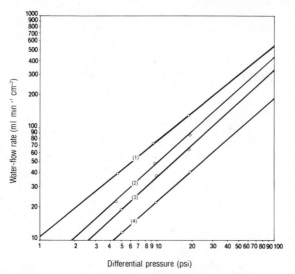

FIGURE 2.20. Water-flow rate versus pressure for various 0.45-μm membranes (1 = highly aniso-tropic; 2 = conventional mixed ester; 3 = Nylon 6,6; 4 = PVF₂) (from Kesting et al.[97]).

layer—, the coarse prefilter—comprises the balance of the membrane. If the ani-sotropic membrane is positioned such that the coarse side is adjacent to the feed, significant increases in throughput result (Fig. 2.21; Table 2.12). If, on the other hand, the fine side is adjacent to the feed, little difference can be seen between the throughput of anisotropic and isotropic membranes. In order to accommodate the fact that the final filter layer of the anisotropic membrane is thinner than a corre-sponding isotropic membrane of the same rated pore size and can, therefore, be expected to have a slightly lower LRV (log reduction value), the pore size of the fine layer of the anisotropic membrane is designed to be slightly smaller than that of isotropic membranes of the same rated pore size. The slight adjustment re-quired—for example, an anisotropic membrane with a degree of anisotrophy (DA) of 5 (coarse-side/fine-side pore-size ratio) should have a kerosene bubble point of 28 psi (0.18-μm pore size), compared to 25 psi for the isotropic membrane—has little effect on the increased flow rates and throughputs achieved by the anisotropic membrane. Of course as the DA increases and the relative thickness of the fine-celled layer decreases, the bubble point must be proportionately larger. A 0.2-μm-rated commercial anisotropic microfilter is available which has a DA of 100 and a fine layer which is only $\frac{1}{25}$ of the membrane thickness. Although this membrane is characterized by a kerosene bubble point of 50 psi (0.1-μm-rated pore size), it, nevertheless, exhibits permeabilities and throughputs which are greater than those for corresponding 0.2-μm-rated isotropic membranes.

Pseudomonas diminuta is the 0.3-μm-diam bacterium which is utilized as the standard microorganism with which *sterile* (actually, *sterilizing*) 0.2-μm-rated membrane microfilters are challenged. Since bubble point is inversely related to pore size, bacterial retention increases with the bubble point (Fig. 2.22). It is ap-

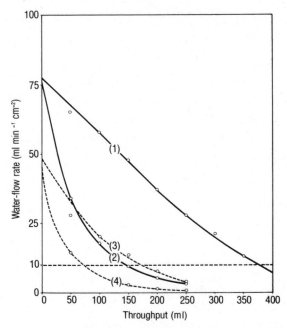

FIGURE 2.21. Effect of anisotropy upon flow decay of 0.45-μm highly anisotropic and isotropic mixed-ester membranes (from Kesting et al.[97]).

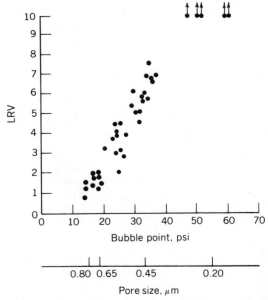

FIGURE 2.22. Effect of bubble point on the passage of *Pseudomonas diminuta*. A point with an arrow indicates no passage. The ordinate depicts the log reduction value (LRV) (from Sladek and Leahy[98]).

TABLE 2.12 AIR AND WATER PERMEABILITIES AND FILTRATION CAPACITIES OF VARIOUS MICROFILTRATION MEMBRANES[a]

Membrane	Bubble Point (psi)	Air Permeability $(L/min\ cm^2)$		Water Permeability $(mL/min\ cm^2)$		Filtration Capacity (throughput)[b]
0.2 μm						
Highly anisotropic mixed ester	59	6.28	2.75[c]	39.6	2.54[c]	4.40
Conventional mixed ester	46	2.28	1.00	15.6	1.00	1.00
0.45 μm						
Highly anisotropic mixed ester	35	12.3	1.82	76.3	1.55	2.14
Conventional mixed ester	32	6.74	1.00	49.1	1.00	1.00
0.8 μm						
Highly anisotropic mixed ester	16	40.4	1.64	230	1.95	1.50
Conventional mixed ester	16	24.5	1.00	118	1.00	1.00

[a]From Kesting et al.[97]
[b]Normalized relative to the values for conventional mixed-ester membranes after 80% flow decay.
[c]Values in this column are normalized relative to the value for conventional mixed-ester membranes at 10 psid (corrected).

parent that microfilters with rated pore size greater than those of the bacteria to be filtered can be used effectively, provided that the bacterial content of the material to be filtered is sufficiently low. The principal reason for this is the large number of cell layers most of whose apertures are considerably smaller than the rated pore size. The less critical 0.45-μm-rated membranes are challenged by *Serratia marcesens,* a 0.5-μm bacterium. Because the ability of a membrane filter to sterilize, that is, to provide sterile filtrate, increases with decreasing feed concentration and membrane thickness, and decreases with increased pressure, loading, and membrane area, it was necessary to establish standard conditions for qualifying the efficiency of sterilizing membrane microfilters. A challenge of 10^7 colony forming units (CFUs) of *Pseudomonas diminuta*/cm^2 of membrane area, at a pressure differential of 30 psi, must result in a sterile filtrate. Filtrate sterility is established by the absence of turbidity in the nutrient medium filtrate after three days incubation at 37°C. The passage of a single CFU is sufficient to effect turbidity under these conditions because of the exponential growth rate of these bacteria.

A membrane which has been utilized for filtering a large volume of solution and/or for a long period of time will become loaded with bacteria which can multiply and "grow through" the membrane even as filtration is under way. For this reason, in critical applications, such as the sterilization of pharmaceuticals and large-volume parenteral (LVP) (for injection) solutions, the filters are routinely replaced after 12 h of operation, regardless of how great the residual filtration capacity may be.

The broader the pore-size distribution, the greater will be the effect of membrane area upon challenge results. It is statistically more probable that a cartridge containing a square meter (10^4 cm^2) of membrane surface will contain both occasional pores and the necessary tortuous conduits which must accompany them to permit the passage of a single CFU than it is that a 1-cm^2 disk will. For all these reasons, the efficacy of a given filter combination must be established empirically under actual operating conditions. Filter characteristics such as bubble point, forward flow rate (air diffusion at sub bubble-point pressure), and the like should be considered as no more than effective indices of sterilization efficiency. This is particularly true for systems which are operating under conditions of high stress. Ironically, although the challenge organism is the final arbiter of sterilization efficiency, its size is not constant but will vary from strain to strain. It should be borne in mind, however, that the undeniable success of membrane filters at achieving sterile filtration is not due to the existence of "absolute" filters (a meaningless concept which should be discarded), but rather to the design of filters with a large statistical overkill. Filters with LRVs of > 10 are routinely utilized where LRVs of only 7 are required.

2.3.2 Ultrafiltration

The term *ultrafiltration* (UF) has changed its meaning over the years. At the time of Ferry's[99] 1936 review article, UF referred to filtrative separation of particles in the colloid-size range and included both UF and what we now designate as MF.

In ultrafiltration the dispersed phase ("solute" in its most general sense) passes through the membrane less readily than the "solvent" for one of several reasons[99, 100]:

1. It is adsorbed in the surface of the filter and its pores (primary adsorption).
2. It is either retained within the pores or excluded therefrom (blocking).
3. It is mechanically retained on top of the filter (sieving).

Inasmuch as the desired effect in ultrafiltration is sieving, primary adsorption and blocking should be eliminated as completely as possible.

For the ideal case, in which adsorption and blocking are absent, Manegold and Hofmann[101] defined as sieving constant ϕ such that

$$\phi = C_f/C_s$$

where C_f is the concentration of a small volume of filtrate at a given moment and C_s is the simultaneous concentration of the filtering solution. In a closed system, both C_f and C_s increase with time when $0 < \phi > 1$. The amounts of solute and solvent which permeate through the membrane depend on statistical considerations, even when there is a large size difference between the two. ϕ can be calculated as a function of the ratio of pore size to particle size[99] assuming that:

1. Adsorption and blocking are absent.
2. The membrane structure is represented by cylindrical pores penetrating the membrane at right angles to both surfaces (not valid except for irradiation track membranes).
3. Solute particles are constant in size and shape, travel vertically toward the membrane surface, and permeate only when they are wholly within the pore walls.
4. The vertical velocity has a parabolic distribution across the capillary in accordance with viscous flow.
5. The solution above the membrane is homogeneous.

In such a case

$$\phi = 2 \left(1 - \frac{R}{r}\right)^2 - \left(1 - \frac{R}{r}\right)^4$$

When ϕ is plotted versus the logarithm of the ratio r/R (Fig. 2.23), a theoretical curve is obtained which resembles the experimental end-point curves for normal filtration (Fig. 2.24a). When primary adsorption or blocking is present, the curves assume an abnormal shape (Fig. 2.24b). The extent to which primary adsorption

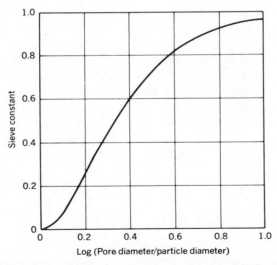

FIGURE 2.23. Theoretical curve for the sieve constant as a function of the logarithm of the ratio of pore diameter (from Ferry[99]; reprinted with permission from *Chemical Review*, © 1936 American Chemical Society).

and/or blocking occur depends on such factors as solution concentration, volume filtered, and the strength of membrane–solute interactions. Primary adsorption capacity is met most rapidly for solutions of high initial concentration, high filtration pressures, thin membranes, and in the presence of surfactants. When the pore diameter is much wider than that of the solute particles, curves of type I result. If the pore size is of the same order of magnitude as the particle size, curves of types II to IV may result. Initial low values of C_f are due to primary adsorption of the solute on the membrane or within its pores. After the membrane's exterior and interior surfaces have been coated with a layer of solute, the solute either appears in the filtrate in undiminished concentration (curve I) or C_f levels off or slowly increases (Fig. 2.24 a). In abnormal filtration where blocking occurs, C_f will reach a maximum, after which it will decline (Fig. 2.24 b).

The concentration of the residual solution, of course, remains unchanged where pore size greatly exceeds that of the solute. Insofar as sieving occurs, the concentration in the residue will increase. This also occurs with blocking, particularly if blocking is occasioned by the presence of foreign particles. Blocking can also result from the precipitation of solute at the membrane–solution interface. When there is a marked primary adsorption from a small amount of a dilute solution, residue concentration may be diminished.

As filtration pressure is increased, the zone of primary adsorption is narrowed by shearing away all but the most tightly sorbed layers.[99] Another means of limiting primary adsorption is to include surfactants whose surface activity causes them to preferentially coat the membrane surface (Fig.2.25).

Blocking (Fig.2.25 c) occurs most readily at high solution concentrations and high pressure; blocking is much less pronounced in dialysis, where there is no applied pressure. It increases with increasing membrane thickness and the presence of foreign particles. As with primary adsorption, the presence of surfactants limits blocking. The effect of surfactants in increasing permeability in ultrafiltration is to be contrasted with their opposite effects in hyperfiltration. Whereas they coat pore walls in ultrafiltration, thereby increasing lubricity, they fail to permeate the tighter HF membranes and by remaining at the membrane–solution interface in effect function as surfactant liquid membranes in series with the underlying solid (gelatinous) membranes (Chapter 9).

Although the pressure-driven processes which are utilized to separate particles

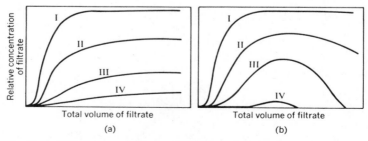

FIGURE 2.24. Typical filtration curve: (a) normal filtration and (b) abnormal filtration (from Ferry[99]); reprinted with permission from *Chemical Review*, © 1936 American Chemical Society).

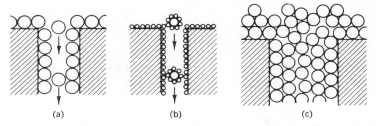

(a)   (b)   (c)

FIGURE 2.25. Schematic representation of normal and abnormal filtration (from Ferry[99]; reprinted with permission from *Chemical Review*, © 1936 American Chemical Society).

on the basis of size are viewed as a continuum, they may usefully be distinguished from one another. Ultrafiltration occupies that portion of the spectrum which separates MF from HF. Ultrafiltration begins (MF ends) at the point where the size of the solute particle is approximately 10-fold greater than the size of solvent molecules. Perhaps the most significant difference between HF and UF is that the former treats solutions of low MW and high concentration where osmotic pressure is appreciable, whereas the latter treats solutions of high MW and low concentrations so that osmotic pressure is negligible. The dividing line between UF and MF is also arbitrary. Owing to improvements in scanning electron microscopy the "dense skin" layer of UF membranes has recently been shown to be porous (Fig. 2.26). The tighter character of UF relative to MF is evident in the smaller pore size, higher pore density, and lower porosity of the former (Table 2.9). Operating pressures also tend to be intermediate between those utilized for MF and HF.

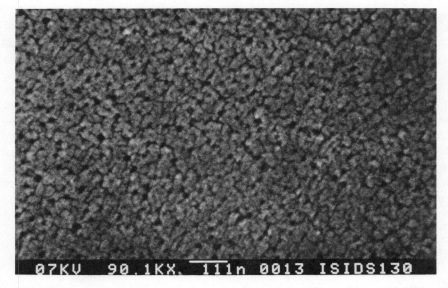

FIGURE 2.26. SEM photomicrographs of pores in skin surface of UF membranes with MW cutoff 10,000 ($\times 10^5$) (from Zeman[102]).

Ultrafiltration is the realm of macromolecular solutes of both synthetic and biological origin. Purification and, to a lesser extent, fractionation of macromolecules can be accomplished by UF. In the former, impurities of low MW (microsolutes) are permitted to permeate the membrane while the macromolecules are retained. In this case, UF, although pressure-driven, fulfills a role similar to that of dialysis which is concentration-driven. Fractionation involves the separation of macromolecules of various sizes.

The fractionation efficiency of UF is hindered by a number of factors such as interaction between macromolecules, the formation of a boundary layer of increased concentration at the interface between the membrane and the feed solution, and membrane–solute interaction. Boundary-layer buildup is a natural consequence of concentration polarization which results from the preferential loss of solvent from the solution at its interface with the membrane. Because of solute–solute and membrane–solute interaction, this boundary layer is sometimes irreversible and constitutes an insoluble *gel layer* which dramatically alters the surface of the original membrane. In such cases, the functional UF membrane is actually a composite consisting of the original membrane and the gel layer in series. An important consequence of this phenomenon is that the high hydraulic permeabilities which are obtained with prefiltered solvent are diminished 10-fold in operation with solutions to values which are not significantly greater than those obtained with the much tighter HF membrane. Another is that the fractionation efficiency of UF membranes is generally low. In the case of protein separations, for example, it has been suggested[103] that a minimum of an order-of-magnitude difference in MW is necessary to separate two protein fractions from one another. Even granted this size difference—as in the case of albumin and globulin—no fractionation may occur. Fractionation of polyethylene glycols in the MW range 600–6000 has been achieved utilizing cellulose acetate, polysulfone, and aromatic polyamidehydrazide membranes.[104] However, this was done utilizing very dilute (100 ppm) solutions. When concentrated solutions are employed, the results are much less promising.

The primary applications of UF are and will probably remain in the purification of macromolecular solutions and colloidal suspensions by the selective permeation of microsolutes. Examples are the recovery of colloidal paint particles from spent electrolytic paint particle suspensions, the recovery of lactose-free protein from whey, and the filtration of water for removal of colloidal foulants prior to its demineralization by hyperfiltration or ion exchange. In the first of these applications, it was found that nonionogenic polysulfone UF membranes, although effective when applied to suspensions of negatively charged colloids, fouled rapidly upon exposure to suspensions containing positively charged particles. The solution to this problem was the development of large-diameter capillary polycationic membranes which, by virtue of their repulsion of particles of like charge, successfully resisted fouling (Fig. 2.27).

2.3.3 Hyperfiltration

Hyperfiltration or *reverse osmosis* lies at the tight end of the pressure-driven membrane separation process spectrum. As a result, membrane pore size is smaller,

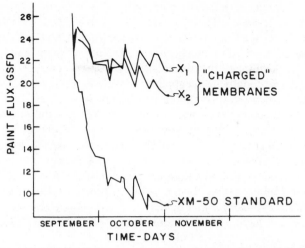

FIGURE 2.27. Comparative data showing the flux versus time performance of the XM-50 membrane and that of the newly developed "charged" membranes on cathodic paint. The designations X_1 and X_2 refer to different "charge density" levels (from Breslau et al.[105]).

porosity lower, and pore density higher than for UF and MF membranes (Table 2.9). These properties enable HF membranes to retain microsolutes (including ions) whose size is less than 10 Å. The small size and formula weights of these particles in turn means that their concentrations, even if modest when expressed as percent by weight, tend to be appreciable in molar terms.

According to van't Hoff,[2] the osmotic pressure of a dilute solution can be described with an equation which is analogous to the ideal gas equation:

$$\pi V = nRT$$

where π is the osmotic pressure, V is the solvent volume, n is the number of moles of solute, R is the gas constant, and T is the absolute temperature. Since $n/V = C$, the molar-concentration-of solute equation may be rewritten

$$\pi = RTC$$

However, certain effects, such as solute–solute interaction, reduce the activity of the individual solute particle below unity. To correct for this, C is multiplied by the osmotic coefficient g which yields

$$\pi = gRTC$$

In other words, at a given temperature, π is proportional to the molar concentration of solute.

The osmotic pressure of seawater ($\sim 3.5\%$ NaCl) is approximately 350 psi at 25°C. Therefore, if a semipermeable membrane separates seawater from a freshwater reservoir, the concentration gradient will cause water to permeate the membrane to dilute the seawater. This migration of a solvent across a semipermeable barrier from a less concentrated to a more concentrated solution is known as *osmosis*. If pressure is applied to the concentrated solution, the passage of water will be diminished. If the applied pressure equals the osmotic pressure, no net flow of water will take place. Finally, if the applied pressure exceeds the osmotic pressure, the osmotic flow will be reversed and a net flow of water from the more concentrated to the less concentrated solution will occur. This is the process known as *reverse osmosis** or hyperfiltration. Any pressure in excess of the osmotic pressure is known as the *effective* pressure and constitutes the driving force for solvent (water) transport. Permselectivity in hyperfiltration is usually expressed in one of the two ways: percent solute rejection (or retention), that is, 100 × [(solute concentration in feed minus solute concentration in product)/solute concentration in feed)]; or the *solute-reduction factor,* that is, solute concentration in feed/solute concentration in product. The latter description offers greater sensitivity when working with highly permselective membranes.

The tremendous impetus given to desalination with the invention of the integrally-skinned cellulose acetate membrane by Loeb and Sourirajan,[106] moved hyperfiltration to the forefront of interest in membrane separation processes during the period 1960–1980. It is not surprising, therefore, that a considerable amount of emphasis has been placed upon functional descriptions of this process. Among the many models which have been proposed are the preferential-sorption–capillary-flow model of Sourirajan,[107] and the solution–diffusion model of Lonsdale *et al.*[108]

The interfacial tension σ of a solution and the adsorption Γ of a solute at an interface are related[109,110]

$$\Gamma = -\frac{1}{RT}\left(\frac{d\sigma}{d\ln a}\right)$$

where R is the gas constant, T is the absolute temperature, and a is the activity of the solute. For aqueous sodium chloride solutions, there is a negative adsorption of the solute resulting in a monomolecular layer of pure water at the air–solution interface.[109-112] The thickness t of this layer at the air–solution interface is[109]:

$$t = -\frac{1000\alpha}{2RT}\left(\frac{d\sigma}{d(\alpha m)}\right)$$

where α is the activity coefficient of the salt in solution, and m is the molality of the solution. The thickness of the pure water layer at other interfaces, such as a membrane–solution interface, will vary with the chemical natures of both solute and interface (Chapter 4). If the interface contains pores whose diameter $\leq 2t$ (Fig.

---

* Reverse osmosis (RO) is the preferred term in the United States. However, the author is using the term hyperfiltration (HF) in this volume to stress the relationship between it and the other closely related pressure-driven processes.

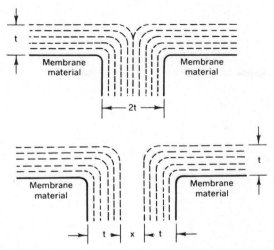

FIGURE 2.28. Diagramatic representation of ordered water sheath (thickness t) in the vicinity of a pore in a hydrophilic membrane (from Sourirajan[107]; reprinted with permission from *Industrial Engineering Chemistry (Fundamentals)*, © 1963 American Chemicals Society).

2.28), then the application of pressure in excess of the osmotic pressure will cause the freshwater layer to permeate the membrane, leaving behind a more concentrated salt solution. The strength of this model is its ability to encompass solution properties as well as the chemical and physical properties of the membranes. Variations in permeability and selectivity can be accommodated on the basis of the existence of pores of various sizes. In fact, it is now held[113,114] that two size distributions of pores exist: Numerous small pores which approximate the $2t$ value (≤ 10 Å) thought to exist in perfect or ideal membranes and occasional large (≥ 100 Å) pores which are attributable to the inevitable existence of defects in the skin layer of integrally-skinned membranes (Chapter 7).

There are two separate but related physical explanations for the existence of a solute-free layer of water within the membrane pores. The first focuses upon the direct repulsion of ions by the membrane polymer[113,115,116] and the second upon the indirect repulsion of ions by the influence of pore (capillary) walls upon the water which they contain.[117-120] Glueckauf and Sammon,[113] Bean,[115] and Yu et al.[116] have analyzed the electrostatic force which causes ions to avoid a region of low dielectric constant. The dielectric constant ϵ is a macroproperty which is derived from the various polar and nonpolar groups in the bulk polymer and bears a simple relationship to the solubility parameter δ; $\epsilon = \delta/3.5$. The separation of ionic solutes is inversely related to the dielectric constant of the membrane polymer and to pore size, and directly related to pressure. According to Bean,[115] for a membrane with a dielectric constant of 3 and a dilute solution of 1–1 electrolyte, solute separations of 90, 99, and 99.9% could be obtained from 27-, 13.5-, and 9-Å-diam pores. For 99% solute separation through the same membrane, the pore size could be 27 Å for a 2–1 electrolyte and 40 Å for a 2–2 electrolyte. On the other hand, for a membrane polymer with a dielectric constant of 10- and 27-Å pores, the solute separations for 1–1, 2–1, and 2–2 electrolytes are 60, 83, and 97%, respectively.

Eisenman[117] explained the selectivity of membranes for different ions on the basis of the relative free energy of interaction between water and membrane sites and between water and dissolved ions. Schultz and Asunmaa[118] considered the special case of Poiseuille flow through membrane pores which have the ideal critical diameter (2t) for desalination given by

$$J = \frac{n\pi P(r_{\text{eff}})^4}{8\eta L\tau} \tag{5}$$

where J = permeability of ordered water,
   n = pore density,
   P = effective pressure,
   r_{eff} = effective pore radius,
   η = average viscosity of ordered water,
   L = thickness of skin (resistance) layer of membrane, and
   τ = tortuosity factor.

On the basis of electronmicrographs of the skins of cellulose membranes, the pore radius was estimated as 18.5 Å (too high by a factor of ~4) and the pore density as ~6.5 × 10^{11}/cm² (too low perhaps by a factor of 4). These inaccuracies tend to offset one another and do not detract significantly from their argument.

The tortuosity factor τ of 2.5 is equivalent to the random mesh of noncircular, tortuous, interconnecting pores whose flow properties were analyzed by Carman.[121] If Eq. (5) is solved for η, a value of 0.35 P is obtained for the viscosity of the ordered water in a "tight" cellulose acetate membrane at 23°C. This viscosity is about 37 times greater than the value of 0.00936 P for ordinary water at this temperature. (If it is assumed that the ϵ of water is proportional to its viscosity, this leads to a value of 2.12 for the ϵ of ordered water.) The activation energy for viscous flow of supercooled water with this viscosity, calculated from the empirical equation

$$E_v = 1020 \left(\frac{T'}{T' - 150} \right)^2 \frac{\text{cal}}{\text{mole}}$$

equals 6015 cal/mole, almost exactly matching the asymptotic experimental value of 5959 cal/mole reported by Vincent[122] (Fig. 2.29) for permeation of water through an integrally-skinned cellulose acetate membrane with 53.5% water content and 99.5% salt retention. As the pore radius decreases as the result, for example, of annealing, the average viscosity of water increases and with it resistance to flow. Vincent et al.[123] observed that the capacity of the water within such a membrane to solvate cobalt chloride, a salt whose various hydrates have different colors, increases with increasing water content. Thus as the pore radius exceeds a certain value, some ordinary (ion-solvating) water penetrates into the center of the pore channel.

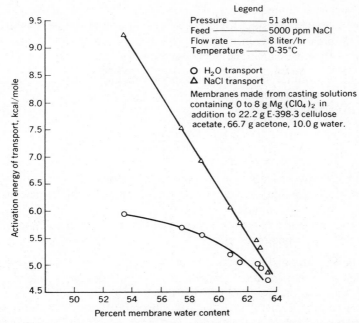

FIGURE 2.29. Activation energy of transport as a function of membrane-water content (from Vincent[122]).

As stated above, the efficiency of hyperfiltration varies with the nature of the ionic solute. Ions which have a high charge density (small radius and/or high charge) tend to be more strongly hydrated than those with a low charge density. Moreover, as crystal radii increase, a point is reached beyond which hydration no longer takes place. This accounts for the fact that selectivity increases in the series $K^+ < Na^+ < Li^+$, whereas

$$Cs^+ > Rb^+ > K^+$$

The solution–diffusion model was originally employed to explain the permeation of gases, vapors, and organic liquids through dense but defect-free membranes. It was subsequently extended to hyperfiltration by Lonsdale et al.[108]

Water permeability J is a function of both diffusion and solubility factors:

$$J_1 = -D_1 C_1 \frac{\overline{V}_1}{RT} \frac{\Delta P - \pi}{d} \tag{6}$$

where J_1 = flux of water (cm/s),
  D_1 = diffusion coefficient of water in the polymer (cm^2/s),
  C_1 = average concentration of water solvent in the polymer (g/cm^3),
  $\overline{V}_1$ = partial molal volume of water in the polymer,
  R = gas constant,

T = absolute temperature,

$\Delta P - \pi$ = applied pressure difference across the membrane minus the osmotic pressure difference, and

d = effective membrane thickness (cm).

$D_1 C_1$ can be determined from osmosis experiments and C_1 from sorption data of water in the dense membrane.

The permeability of solute

$$J_2 = D_2 K \frac{\Delta C_2}{d} \tag{7}$$

where D_2 = diffusion coefficient for salt in the membrane,

K = distribution coefficient for solute between the membrane and external solution, and

ΔC_2 = difference in concentration of salt in the exterior solution and in the membrane.

K is to a first approximation, independent of pressure. The diffusion coefficients are obtained by immersing dense membranes of known thickness in a solution and measuring solute sorption as a function of time.

With the help of Eqs. (6) and (7), the "theoretical" solute rejections at any given pressure and solute concentration should be calculable. The values so obtained rarely, if ever, match the values obtained under hyperfiltration conditions and this is attributable to the nonequivalence of the structure of dense membranes and the skin layers of integrally-skinned hyperfiltration membranes.

In contrast to integrally-skinned membranes which *inevitably* contain defects owing to occasional incomplete coalescence of micelles in the skin layer (Chapter 7), dense films can be made free of defects. Furthermore, the glassy state is not an equilibrium condition and defect-free films of varying density can be prepared in which the salt diffusion coefficients vary by an order of magnitude.[124] Among the more obvious factors which can influence the density of films in the glassy state are the relative strengths of polymer–solvent and polymer–polymer interactions within the solutions from which the films are cast, the rate of desolvation, and environmental conditions (Chapters 5–7). Therefore, there can be no "theoretical" salt rejection for any integrally-skinned membrane and the solution–diffusion model can be applied quantitatively only to defect-free dense films prepared under set conditions.

Various attempts have been made to salvage the solution–diffusion model by accommodating imperfections. The equations of Sherwood et al.[125]:

$$J_{H_2O} = k_1 (\Delta P - \Delta \pi) + k_3 \Delta P$$

$$J_{salt} = k_2 \Delta \pi + k_3 \Delta P C_0$$

where k_3 is a constant associated with flow through imperfections, were found by Applegate and Antonson[126] to adequately describe the transport behaviors of integrally-skinned cellulose acetate and aromatic polyamide membranes. However, the retention of even such a modified solution–diffusion model for hyperfiltration appears inadvisable since not only are its basic premises false, but its predictions lead to false conclusions and it provides no direction for future research.

In contrast, the preferential sorption–capillary-flow mechanism leads to the following specific conclusions with respect to the physical and chemical prerequisites for hyperfiltration membranes and membrane polymers.

1. There must exist a maximum number of pores with a narrow size distribution centered around a value of $2t$. This value will, however, vary with the nature of the ionic solute. To a first approximation $2t$ will increase as the diameters of hydrated ions increase.

2. The number of pores with a diameter $>2t$ must be minimal.

3. The permeation of ionic solutes is inversely proportional to the dielectric constant ϵ of the water contained within the dense skin layer of the membrane. Since the ϵ of the membrane polymer undoubtedly is an important contributing factor to the value of ϵ of the water within the skin layer, it too is of importance to the permeation of ionic solutes. Note too that the effects of pressure and temperature are consistent with this hypothesis. The rejection of ionic solutes increases with pressure. Water content decreases with pressure and thus results in an effective decrease in the ϵ of membrane water as well. Increasing temperature, on the other hand, acts to increase the ϵ of capillary water and hence to decrease solute rejection.

4. The affinity of water for the walls of the pores and capillaries in the skin layer and with it water permeability will tend to decrease as the dielectric constant of the membrane polymer decreases. Furthermore, the size of water clusters will also tend to increase with decreasing polymer ϵ, and this will act to further decrease permeability. Both solute and water permeation decrease with the ϵ of polymer. However, since water permeability decreases at a faster rate, an optimum value of ϵ will exist for a membrane with given pore statistics.

Several additional requirements for hyperfiltration membranes and/or polymers may be added as a result of empirical observations:

5. The skin layer should be in the glassy amorphous state. Crystallites are impermeable and will serve only to reduce effective pore density. Small, that is, $<2t$, pores are probably slightly skewed displacements or kinks between neighboring polymer chain segments. Large, that is, $>2t$, pores are domains where micelles have failed to coalesce completely (Chapter 7).

6. The polymer must have a high modulus to resist creep or compaction and attendant water loss with time. This in turn requires a high value of T_g which in turn suggests the need for ring structures, such as the anhydroglucose units of cellulose acetate or the aromatic moieties of aromatic polyamides. The purpose of this requirement is to guarantee long-term operation with minimal loss of permeability.

7. The polymer must have a narrow MW distribution and, consistent with processing requirements, a high MW. These parameters bear strongly upon membrane integrity (skin *and* substructure) and mechanical properties such as toughness and flexibility (Chapter 4).

8. The membrane should exhibit sufficient hydrolytic stability as to permit continuous operation for three to five years. For cellulose acetate this requirement is met by acidification of the feed so as to control the pH between 4.5 and 6.0. Several noncellulosic polymers, notably the aromatic and fully aromatic polyamides, and the sulfonated polysulfones, exhibit the requisite hydrolytic stability without the need for pH adjustment.

9. The membrane must be resistant to attack by microorganisms. For cellulose acetate (CA) this is accomplished by dry storage, prefiltration of chlorinated feed solutions, and the incorporation into CA of ester groups with terminal quaternary ammonium groups.[127,128] Polyamide and other noncellulosic membranes tend to be nonbiogradable.

10. The polymer should be resistant to degradation by chlorine or other oxidants, which are added to the feed to control growth of microorganisms which act to degrade and/or foul the membranes. Cellulose acetate will tolerate the continuous presence of 1-ppm chlorine in the feed. Most polyamides, with the possible exception of those such as the polypiperazinamides (Chapter 4) which are prepared utilizing di-secondary amines, are subject to attack by chlorine either at the —NH— group (N-chlorination) or to halogenation of the aromatic rings.[129] In such cases pretreatment must include a dechlorination procedure.

11. When fouled, the membrane must be capable of undergoing repeated cleaning cycles. Prevention is the best cure for fouling. In hyperfiltration, as for other pressure-driven separation processes, prefiltration of the feed is necessary to guarantee a useful membrane life. Among the principal foulants which are removed by prefiltration are colloidal particles of both inorganic and organic types. Examples are silica, iron, both living and dead microorganisms, and organic debris. Fortunately, the colloidal size of these particles result in solutions whose turbidity is a convenient measure of foulant concentration.

Prefiltration is utilized to lower the turbidity of the feed to ~1 JTU (Jackson turbidity unit). For feeds of high turbidity the cost of pretreatment (including prefiltration) can be as high as the cost of hyperfiltration itself. Deep well waters tend to be clear and are therefore the preferred feeds for hyperfiltration; turbid surface waters are less desirable.

To date, the principal applications of hyperfiltration are desalination of sea and brackish waters for human consumption, the demineralization of substandard tap water for human consumption, and the preparation of high-purity water from tap water for industrial and medical applications.

Many studies have been made of the retention of organic solutes by hyperfiltration membranes. In general polyamide membranes are less permeable to organic solutes than are cellulose acetate types. The reason for this is believed to be the greater solubility of these solutes in cellulose acetate than in polyamide. Since the solubility of organic compounds in a polymer is related to the solubility of the polymer in organic solvents, the reader is referred to Chapter 5 for a discussion of the relevant factors.

Because few if any commercial applications have been found for hyperfiltration of solutions of organic solutes in either aqueous or organic solvents, this subject will not be considered further here.

Consider the case of a hyperfiltration membrane which is perfectly semipermeable. Both water and solute are brought to the membrane, but only water can permeate. Since the solute cannot permeate, its concentration at the membrane–solution interface increases. Theoretical analysis of this concentration–polarization effect at the surface of hyperfiltration membranes have been presented by Brian,[130] Sherwood et al.,[131] Gill et al.,[132] Dresner,[133] and others.

The concentration–polarization (boundary-layer) effect can influence material transport in several ways:

1. The osmotic pressure at the membrane–solution interface is increased because of the increase in solute concentration near the membrane surface over that in the bulk solution. Water permeability therefore decreases since the effective pressure (the applied pressure $P - \pi$) is decreased.

2. Since the fraction of solute retained by the membrane is, to a first approximation, independent of concentration, more solute will permeate because the interfacial concentration is higher than it would be in the absence of a boundary layer.

3. If the interfacial concentration of such solutes as are capable of forming insoluble precipitates exceeds a certain value, surface scaling will occur. Such a development may lead to physical blockage.

Concentration–polarization becomes increasingly significant with increasing solution concentration, permeability, and permselectivity. Although it cannot be completely eliminated, it can be minimized by: (1) agitation of the fluid layer by passage of the bulk solution past the membrane surface at velocities where turbulence (Fig. 2.30) develops; (2) utilizing entrance effects in thin channels in the laminar-flow region; and (3) the insertion of baffles in narrow channels to promote turbulence.

The fraction of the feed which is recovered as usable product is known as the *recovery.* It varies from about 0.33 for seawater to as high as 0.9 for low-salinity brackish waters. Recovery is an important consideration for large plants where the energy costs associated with pressurization and circulation of feed are a significant fraction of the overall cost. It is less significant for operation at tap water (< 100 psi) pressures where circulation is not needed and the raw-water cost is negligible. In such cases recoveries as low as 0.1 are acceptable.

Variations in temperature have a great many largely interrelated effects upon hyperfiltration as well as upon other membrane processes. In addition to the increased motion of solvent and solute molecules and decreasing interactions between the two, membrane structure can undergo temperature-dependent changes on the molecular, supermolecular, and colloidal levels. The net effect of increasing temperature, particularly above the T_g, is to decrease the porosity and, with it, the water content of the membrane gel (Chapter 7). Membrane–solvent and membrane–solute interactions are also affected. It should be noted, however, that solubility can either increase or decrease with increasing temperature. It will decrease for substances (such as noncondensible gases) whose solubility decreases with increasing temperature, and it will increase where the reverse is true.

For dense membranes like those employed in gas separations and, to a first approximation, the skin layer of integrally-skinned hyperfiltration membranes, the temperature dependence of the permeability constant P can be expressed as an Arrhenius relationship in which

$$P = P_0 e^{-E_p/RT}$$

Here E_p is the activation energy, and the P_0 is the factor associated with the overall permeation process. Since

$$P_0 = S_0 D_0$$

where S is the solubility of permeant species in the membrane and D is the diffusion coefficient, S and D can also be expressed as Arrhenius functions with

$$S = S_0 e^{-H_s/RT} \quad \text{and} \quad D = D_0 e^{-E_D/RT}$$

where H_s is the heat of solution and E_D the activation energy for the diffusion process. In most cases permeability increases and activation energy decreases with increasing temperature. Exceptions do, however, occur, as when permeant solubility in the membrane and temperature are inversely related.

In the transport of water and salt across integrally-skinned cellulose acetate membranes in hyperfiltration, activation energies decrease with increasing membrane water content (Fig. 2.30). The effect of increasing temperature is to decrease the size of water clusters (Table 2.13) which in turn acts to increase the ϵ of ordered water. The extent of hydrogen bonding, and with it the proportion of more clustered (less highly permeable) water units, decreases with increasing temperature (Table 2.14).

Drost-Hansen[136] has cited evidence for the existence of thermal anomalies or kinks in the structure of water near 15°C and at approximately 15° intervals thereafter. Similar kinks exist for the properties of aqueous solutions, all or most of which are presumably attributable to decreasing interaction with water with in-

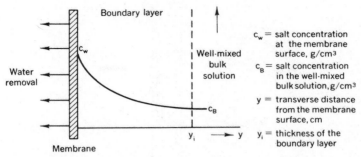

FIGURE 2.30. Concentration profile in the boundary layer for well-developed turbulent flow.

The labels in the figure read:

c_w = salt concentration at the membrane surface, g/cm^3

c_B = salt concentration in the well-mixed bulk solution, g/cm^3

y = transverse distance from the membrane surface, cm

y_i = thickness of the boundary layer

Boundary layer

Well-mixed bulk solution

Water removal

Membrane

TABLE 2.13 AVERAGE SIZE OF WATER CLUSTERS AS A FUNCTION OF TEMPERATURE[a]

	Number of Molecules per Cluster	
$T_1(°C)$	$H_2O(l)$	$D_2O(l)$
0	130	140
20	90	100
90	45	40

[a]From Prengle.[134]

TABLE 2.14 MOLE FRACTIONS OF SPECIES WITH NO, ONE, AND TWO BONDS IN LIQUID WATER[a]

$T_1(°C)$	C_0	C_1	C_2	Hydrogen Bonding (%)
5	0.27	0.42	0.31	52
10	0.29	0.42	0.29	50
20	0.31	0.42	0.27	48
30	0.33	0.42	0.15	46
40	0.34	0.43	0.23	45
50	0.36	0.43	0.21	43
60	0.38	0.42	0.20	41
70	0.40	0.42	0.18	39
80	0.42	0.42	0.16	37
90	0.44	0.41	0.15	36

[a]From Choppin.[135]

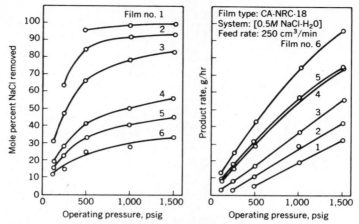

FIGURE 2.31. Effect of operating pressure on the separation and permeability characteristics of different films (from Sourirajan and Govindan[138]).

creasing temperature. The effect of temperature upon membrane separations of aqueous solutions is particularly complex in that membrane, solute, and water molecules themselves may exhibit varying affinities for water, so that a stepwise increase in kinetic energy may cause the departure of water from one species and not another.

Increasing pressure generally acts to increase the permeation rate of a given species through a membrane. The effects of pressure are mitigated by the mechanism of permeability, for example, by membrane structure, the various interactions between membrane and permeant species (and between the permeant species themselves), solute concentration, and the charge characteristics of membrane and solute. In addition, pressure in excess of the membrane's compressive yield point results in compaction and resultant loss of void volume (porosity) and permeability. Not only does the average porosity of the membrane vary with pressure, but the porosity may decrease with distance from the high-pressure side. Under a solution pressure of 1000 psi, for example, a 20-fold change in permeability across the membrane has been observed, with 50% of the pressure drop occurring in the last 20% of the membrane.[137] Pressurization causes a change in void dimension in the plane perpendicular to the surface, whereas heating causes shrinkage in all three dimensions. Both effects synergistically act to decrease porosity (water content).

Although, ideally, the product flux should increase linearly with pressure, in practice this is found only for the tightest membranes (films 1, 2, and 3 in Fig. 2.31 and even for these only transiently, since the membranes tend to compact rather substantially at pressures in excess of 750 psig). While salt permeability does not vary appreciably with pressure, product concentration decreases because of the increase in the flux of water which accompanies a given amount of salt.

Unlike distillation, where it is of relatively minor significance, feed quality is one of the primary factors influencing the economics of hyperfiltration. This comes

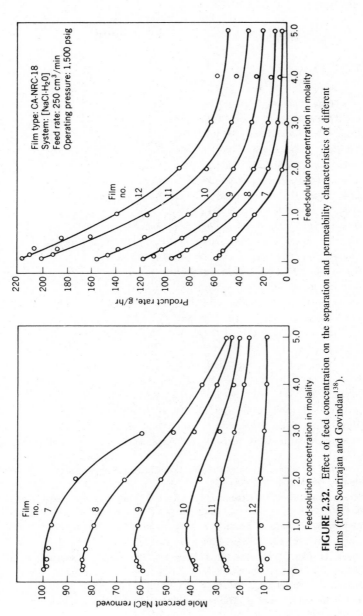

FIGURE 2.32. Effect of feed concentration on the separation and permeability characteristics of different films (from Sourirajan and Govindan[138]).

about both directly, as a result of effects upon permeability and permselectivity (Fig. 2.32), and indirectly, as a result of its influence upon the fraction of feed water recoverable as product.

REFERENCES

1. A. Fick, *Pogg. Ann.*, **94**, 59 (1855).
2. J. H. van't Hoff, *Z. Phys. Chem.*, **4**, 129 (1889).
3. G. Park, *Trans. Faraday Soc.*, **53**, 107 (1957).
4. M. Marel, Y. Aiakawa, and B. Proctore, *Mod. Packaging*, **28**,(8), 213 (1955).
5. D. Davis, *Paper Trade J.*, **123**(9), 33 (1946).
6. A. Lancock and B. Proctor, *TAPPI*, **35**, 241 (1952).
7. D. Brubaker and K. Kammermeyer, *Ind. Eng. Chem.*, **44**, 1465 (1953).
8. H. Todd, *Mod. Packaging*, **17**(4), 124 (1944).
9. D. Brubaker and K. Kammermeyer, *Ind. Eng. Chem.*, **45**, 1148 (1953).
10. C. Rogers, J. Meyer, V. Stannett, and M. Szwarc, *TAPPI*, **39**, 737 (1956).
11. H. Daynes, *Proc. Roy. Soc. London, Ser. A*, **97A**, 286 (1920).
12. R. Barrer, *Trans. Faraday Soc.*, **35**, 628 (1939).
13. A. Michaels and R. Parker, Jr., *J. Polym. Sci.*, **41**, 83 (1959).
14. A. Myers, C. Rogers, V. Stannett, and M. Szwarc, *Mod. Plastics*, **34**(9), 157 (1957).
15. G. van Amerogen, *J. Appl. Phys.*, **17**, 978 (1946).
16. A. Michaels, and R. Parker, Jr., *J. Polym. Sci.*, **41**, 83 (1959).
17. C. Klute, *J. Polymer Sci.*, **41**, 307 (1959).
18. C. Klute, *J. Appl. Polym. Sci.*, **1**, 340 (1959).
19. A. Michaels and H. Bixler, *J. Poly. Sci.*, **50**, 393 (1961).
20. A. Michaels, W. Vieth, and J. Barrie, *J. Appl. Phys.*, **34**, 13 (1963).
21. A. Michaels and H. Bixler, *J. Polym. Sci.*, **50**, 413 (1961).
22. A. Michaels, H. Bixler, and H. Fein, *J. Appl. Phys.*, **35**, 3165 (1964).
23. R. Baddour, A. Michaels, H. Bixler, R. de Filippi, and J. Barrie, *J. Appl. Polym. Sci.*, **8**, 897 (1964).
24. H. Bixler and A. Michaels, American Institute of Chemical Engineering, 53rd National Meeting, Pittsburgh, PA, May 17–20, 1964.
25. W. Krewinghaus, Sc.D.Dissertation, Massachusetts Institute of Technology, Cambridge, MA, April 1966.
26. A. Meyers, V. Tammela, V. Stannett, and M. Szwarc, *Mod. Plastics*, **37**(6), 139 (1960).
27. J. Henis and M. Tripodi, *Science*, **220**(4592), 11 (1983).
28. L. Pilato, L. Litz, B. Harectay, R. Osborne, A. Farnham, J. Kawakami, P. Fritze, and J. McGrath, *ACS Prepr.*, **16**(2), 42 (1975).
29. H. Hoehn, U.S. Patent 3,822,202 (1974).
30. W. Koros, Paper presented at 187th National Meeting, American Chemical Society , St. Louis, MO, April 9–14, 1984.
31. V. Stannett and H. Yasuda, in *Crystalline Olefin Polymers*, R. Raff and K. Doak, Eds., Pt. 2, Interscience, New York, 1965.
32. R. Binning, R. Lee, J. Jennings, and E. Martin, *Ind. Eng. Chem.*, **53**, 45 (1961).

33. R. Binning and F. James, *Petrol. Refiner,* **37,** 214 (1958).
34. R. Binning, R. Lee, J. Jennings, and E. Martin, *Prepr., Div. Pet. Chem., Am. Chem. Soc.* **3**(1), 131 (1958).
35. V. Stannett and H. Yasuda, *J. Polym. Sci. B* **1,** 289 (1963).
36. R. Kokes and F. Long, *J. Am. Chem. Soc.,* **75,** 6142 (1953).
37. E. Steigelmann and R. Hughes, U.S. Patent 4,047,908 (1977).
38. I. Cabasso, *Ind. Eng. Chem. Prod. Res. Dev.,* **22,** 319 (1983).
39. R. Lacey, in *Industrial Processing with Membranes,* R. Lacey and S. Loeb, Eds., Chap. 1 Wiley-Interscience, New York, 1972.
40. W. Pauli, *Biochem. Z.,* **152,** 355 (1924).
41. K. Spiegler, *Principles of Desalination,* Academic, New York, 1966.
42. T. Nishiwaki, Reference 39, Chap. 6.
43. J. Zang, Paper presented at the Membrane Processes for Industry Symposium, sponsored by Southern Research Institute, Birmingham, AL, May 19–20, 1966.
44. P. Kollsman, U.S. Patent 3,025,227 (March 13, 1962).
45. M. Baizer, *Tetrahedron Lett.,* **15,** 973 (1963).
46. F. Helfferich, *Ion Exchange,* McGraw-Hill, New York, 1962.
47. P. Mazur and J. Overbeck, *Rec. Trav. Chim. Pays-Bas,* **70,** 83 (1951).
48. J. Overbeck, *J. Colloid Sci.,* **8,** 420 (1953).
49. A. Stavermann, *Trans. Faraday Soc.,* **48,** 176 (1952).
50. E. Wiebenga, *Rec. Trav. Chim. Pays-Bas,* **65,** 273 (1946).
51. T. Teorell, *Proc. Soc. Exp. Biol. Med.,* **33,** 282 (1935).
52. K. Meyer and J. Sievers, *Helv. Chim. Acta,* **19,** 649 (1936).
53. E. Erickson, *Kgl. Lantbrucks-Högskol. Ann.,* **16,** 420 (1949).
56. R. Keynes, *J. Physiol.,* **114,** 119 (1951).
57. J. Kobatake, *Brusseiron Kenkyn,* **84,** 35 (1955); **87,** 58 (1955); CA **49,** 12918 (1955).
58. P. Lorenz, *J. Phys. Chem.,* **56,** 775 (1952).
59. J. Lorimer, E. Boterenbrood, and J. Hermans, *Disc. Faraday Soc.,* **21,** 141 (1956).
60. D. Mackay and P. Meares, *Kolloid Z.,* **167,** 37 (1959).
61. D. Mackay and P. Meares, *Trans. Faraday Soc.,* **55,** 1221 (1959).
62. D. Mackay and P. Meares, *Kolloid Z.,* **171,** 139 (1960).
63. J. Mackie and P. Meares, *Proc. Roy. Soc. London, Ser. A* **232,** 498, 510 (1955).
64. J. Mackie and P. Meares, *Disc. Faraday Soc.,* **21,** 11 (1956).
65. P. Meares and H. Ussing, *Trans. Faraday Soc.,* **55,** 142 (1959).
66. G. Scatchard, *J. Am. Chem. Soc.,* **75,** 2883 (1953).
67. R. Schlögl, *Z. Phys. Chem. (Frankfurt),* **1,** 305 (1954).
68. R. Schlögl, *Z. Phys. Chem. (Frankfurt),* **3,** 73 (1955).
69. R. Schlögl and U. Schödel, *Z. Phys. Chem. (Frankfurt),* **5,** 372 (1955).
70. G. Schmid, *Z. Elektrochem.,* **54,** 424 (1950).
71. K. Spiegler, *Trans. Faraday Soc.,* **54,** 1409 (1958).
72. H. Davson and J. Danielli, *The Permeability of Natural Membranes,* Chap. 21 and Appendix A, Cambridge University Press, New York, 1943.
73. K. Laidler and K. Schuler, *J. Chem. Phys.,* **17,** 851 (1949).
74. W. Katz and N. Rosenberg, U.S. Patent 2,694,680 (1954).
75. K. Sollner, *Z. Elektrochem.,* **36,** 234 (1930).
76. K. Sollner and A. Grollman, *Z. Elektrochem.,* **83,** 274 (1932).

77. A. Peers, *Disc. Faraday Soc.*, **21**, 124 (1956).

78. O. Kedem, J. Cohen, A. Warshawsky, and N. Kahana, *Desalination*, **46**, 291 (1983).

79. R. Kilburn, *Florida State Hort. Soc. Proc.*, **73**, 216 (1960).

80. R. Smith, C. Hicks, and R. Mosky, Conference on Convenience Foods and Dehydration, Philadelphia, PA, 1964.

81. J. Zang, R. Mosky, and R. Smith, Paper presented at the 57th National Meeting, American Institute of Chemical Engineering, Minneapolis, MN, Sept. 1965.

82. J.-G. Helmcke, *Kolloid Z.*, **135**, 29 (1954).

83. D. Pall and E. Kirnbauer, "Bacterial Removal Prediction in Membrane Filters," paper presented at the 52nd Colloid & Surface Science Symposium, University of Tennessee, Knoxville, TN, 1978.

84. H. Bechold, *Z. Phys. Chem.*, **64**, 328 (1908).

85. M. Cantor, *Ann. Phys. (Newfoundland)*, **47**, 399 (1892).

86. H. Bechhold, M. Schlesinger, and K. Silbereisen, *Kolloid Z.*, **55**, 172 (1931).

87. H. Knöll, *Kolloid Z.*, **86**, 1 (1939).

88. H. Knöll, *Kolloid Z.*, **90**, 189 (1940).

89. H. Knöll, *Zentr. Bakteriol. Abt. 1*, **143**, 479 (1939).

90. E. Honold and E. Skau, *Science*, **120**, 805 (1954).

91. H. Thiele and K. Hallich, *Kolloid Z.*, **163**, 115 (1959).

92. F. Erbe, *Kolloid Z.*, **59**, 32, 195 (1932).

93. W. Elford and D. Ferry, *Biochem. J.*, **28**, 650 (1934).

94. F. Erbe, *Kolloid Z.*, **63**, 277 (1933).

95. P. Grabar and J. Niktine, *J. Chim. Phys.*, **33**, 50 (1936).

96. C. Badenhop, A. Spann, and T. Meltzer, in *Membrane Science and Technology*, J. Flinn, Ed., Plenum, New York, 1970.

97. R. Kesting, A. Murray, K. Jackson, and J. Newman, *Pharm. Technol.*, **5**(5), 52 (1981).

98. K. Sladek and T. Leahy, *Proc. Second World Cong. Chem. Eng. (Montreal)*, **4**, 158 (1981).

99. J. Ferry, *Chem. Rev.*, **18**, 373 (1936).

100. S. Tuwiner, *Diffusion and Membrane Technology*, Reinhold, New York, 1962.

101. E. Manegold and R. Hofmann, *Kolloid Z.*, **52**, 19 (1980).

102. L. Zeman, personal communication (1984).

103. A. Michaels, in *Ultrafiltration Membranes and Applications*, A. Cooper, Ed., Plenum, New York, 1980.

104. K. Chan, T. Matsuura, and S. Sourirajan, *Ind. Eng. Chem., Prod. Res. Dev.*, **21**, 605 (1982).

105. B. Breslau, A. Testa, B. Milnes, and G. Medjanis, cited in Reference 103, p. 109.

106. S. Loeb and S. Sourirajan, UCLA Engineerging Report 60–60 (1960).

107. S. Sourirajan, *Ind. Eng. Chem. Fundam.*, **2**(1), 51 (1963).

108. K. Lonsdale, U. Merten, and R. Riley, *J. Appl. Polym. Sci.* **9**, 1341 (1965).

109. W. Harkins and H. McLaughlin, *J. Am. Chem. Soc.*, **47**, 2083 (1925).

110. J. McBain and R. Dubois, *J. Am. Chem. Soc.*, **51**, 3534 (1929).

111. A. Goard, *J. Chem. Soc.*, **127**, 2451 (1925).

112. J. Langmuir, *J. Am. Chem. Soc.*, **39**, 1848 (1917).

113. E. Glueckauf and D. Sammon, *Proc. Third Int. Symp. Fresh Water from the Sea (Dubrovnik)*, **2**, 397 (1970).

114. K. Chan, L. Tinghui, T. Matsuura, and S. Sourirajan (to be published).

115. C. Bean, OSW Research & Development Progress Report 465 (1969).

116. N. Yu., G. Dytnerskii, G. Polyakov, and L. Lukavyi, *Theor. Found. Chem. Eng.*, **6**(4), 565 (1972), Publication Consultants Bureau, New York, May 1973.
117. G. Eisenman, *Biophys. J.*, **2**, 259 (1962).
118. R. Schultz and S. Asunmaa, *Recent Progr. Surface Sci.*, **3**, 293 (1971).
119. C. Reid and E. Breton, *J. Appl. Polym. Sci.*, **1**, 133 (1959).
120. R. Kesting, A. Vincent, and M. Barsh, *J. Appl. Polym. Sci.*, **9**, 2063 (1965).
121. P. Carman, *Disc. Faraday Soc.*, **3**, 72 (1948).
122. A. Vincent, cited in *Saline Water Conversion Rept. for 1965,* p. 36, GPO, Washington, D.C.
123. A. Vincent, M. Barsh, and R. Kesting, *J. Appl. Polym. Sci.*, **9**, 2363 (1965).
124. C. W. Saltonstall, personal communication.
125. T. Sherwood, P. Brian, and R. Fisher, *Ind. Eng. Chem. Fundam.*, **6**(1), 2 (1967).
126. L. Applegate and C. Antonson, in *Reverse Osmosis Membrane Research,* H. Lonsdale and H. Podall, Eds., Plenum, New York, 1972.
127. R. Kesting, K. Jackson, and J. Newman, *Proc. Fifth Int. Symp. Fresh Water from the Sea,* **4**, 78 (1976).
128. R. Kesting, J. Newman, K. Nam, and J. Ditter, *Desalination,* **46**, 343 (1983).
129. J. Glater and M. Zachariah, 188th National ACS Meeting, Philadelphia, PA, August 26–31, 1984.
130. P. Brian, *Ind. Eng. Chem. Fundam.*, **4**, 439 (1965).
131. T. Sherwood, C. Tien, and D. Zeh, *Ind. Eng. Chem. Fundam.*, **4**, 113 (1965).
132. W. Gill, C. Tien, and D. Zeh, *Ind. Eng. Chem. Fundam.*, **5**, 367 (1966).
133. L. Dresner, ORNL Report 3621 to OSW (1964).
134. H. Prengle, cited in Reference 122, p. 16.
135. G. Choppin, cited in Reference 122, p. 15.
136. W. Drost-Hansen, *Ann. N.Y. Acad. Sci.*, **125**(2), 471 (1965).
137. P. Harriot and D. Michelson, OSW Research & Development Report 330 (April 1968).
138. S. Sourirajan and T. Govindan, Paper presented at the First International Symposium on Water Desalination, Washington, D.C., October 3–9, 1965.

3 MISCELLANEOUS USES OF MEMBRANES

The division of membrane applications into separation processes and miscellaneous uses does not imply the existence of nonseparative roles for the latter. Indeed all membranes are by their very nature semipermeable barriers. Instead, the distinction is made on the basis of generally perceived differences between the utilization of membranes in situations in which the separation itself is the principal focus and their uses where a separation is more or less incidental to the primary objective which may instead be of an analytical, containment, or metering nature. In some cases membranes which were developed for use in separation processes have been employed with little or no modification for these miscellaneous uses. In others the membranes have been specially designed for their roles. Taken together, miscellaneous uses constitute the rapid growth segment of membranology, and are expected by some to surpass separation processes in economic importance by 1990.

3.1 SELECTIVE MEMBRANE ELECTRODES

Selective membrane electrodes are devices that sense chemical species and experience the internal transport of electronic charge as a consequence. The electrode can be viewed as one-half of a concentration cell in which a selective membrane separates the reference electrode from the test solution. A measureable electropotential is generated by concentration (activity) differences at opposite sides of the membrane. If the concentrations (activities) of the chemical species are m_1 and m_2, they will have emfs (electromotive forces) of E_1 and E_2. The emf E of the cell will be given by

$$E = E_1 - E_2$$

where

$$E_1 = E^0 - (2RT/\mathcal{F})\ln a_1$$

$$E_2 = E^0 - (2RT/\mathcal{F})\ln a_2$$

so that

$$E = (2RT/\mathcal{F})\ln a_2/a_1$$

The diffusing chemical species can be gaseous, ionic, or even nonionic in nature. It is only necessary that this species participate in an equilibrium reaction that results in an ion which can be directly measured by an ion-selective electrode. The membrane itself can be on any material: glass, crystalline solid, mixtures of finely divided crystalline solids and polymeric film formers, microporous solid, liquid in microporous solid, electroactive compounds in plasticized polymers, and enzymes immobilized within a microporous solid.

Gas-sensing electrodes utilize the fact that gases can be involved in equilibria in which they can be converted to ions (Table 3.1). Because the permeability of gases through dense films tends to be low, microporous as well as dense membranes have been investigated (Table 3.2). Microporous membranes must remain unwetted by the aqueous phase so that the diffusing species can permeate as a gas. Such hydrophobic microfiltration *air-gap* membranes are also commonly employed as sterile air vent filters. The data clearly show a 10^4 superiority of air-gap membranes over homogeneous thick dense films. However, since even the most hydrophobic air-gap membranes will wet eventually, future consideration will almost certainly be given to integrally-skinned membranes of the type now employed for commercial gas separations (Chapter 2).

The performance of a number of gas-sensing electrodes is summarized in Table 3.3. The electrodes employ air-gap membranes and are made using the construction shown in Figure 3.1. the lower limit is the apparent concentration indicated by the

TABLE 3.1 POSSIBLE EQUILIBRIA ASSOCIATED WITH GAS-SENSING ELECTRODES[a]

Diffusing Species	Equilibria	Sensing Electrode
NH_3	$NH_3 + H_2O \rightleftharpoons NH_4^+ + OH^-$	H^+
	$xNH_3 + M^{n+} \rightleftharpoons M(NH_3)_x^{n+}$	$M = Ag^+, Cd^2, Cu^{2+}$
SO_2	$SO_2 + H_2O \rightleftharpoons H^+ + HSO_3^-$	H^+
NO_2	$2NO_2 + H_2O \rightleftharpoons NO_3^- + NO_2^- + 2H^+$	H^+, NO_3^-
H_2S	$H_2S + H_2O \rightleftharpoons HS^- + H^+$	S^{2-}
HCN	$Ag(CN_2)^- \rightleftharpoons Ag^+ + 2CN^-$	Ag^+
HF	$HF \rightleftharpoons H^+ + F^-$	F^-
	$FeF_x^{2-x} \rightleftharpoons FeF_y^{3-y} + (x - y)F^-$	Pt(redox)
HOAc	$HOAc \rightleftharpoons H^+ + OAc$	H^+
Cl_2	$Cl_2 + H_2O \rightleftharpoons 2H^+ + ClO^- + Cl^-$	H^+, Cl^-
CO_2	$CO_2 + H_2O \rightleftharpoons H^+ + HCO_3^-$	H^+
X_2	$X_2 + H_2O \rightleftharpoons 2H^+ + XO^- + X^-$	$X = I^-, Br^-$

[a] From Ross et al.[1]

TABLE 3.2 THE DIFFUSION AND PARTITION PARAMETERS OF CARBON DIOXIDE AND OXYGEN FOR THREE DIFFERENT MEMBRANES[a,b]

Gas	Air $D(cm^2/s)$	k	Silicone Rubber $D(cm^2/s)$	k	Low-Density Polyethylene $D(cm^2/s)$	k
CO_2	1.3×10^{-1}	1.2	1.1×10^{-5}	2.6	8×10^{-7}	4.8×10^{-1}
O_2	1.8×10^{-1}	32	1.6×10^{-5}	9.9	9×10^{-7}	2.2×10^{-1}

[a]From Ross et al.[1]
[b]The k values were calculated wherever possible from solubility data in *Solubilities of Inorganic and Metal-Organic Compounds* 4th Ed. W. R. Link, Ed. *American Chemical Society*, Washington, D.C. 1958. Other values were estimated using free-energy data in *Selected Values of Chemical Thermodynamics Properties*, Circular 500, National Bureau of Standards, Washington, D.C. 1952.

electrode in a zero concentration solution after 5 min, coming from a solution with a concentration fo $10^{-2}\ M$ of the gas sensed. An adequate electrode for laboratory use should reach 99% of the final reading in 2 to 3 min for a 10-fold increase in concentration of the species sensed. Time–response characteristics are less critical for continuous monitoring. In such cases time lags of up to 30 min can be tolerated. The slope in Table 3.3 is the approximate potential change at 29°C for a 10-fold increase in gas concentration within the range in which the electrode is Nernstian. Although the kD value for water is lower than for most gases, its high concentration ($\sim 55\ M$) makes water vapor a potential electrode interference. If water transport results in dilution of the internal electrolyte and this alters parameters in the equilibrium equation, electrode instability and drift will be observed. Optimal electrode performance is obtained by adjusting the sample osmolality to that of the internal electrolyte, and by minimizing any temperature difference between electrode and sample.

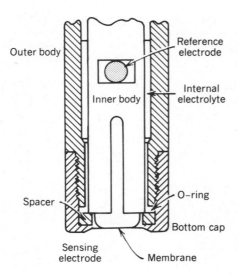

Outer body
Reference electrode
Internal electrolyte
Inner body
Spacer
O–ring
Bottom cap
Sensing electrode
Membrane

FIGURE 3.1. Construction of Orion series 95 gas-sensing electrode (from Ross et al.[1]).

TABLE 3.3 SOME SPECIFICATIONS OF GAS-SENSING ELECTRODES[a]

Species Sensed	Sensor	Internal Electrolyte	Lower Limit (M)	Slope	Sample Preparation	Interferences	Applications
CO_2	H^+	0.01 M $NaHCO_3$	$\sim 10^{-5}$	$+60$	$<pH\ 4$		Blood, fermentation vats
NH_3	H^+	0.01 M NH_4Cl	$\sim 10^{-6}$	-60	$>pH\ 11$	Volatile amines	Soil, water, Kjeldahl analyses
Et_2NH	H^+	0.1 M Et_2NH_2Cl	$\sim 10^{-5}$	-60	$>pH\ 11$	NH_3	b
SO_2	H^+	0.01 M $NaHSO_3$	$\sim 10^{-6}$	$+60$	HSO_4 buffer	Cl_2, NO_2 must be destroyed (N_2H_4)	Stack gases, foods, wines, S in fuels
	H^+	0.1 M $NaHSO_3$	$\sim 10^{-4}$				
NO_2	H^+	0.02 M $NaNO_2$	$\sim 5 \times 10^{-7}$	$+60$	Citrate buffer	SO_2 must be destroyed, (CrO_4^{2-}) CO_2 interferes	Stack gases, ambient air (after scrubbing), nitrite in foods
H_2S	S^{2-}	Citrate buffer (pH 5)	$\sim 1^{-8}$	-30	$<pH\ 5$	O_2 (ascorbic acid must be added to samples)	Pulping liquors anaerobic muds, fermentation
HCN	Ag^+	$KAg(CN)_2$	$\sim 10^{-7}$	-120	$<pH\ 7$	H_2S (add Pb^{2+})	Plating baths, plating wastes
HF	F^-	1 $M\ H^+$	$\sim 10^{-3}$	-60	$<pH\ 2$		Etching baths, steel pickling
HOAc	H^+	0.1 M NaOAc	$\sim 10^{-3}$	$+60$	$<pH\ 2$		c
Cl_2	Cl^-	HSO_4 buffer	5×10^{-3}	-60	$<pH\ 2$		Bleaching

[a]From Ross et al.[1].
[b]Example of volatile weak-base electrode.
[c]Example of volatile weak-acid electrode.

Several types of ion-selective electrodes based upon polymeric membranes are available: heterogeneous membranes in which finely divided salts such as Ag_2S and AgSCN are mixed with a film former such as polyethylene and then pressed and sintered[2]; ionomers which contain fixed cationic or anionic groups; and plasticized polymers containing highly selective electroactive compounds such as valinomycin.[3] The last mentioned is related to facilitated transport in internally supported liquid membranes (Chapter 9). Because facilitated transport through liquid membranes is such a versatile and highly selective process, the factors affecting its utility will be considered here.

The K^+-ion-selective membrane based on poly(vinyl chloride) (PVC), a plasticizer such as dioctyl phthalate (DOP) or dioctyl adipate (DOA), and the electroactive compound valinomycin, is an excellent prototype of an ion-selective membrane.[3] Valinomycin is a cylinder with a diameter of 15 Å and a height of 12 Å. Its polar groups are oriented towards the center of the molecule where K^+ is held and its lipophilic groups are turned outwards. Valinomycin forms complexes with alkali-metal ions in decreasing order of stability: $Rb^+ > K^+ \gg Na^+ > Li^+$. It is 10,000 times more selective for K^+ than for Na^+. Poly(vinyl chloride) is a polymer with a fairly high T_g ($\sim 81\,°C$) but which can be plasticized to a T_g below room temperature by the addition of compounds such as DOP or DOA. Both PVC and the plasticizer are compatible with valinomycin. To prepare a membrane, PVC, the plasticizer, and valinomycin are dissolved in THF (tehahydrofuran) and the solution is cast and allowed to evaporate to dryness. The resultant membrane may be visualized as a swollen gel containing mobile solutions of valinomycin in the plasticizer. All components are compatible, hydrophobic, and nonvolatile and hence not prone to exudation or leaching by water. Calcium ion selectivity is accomplished by the addition of dioctylphenyl phosphonate or tributyl phosphate which serve as plasticizers and electroactive materials.

When the electrode is immersed in the test solution the specific ion forms a complex with the electroactive compound at the membrane–solution interface. This complex is then shuttled across the membrane between polymer chains whose flexibility has been increased by the presence of the plasticizer.

Nonionic solutes can also be analyzed by selective electrodes, most commonly through the conversion of the solute to an ionic form by a specific enzyme. For example, glucose can be converted to gluconic acid by the action of glucose oxidase:

$$\text{glucose} + O_2 \longrightarrow \text{gluconic acid} + H_2O_2 \tag{1}$$

When oxygen is in non-rate-limiting excess and the glucose concentration is well below the apparent K_m for the immobilized glucose oxidase, there is a linear relationship between glucose concentration and decrease in oxygen pressure.[4] Glucose concentration is then determined by the amperometric measurement of O_2 depletion. A dual (differential) electrode is used so that only the O_2 consumption due to glucose is measured (Fig. 3.2). this cancels out background noise due to species other than O_2 and O_2 consumption unrelated to the oxidation of glucose. The en-

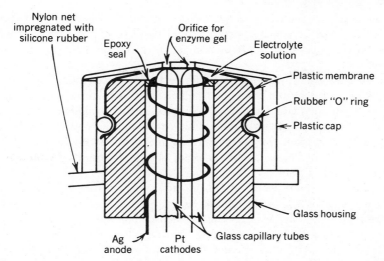

FIGURE 3.2.   Dual cathode enzyme electrode (from Updike and Hicks[4]; reprinted with permission from *Nature*, Vol. 214, p. 986, © 1967 Macmillan Journals Limited).

zyme is immobilized within the membrane by the simple expedient of copolymerizing acrylamide and N, N'-methylene bisacrylamide in a solution which also contains the enzyme. The whole process is accomplished on a microscale. An orifice ~ 30 μm in diameter within an impermeable shield which separates the glucose solution from the Pt cathode is filled with a droplet of monomer–enzyme solution. Polymerization is then effected in an oxygen-free atmosphere. The nature of the polymer matrix has a considerable effect upon enzyme efficiency and stability. Glucose oxidase has also been immobilized within a cellulose acetate membrane directly on the platinum tip of an O_2 electrode.[5] An alternate approach to the determination of glucose by amperometric measurement of oxygen consumption is to convert glucose to lactic acid whose concentration can be determined with a pH electrode. Ordinary dental plaque can provide the enzymes for a bacterial electrode which responds to several hexoses and pentoses.[6]

3.2   COLLECTOR MEMBRANES

A large and growing number of applications has been developed in which microfiltration membranes are utilized as filters for the collection of particles which are then subjected to analysis. Among these are the collection of bacteria for direct counting by transmitted light microscopy, particle counting in beverages, phytoplankton counting, measuring and counting inert particles, counting asbestos fibers, counting of cancer cells, virus adsorption and elution, and viability counting of water-borne bacteria.[7] Electrophoresis is an electrically driven process in which protein fractions are separated in either microporous cellulose acetate[8] or ultraporous agarose or polyacrylamide gels. The active field of nucleic acid hybridization is aided by the propensity for cellulose nitrate to strongly sorb DNA so that both DNA:DNA- and DNA:RNA- hybridizations can be carried out on the membrane filter.[7]

Most uses for collector membranes derive from the fact that the majority of particles which are removed from gas or liquid streams by filtration are deposited at or near the surface of the filter. These particles are then available for microscopic (or other) analysis. Every class of microfilter can be used for collection when dead-end filtration if employed. However, where, as in electrophoresis, lateral movement in a plane parallel to the surface is required, only high-porosity filters consisting of open cells are applicable.

The examination by transmitted light microscopy of particles on the surface of a membrane requires that the membrane be transparent to visible light. Irradiation track membranes (Chapter 8) are essentially dense films with occasional cylindrical pores, so this requirement is fulfilled by a membrane as it stands. For opaque filters, such as phase-inversion membranes of cellulose acetate, two techniques are available for rendering the membrane transparent. The first is the immersion of the dry membrane in a liquid of the same refractive index (~ 1.5) as the membrane polymer. The second is the process of *clearing* in which the membrane is densified by immersion in a solution consisting of a relatively volatile nonsolvent such as ethanol and a relatively nonvolatile solvent such as cyclohexanone. Because of the high concentration of nonsolvent, a clearing solution will not affect the membrane immediately upon immersion. However, as evaporation proceeds, the immersion medium becomes progressively richer in solvent owing to the loss of the volatile nonsolvent. Plasticization and gravity then act in concert to reduce porosity, thereby achieving transparency. This technique is widely utilized in cellulose acetate electrophoresis.[8] It will become apparent (Chapter 7) that the clearing solution is essentially the opposite of the solvent system which is utilized in the dry casting process for the preparation of microporous membranes. The latter employs volatile solvents and less volatile nonsolvents.

Direct microscopic counting of bacteria produces values which are many times higher than viable counts. In direct counting the bacteria are fixed with formalin and stained with a solution of erythrosin. *Viability counting* is a test of water quality based upon the presence of viable indicator microorganisms such as *Escherichia coli*.[7] Colonies are grown from individual bacteria or colony-forming units (CFUs). A given volume of water is allowed to pass through a 0.45-μm cellulose mixed-ester filter, catching any viable bacteria at or just beneath the surface of the membrane which is adjacent to the feed solution.[9] The filter is then transferred in such a manner that the downstream side of the membrane faces a piece of blotter which has been saturated with a liquid-nutrient medium. The growth medium is conveyed to the top surface of the membrane filter by capillary action. With incubation at 37°C, the growth medium nurtures each viable bacterium into a colony of bacteria large enough to be seen by the naked eye. The membrane filter technique is far less cumbersome than the agar spread-plate technique which preceded it and which together with the most probable number technique still serve as the primary standards against which membrane results are compared.[10] There are substantial differences between the various membrane filters used for viability counting. For example, the degree of anisotropy or inhomogeneity in depth (Chapter 7) strongly influences the total viability count.[11,12] The existence of large pores and cells at and below the membrane surface at which the bacteria are collected results in large pools of nutrient medium which enhance the probability that each bacterium will survive and grow into a colony. This phenomenon has been entitled *cradling* by Sladek.[11] Minor differences in the nature of the bacteria, growth media, filter wetting agents, and so on, can also significantly influence both the total count and colony characteristics, such as uniformity, shape, size, and sheen, and the background color of the membrane itself. Although no single membrane filter can be considered superior in every respect, highly anisotropic filters tend to yield the closest correspondence between their values and those obtained by primary standard techniques.[7] Furthermore, because highly anisotropic filters are much less prone to fouling than are isotropic filters, the former should always be employed where turbid feeds are encountered.

Viruses are much smaller than bacteria and are also present at much lower concentrations in water from which they must be assayed. They will therefore tend to pass through microfilters unless the latter are treated with acid or some other reagent which will introduce a positive charge onto the filter surface. Under such conditions, the negatively charged virus particles will be adsorbed.[7] Subsequent elution of the viruses is effected by increasing the pH, thereby decreasing virus–membrane interaction. Positive surface charges also effect the sorption of *pyrogens*—the fever-producing endotoxins which originate as bacterial cell wall fragments. However, it should be kept in mind that surface charges can be rapidly neutralized by any oppositely charged solute or suspended particle. Once the charged sites have been thus preempted, their capacity to selectively sorb contaminants will have been lost, and retention or passage will depend only on the steric characteristics of permeant and membrane.[13]

Recombinant DNA technology (gene engineering) has been greatly facilitated by the fact that cellulose nitrate and certain nylon filters selectively bind denatured DNA and RNA/DNA hybrids, while allowing free RNA to pass through.[7] Cellulose nitrate filters bind 50–80 μg/cm^2 of single-stranded DNA. By contrast, cellulose acetate filters bind only about 1 μg/cm^2. Gillespie and Spiegelman[14] developed a technique for effecting hybridization of both DNA:DNA and DNA:RNA types directly on the cellulose nitrate filter. In both cases, single-stranded DNA is filtered onto 0.45-μm cellulose nitrate membranes and fixed in dry heat at 80°C for 2 h. At this point the DNA:DNA membranes are treated with Denhardt's[15] reagent, a dilute solution of ficoll, polyvinylpyrollidone, and bovine serum albumin, which blocks any surplus DNA binding sites on the cellulose nitrate. Now when the radioactive single-stranded DNA is allowed to contact the membrane, it will only bind to the DNA which has been previously bound to the membrane. The membrane is washed to remove any unbound DNA and then placed in a scintillation counter. In the case of DNA:RNA hybridization, Denhardt's reagent is not required. After the addition of radioactive RNA, surplus RNA is removed by treatment with ribonuclease.

In recombinant DNA technology, nucleic acid fragments which are produced by the treatment of double-stranded DNA with restriction endonucleases are separated electrophoretically on agarose or polyacrylamide gels. Staining with ethidium bromide renders the separated bands visible. The DNA is then denatured by treatment with alkali and then placed in a buffer solution. The DNA bands are transferred to cellulose nitrate membranes by a wicking or blot transfer procedure known as a "Southern transfer."[16] After the transfer has been completed and the nucleic acids have been thermally fixed, the hybridization procedure is carried out as described.

Macromolecules such as proteins contain fixed charges. They can be separated from one another by the imposition of an electric field in the process known as *electrophoresis*. The rate of migration of a given macromolecule is governed by such factors as net charge and molecular weight. Although in gravity-free (i.e., space) environments, thermal convection currents do not act to rapidly remix the separated macromolecules, on earth this problem necessitates carrying out electrophoresis within open-celled matrices to frustrate convection. These matrices are essentially of two types: microporous cellulose acetate membranes containing cells 1–2 μm in diameter with surface pores of between 0.35–0.45 μm; and ultraporous agarose or polyacrylamide gels with cells of ~ 0.1 μm in diameter and correspondingly small surface pores.[8] The larger cell size of the cellulose acetate membranes leads to protein fractionation which is primarily based upon differences in net charge so that a small number of fractions results. The much larger number of fractions which result from separations in finely porous agarose gels is attributable to a combination of sieving and net charge effects. Fractionation by electrophoresis is the net result of many factors some of which act in opposition to one another. The presence of fixed charges on the membrane or gel matrix can cause a net flow of water in a direction opposite to that of the migrating proteins in the phenomenon known as *electroendoosmosis*. Depending on the isoelectronic point of a given protein and the composition of the buffer, the protein may be neutral, net cationic, or net anionic. The net charge of a protein will of course determine how fast and in what direction it travels. In view of the ability of electrophoresis to successfully separate materials of biological origin which are of interest to bioengineering studies, it is anticipated that it will grow substantially beyond its already significant analytical usage. In particular it may be anticipated that electrophoretic separations on a large scale and in three dimensions will be realized within this decade.

3.3 CONTROLLED-RELEASE DEVICES

One of the most promising and potentially largest-scale applications of membranes is as containment and metering devices for the controlled release of *effectors* such as drugs, fragrances, insecticides, herbicides, and so on. A variety of stratagems and containment devices have been devised to ensure delivery: imbibition of effectors within *preformed* microporous solids; *monolithic* encapsulation within polymer solutions or melts; establishment of a *reservoir* of effector behind a metering device; and the utilization of *mechanical devices* or *osmotic pumps*. The metering device can be a semipermeable membrane or simply an orifice in an otherwise impermeable film.

The availability of microporous solids in film, rod, tube, cast shapes, pellets, or powder forms, via the thermal variation of the phase-inversion process (Chapter 7), has greatly enhanced the scope of controlled-release applications. This approach is particularly applicable to the containment of liquids and to compounds which are susceptible to degradation if encapsulated by inclusion in polymer melts or solutions. It also has the advantage that porosity and hence effective concentration is greater, and the pore and cell characteristics more precisely defined, than is the case with other loading techniques. These advantages suggest that this approach should be the technique of choice whenever controlled release of vapors is intended.

As an example of this approach, slabs of microporous polymer (50 × 50 × 5 mm) were prepared by pouring a 20% solution of polypropylene in bishydroxyethyl tallow amine (DHTA) at 210°C into molds for cooling and solidification. The slabs were then extracted with acetone to remove the DHTA.[17] The open-celled structure with 80% porosity consisted of voids 1-30 μm in diameter interconnected by pores 0.10 to 10 μm in diameter. The rate of uptake of methyl nonyl ketone (MNK) by the porous slab is shown in Table 3.4. The release of MNK from the microporous PP (polypropylene) slab at three concentrations demonstrates that evaporation rates are equal and constant until approximately 70% of the MNK has been released (Fig. 3.3). A plot of half time versus vapor pressure of MNK, dimethylphthalate (DMP), and N-diethyl-m-toluamide (DEET) demonstrates that release time is inversely proportional to vapor pressure (Fig. 3.4).

TABLE 3.4   RATE OF UPTAKE OF MNK BY POROUS SLAB[a]

Time (min)	Concentration (g/g)	Concentration (g/cm²) (total area)
0	0.00	0.000
1	0.49	0.099
2	0.58	0.117
4	0.60	0.121
10	0.62	0.125
20	0.63	0.127
45	0.65	0.132
90	0.67	0.136
140	0.70	0.142
1080	0.79	0.160

[a]From Brade and Davis[17]; © 1983 Technomic Publishing Company, Inc. Lancaster, PA.

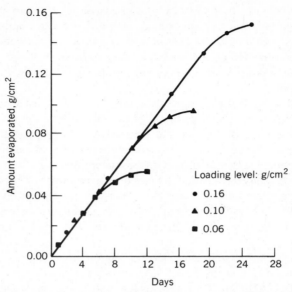

FIGURE 3.3. Release of MNK from porous slab (from Brade and Davis[17]; © 1983 Technomic Publishing Company, Inc., Lancaster, PA).

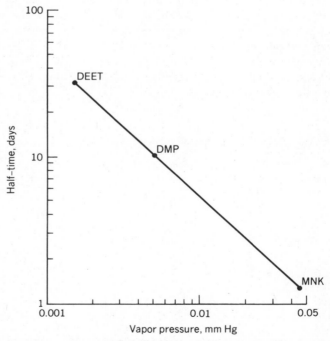

FIGURE 3.4. Relationship between half-time and vapor pressure (from Brade and Davis[17]; © 1983 Technomic Publishing Company, Inc., Lancaster, PA).

TABLE 3.5 TRIPELENNAMINE·HCL LOADINGS[a,b]

Solution Concentration (%)	Drug Loading (%)
5.0	12.54
10.0	23.17
20.0	35.59

[a]From Good.[18]

[b]Loadings achieved by soaking poly(2-hydroxyethylmethacrylate) hydrogels in various aqueous solutions of the drugs. Drug loadings are given as percent (w/w) based on the dry specimens.

In addition to the loading of neat liquids into preformed microporous solids, solutions of solids can also be so incorporated.[18] Slightly cross-linked (1.5% tetraethyleneglycol dimethacrylate) microporous 2-hydroxyethyl methacrylate (HEMA) hydrogels with a porosity of 36% at 37°C were equilibrated in aqueous solutions containing 5, 10, and 20% of tripelennamine · HCl (Table 3.5). The hydrogels were then dried in vacuum at 80°C. Similar loadings can undoubtedly be obtained with organosoluble drugs as well, provided only that the porous matrix is inert to the organic solvent.

In contrast to the containment of liquids for controlled release, solids are best encapsulated in so-called *monolithic* devices by their addition to a polymer solution or melt. In some cases, the solid will be compatible with the polymer in its fluid phases. This is true of some solid-state drugs and, occasionally, even true of some resistant thermophilic enzymes. In such cases, the melt or solution is simply extruded or cast to yield the monolithic solid. The disadvantage of this approach is that the permissible loading will vary significantly depending on the solubilities of solid in the melt or solution. Likewise, the fineness and uniformity of dispersion of the solid in the polymer matrix will depend on the compatibilities of solid and polymer if a melt is involved, and on the interrelated compatibilities of solid, polymer, and solvent if a solution is utilized. Any change of solid type or concentration will require a complete reassesment of release characteristics and may even necessitate the choice of alternate polymers and/or solvents. Nevertheless, despite its many disadvantages, the inherent simplicity of this approach warrants its application in certain instances. If the polymer matrix is biodegradable, is well tolerated by the body, and does not produce toxic decomposition products, it is a candidate for an implantable drug-delivery device. Such is the case for lactic acid (*dl,* L(t)) and glycolic (G) acid homopolymers and copolymers which were dissolved in tetrahydrofuran containing also the hormonal steroid, *d*-norgestrel. After casting into films they were formed into 1.5-mm thick, 3-cm long cylindrical shapes for subdermal implantation.[19] The homopolymer and copolymer composition, molecular weights, and hormonal loadings are listed in Table 3.6. The results certainly demonstrate the feasibility of this approach. The homopolymer L(+)-lactic acid released hormone at a uniform rate of about 4 μg/day-cm^2, which was considered too stable, since a 6-month implant lifetime was the goal (Fig. 3.5). Copolymers from 75 *dl/*

TABLE 3.6 POLYMER CHARACTERISTICS AND HORMONAL LOADINGS OF UNCOATED CYLINDRICAL SUBDERMAL IMPLANTS OF LACTIC/GLYCOLIC ACID POLYMERS[a]

Polymer Composition (% by weight)				Fraction d-Norgestrel in
dl-Lactic	L(+)-Lactic	Glycolic	MW	Implant (by weight)
75		25	56,000	27.5
75		25	78,000	27.5
	75	25	56,000	50.0
90		10	59,000	50.0
50	50		145,000	20.0
50	50		240,000	20.0
100			69,000	50.0
100			135,000	20.0
	100		260,000	20.0

[a]From Wise et al.[19]

25 G, 75 L(+)/25G, and 90 dl/10 G appeared to be *too* unstable. However, polymers 100 dl and 50 dl/50 L(+) were considered worthy of further study (Fig. 3.6). The last mentioned represents a high-MW polymer with a modest hormonal loading. In genera, stability varied directly with MW and inversely with hormonal loading. Stability was greater for homopolymers, which was attributed to the higher crystallinity of the former. In a related study[20] the rate of hydrolysis of a 90/10 poly(L(+)-lactic-*co*-gylcolic acid) implant was followed by gel permeation chromatography (Table 3.7). The MW was halved after about 50 days of implantation.

Where the thermal resistance of the effector is high and/or the melting range of the encapsulating polymer is low, polymer and effector can be combined in the melt to produce a monolithic controlled-release device. Polymer biodegradability is a desirable feature when the device is to be utilized to dispense insecticides or herbicides. This combination of properties is realized in melt spun fibers of polycaprolactone (PCL) mp 60°C, which contain 26% of the herbicide fluoridone.[21] The application of controlled-release herbicides in fiber form has the following advan-

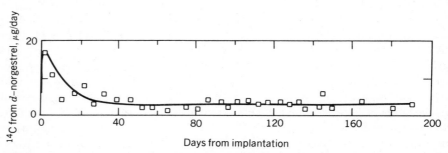

FIGURE 3.5. Recovery of excreted ^{14}C from dog 4B with implants of 100 L(+) polymer of 260,000 $\overline{M}_\eta$ and 20% d-norgestrel. Implant area was 5.0 cm^2 (from Wise et al.[19]).

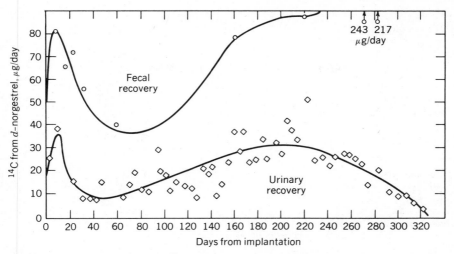

FIGURE 3.6. Recovery of excreted ^{14}C from dog 5B with implants of 50 dl/50 L(+) copolymer of 240,000 $\bar{M}_v$ and with 20% d-norgestrel. Implant area was 6.6 cm^2 (from Wise et al.[19]).

tages: (1) the fibers become entangled with the aquatic weeds which they are to control; and (2) fiber diameter can be utilized to control the duration of the herbicide release. Pelletized versions of the fibers may also be of interest, in that they would permit rapid sinking in flowing water. The herbicide EPTC has been encapsulated about corn seed in a starch–borate complex. This permits the corn plant to grow in a weed-free environment. Biological as well as chemical herbicides have been encapsulated for controlled release. Mycoherbicides, that is, fungi such as *Fusarium lateritium,* a pathogen against sicklepod, have been encapsulated with sodium alginate. The advantage of this approach is specificity, which is often lacking in the broad-spectrum chemical herbicides.

The development of monolithic devices containing bioactive polymers has apparently not followed the tradition of loading preformed microporous solids. The

TABLE 3.7 **MOLECULAR WEIGHT OF THE 90/10 POLY (L(+)-LACTIC-CO-GLYCOLIC ACID^{a,b}**

Removal Schedule from Monkeys	Molecular Weight, $\bar{M}_w$ (material from all recovered beads)
Prior to implantation	37700
7 days (in 1 monkey)	27100
14 days (in 1 monkey)	22100
21 days (in 1 monkey)	19900
28 days (in 1 monkey)	1800
54 days (in 2 monkeys)	18200, 14900

[a]From Wise et al.[20]
[b]Used in the 70% w/w naltrexone base beads removed from monkeys.

TABLE 3.8 INFLAMMATION IN THE RABBIT CORNEA DUE TO
POLYMER IMPLANTS[a,b]

Polymer	Number of Tests	Inflammation (%)		
		None	Mild	Significant
Poly(acrylamide)	10	20	30	50
Poly(vinylpyrrolidone)	4	0	0	100
Poly(vinyl alcohol)	20	15	70	15
Ethylene–vinyl acetate copolymer	20	40	60	0
Hydron-S	15	100	0	0
After alcohol wash:				
Ethylene–vinyl acetate copolymer	20	100	0	0
Poly(vinyl alcohol)	6	67	33	0

[a]From Langer and Folkman.[22]
[b]The criteria for inflammation were the presence of edema, white-cell infiltration, and neovascularization. Inflammation was observed by stereomicroscopy and evaluated as mild if any one of the three characteristics was detected and significant if corneal opacity resulted.

kinetics of macromolecular loading and the low solubility of bioactive macromolecules are probably responsible for this. Furthermore, the solubility and stability of such macromolecules in membrane polymer solutions or melts will only rarely suffice to permit incorporation via these pathways. The alternatives are to place the polymer in a reservoir with a metering device and to add the bioactive polymer as a finely powdered insoluble solid to a polymer solution or melt. The compatibility in rabbit corneas of a number of hydrophilic polymeric encapsulants was tested (Table 3.8). Hydron-S (HEMA), and alcohol-washed Elvax-40, an ethylene–vinyl acetate (EVA) copolymer were found to be excellent membrane candidates. The latter was chosen for further evaluation because of its solubility in methylene chloride. The solid bioactive polymers were prevented from settling out of the membrane polymer solutions by casting the mix at $-80°C$ so that the matrix rapidly froze. Desolvation of methylene chloride solution was effected at $-20°C$ for 2 days followed by final drying in vacuum at room temperature. For an EVA matrix with a 25% loading of bovine serum albumin (BSA) (MW 68,000), the rate of release increased with increasing size of BSA particles. Since the exposed surface area and hence the release rate would be expected to be inversely proportional to the product size, higher release rates of the larger particles suggest the presence of larger-diameter pores and subsurface channels through the membrane polymer matrix. As expected, the release rate increases greatly as the loading increases (Fig. 3.7). Increased porosity necessarily leads to greater accessibility to incoming water and decreased resistance to outgoing polymer. A zero-order release system can be developed by controlling the geometry of the encapsulant (Fig. 3.8). Constant delivery of effector macromolecules is obtained by placing them in a matrix which is hemispheric and coated with an impermeable barrier containing a single orifice in the center of the flat circular surface. This ensures that progressive depletion frees

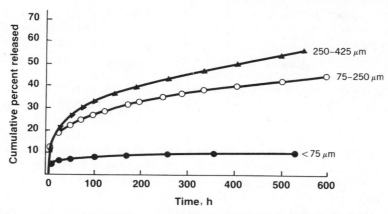

FIGURE 3.7. Effect of particle size on the cumulative release of bovine serum albumin (MW = 68,000); the loading is 25% (from Langer[23]; reprinted with permission from *CHEMTECH*, © 1982 American Chemical Society).

an ever larger surface area from which the effector can be leached, thereby offsetting the greater distance which it must travel. Where controlled—but not constant—release is desired (as in the administration of insulin to diabetics who require higher doses near mealtime), mechanical "stirring" can be employed. The release rates of drug from polymer matrices can be controlled by the imposition of an oscillating external bar magnet. An increase in release rate of up to 30-fold is attainable with this device. Rates return to normal when the magnetic field is discontinued.

A commonly employed concept which is utilized for transdermal drug delivery both in the sclera and on the neck or trunk of the body is the metering of drug

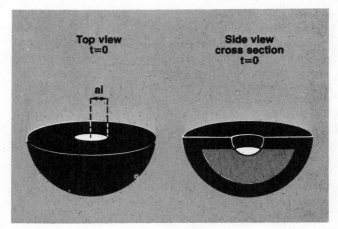

FIGURE 3.8. Schematic diagram of an inwardly releasing hemisphere; a = inner radius. Black represents laminated regions through which release cannot occur. Gray represents drug dispersed in polymer (from Langer[23]; reprinted with permission from *CHEMTECH*, © 1982 American Chemical Society).

from a reservoir which is separated from the skin by a semipermeable membrane, such as a biocompatible EVA copolymer. The reservoir may be a microporous polymer matrix or it may simply contain viscosity and osmotic controlling agents such as lactose, colloidal silica, and silicone fluid.[24] A pilocarpine delivery device has been developed for the treatment of glaucoma, and skin patches have been developed for dramamine and nitroglycerine treatments of motion sickness and angina, respectively. Skin-patch devices also contain a layer of permeable silicone skin contact adhesive which joins the rate-controlling membrane to the skin.

One final controlled-release drug-delivery concept is worthy of mention. The osmotic pump,[25] which is being utilized to study drug dosages in laboratory animals, is a cylinder which contains a collapsible reservoir of flexible, impermeable material, surrounded by a sealed layer containing a saturated solution of osmotic agent, all contained within a semipermeable membrane (Fig. 3.9). After implantation, water penetrates the semipermeable (EVA) membrane into the osmotic solution, thereby generating hydrostatic pressure on the flexible lining of the reservoir. This compresses the reservoir and results in a constant flow of drug through the exit orifice.

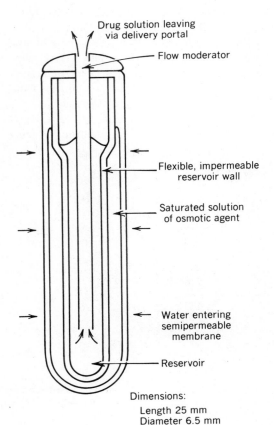

Drug solution leaving
via delivery portal

Flow moderator

Flexible, impermeable
reservoir wall

Saturated solution
of osmotic agent

Water entering
semipermeable
membrane

Reservoir

Dimensions:
Length 25 mm
Diameter 6.5 mm

FIGURE 3.9. Cross section of a functioning osmotic pump (from Capozza[25]).

3.4 MEMBRANE REACTORS

Capillary ultrafiltration membranes can be used as reactors for the growth of microorganisms or even mammalian cells. When viable cells are placed in the shell-side space of a bundle of interiorly skinned capillary fibers and a nutrient solution is passed through the fiber lumina, it is possible to maintain a culture of these cells. If the biochemical products which the cells secrete permeate the skins of the capillary fibers, then this scheme constitutes a continuous process for the production and separation of chemicals of biological origin. Cells grow to tissue like densities of (10^6–10^7 cells/cm^2 of fiber area) an order of magnitude greater than is achieved in roller bottles.[25] Products such as interferon, monoclonal antibodies, antigens, viruses, and hormones may be extracted either continuously or intermittently. Membrane reactor processes have been investigated for the use with pancreatic cells for the synthesis of insulin as well as for the production of other peptide hormones and enzymes. Batch processing has long been utilized for the growth of bacteria and yeast cells in fermentation. The key to continuous fermentation is the removal of toxic products. This is effected most conveniently by the recirculation of cells through a hollow-fiber unit.[26]

In addition to their application to the separation of biochemicals from whole cells, skinned capillary fibers can also be utilized as matrices for the immobilization of enzymes or other catalysts. If the lumen of the capillary is skinned and the porous substructure of the fiber contains the catalyst, the substrate molecules will pass through the skin into the porous substructure. The products of the reaction, once formed, will then diffuse back into the fiber bore for eventual collection and separation. It is worthy of note that the product builds up most rapidly on that side of the membrane (the fiber lumen) where the substrate is present in the higher concentration.[27]

In some cases separation and reaction can be combined. The enzymatic hydrolysis of an amino ester has been combined with the resolution of the resultant racemic mixture.[28]

3.5 SOLID-STATE ELECTROLYTES FOR BATTERIES AND FUEL CELLS

The charged nature of ion-exchange membranes is responsible for their low resistance and high conductivity, characteristics which have prompted their use as separators in cells containing several electrolyte compartments,[29] and as reaction-product-removing depolarizers.[30, 31] They have also been employed to supply a reacting ion to the cell reactions.[32] In the present section, their application as solid-state electrolytes for batteries and fuel cells will be considered.

3.5.1 Electrolytes for Secondary-Cell Batteries

An ion-exchange battery is a galvanic cell in which the electrolyte is a somewhat hydrated solid ion-exchange membrane. When utilized as the electrolyte in a secondary (rechargable) cell, the membrane not only conducts the current but also maintains the physical structure of the battery. Its role is similar to that of a silver halide in the conventional type of solid-electrolyte cells except that its conductance is higher. Another advantage of the ion-exchange membrane battery is that the mobile ion can be varied at will. The resistivities of solid-state electrolytes lie between those of aqueous and solid inorganic electrolytes. Cells containing certain pairs of metal electrodes behave reversibly. Their emf's form an additive series and are only slightly lower than those of their aqueous analogs.

Grubb[33] described the use of two types of ion-exchange membranes for this application: a heterogeneous sulfonated polystyrene (Amberplex C-1) and a homogeneous sulfonated phenolformaldehyde (Neptone Cr-51). Silver and zinc foils were employed as the electrodes. The cell initially contained a membrane which had been equilibrated with zinc ion and washed to remove the free electrolyte before being placed between silver- and zinc-foil electrodes (Fig. 3.10). Charging resulted in a silver-ion region adjacent to the silver electrode.

The emf was found to be a linear function of the current for both the Amberplex and Nepton membranes. Concentration/polarization, possibly the result of solvent movement with ions as hydration cells, occurred when the cells were placed on heavy loads for short time intervals. The substitution of ethylene glycol for a portion of the water had little effect on the emf, indicating that the activity ratio of the Ag^+ and Zn^{++} ions was largely unchanged. However, cell stability improved and conductivity decreased in the presence of ethylene glycol, presumably because of stronger association between fixed and mobile ions as the dielectric constant of the solvent was lowered. When an Amberplex membrane cell was charged for 5 h at 1 mA and placed on a fixed load resistance immediately after charging, approximately 80% of the charge input was recovered after 5 h. Nepton cells were less efficient under the same conditions, permitting the recovery of only 40% of the charge input.

However, the charge recovery for both cells after being allowed to stand for 24 h before discharging was very low. This phenomenon was caused by the cross diffusion of ions to the opposite electrodes and the resultant destruction of the cells. Therefore, before membrane cell batteries can be considered practical, some means of inhibiting such cross diffusion must be found. Several real possibilities exist:

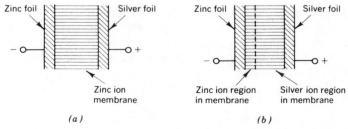

FIGURE 3.10. Schematic diagram of silver–zinc cell: (*a*) uncharged cell, emf 0.8–1.2 V and (*b*) charged cell, emf 1.4–1.5 V (from Grubb[33]; reprinted by permission of the publisher, The Electrochemical Society, Inc.).

the utilization of highly selective ion-exchange membranes to bind the counterions more strongly and the use of complex ions. Recent emphasis has been on cells made of poly (p-phenylene) or polyacetylene electrodes and an organic electrolyte.[34] In their undoped states both polymers are insulators (conductivities $10^{-15}/\Omega$ cm). When they are doped with cations such as Li^+ or anions such as AsF_6^-, conductivity increases by a factor of 10^{16}. In the charged state, the electrolyte's anions serve as counterions for the oxidized or positively charged polymer (the cathode). The electrolyte's cations play the same role for the reduced or negatively charged polymer (anode). Alternatively, oxidized and reduced membrane layers separated by polypropylene spacers are immersed in an organic electrolyte such as propylene carbonate. During discharge, electrons flow from the anode to the cathode through a circuit. As the membranes return to their neutral state, the counterions diffuse back into solution.

3.5.2 Electrolytes for Fuel Cells

The electrochemical generation of electricity in fuel cells, that is, primary cells in which the oxidation of the fuel generates electricity, offers the possibility of achieving higher efficiencies than are obtainable with electromechanical generators driven by heat engines. Fuel cells are just the opposite of electrochemical processes which utilize electricity to generate chemicals. In electrolysis, for example, water is decomposed by electricity into hydrogen and oxygen. In a fuel cell, hydrogen reacts with oxygen to form water, heat, and a direct current. The efficiency of the chemical reaction yielding electric energy can approach 100% whereas that of the heat engine is typically 30%. Fuel cells can be classified as direct or indirect, and reversible or irreversible. Direct fuel cells are those utilizing carbons or hydrocarbons, that is, directly available raw materials, as the fuel together with oxygen as the oxidizer. Indirect fuel cells utilize indirect (prepared from other raw materials) fuel such as water gas, producer gas, or hydrogen. Reversible fuel cells are those in which fuel and oxygen can be produced by reversing the supply of electrons to the cell so that electrons are removed from the oxidizing electrode and fed into the fuel electrode. The hydrogen-oxygen cell is an example of an indirect, reversible cell where $2H_2 + O_2 \rightarrow 2H_2O$, $E_r = 1.23$ V at 20°C and 1 atm or when reversed,

$$2H_2O \longrightarrow 2H_2 + O_2$$

The carbon–oxygen cell, on the other hand, is direct but irreversible because carbon dioxide is not readily converted back into elemental carbon and oxygen. The fuel electrode produces electrons from its supply of fuel while the oxidizing electrode consumes electrons. Work is done when a charge-removing load is placed in the circuit.

Overall cell efficiency is the product of the current efficiency and the voltage efficiencies. The former decreases insofar as the fuel or oxidant reacts with the cell materials; the separation of fuel and oxidant by the electrolyte is imperfect; and fuel is lost in the cell exhaust. Voltage efficiency decreases with increasing cell

TABLE 3.9 EFFICIENCIES OF SEVERAL PROPOSED FUEL CELLS[a]

Cell and Reaction	Current Density (mA/cm^2)	Output, (mw/cm^2)	Voltage Efficiency, $\eta_v(\%)$	Current Efficiency, $\eta_c,(\%)$	Relative Size of Unit[b]
Bischoff–Justi:					
carbon air	1	0.75	75	Not quoted	440
	2	1.0	50		330
Davtyan:					
water					
gas–air	10	8.2	84	Not quoted	40
	20	15.7	80		21
	30	22.8	77		~ 15
Bacon:					
hydrogen–oxygen	162	145.8	77	Approaches	~ 2
	413	330.4	68	100	1
	1076	645.6	51		~ ½
Davtyan:					
hydrogen–oxygen	5	4.1	67	Not quoted	80
	10	7.4	60		45
	20	13.6	55		24

[a]From Watson.[35]
[b]Assuming that each cell has approximately the same thickness.

resistance, increasing concentration/polarization, and increasing the extent of electrode reactions (chemical polarization).

The efficiencies of recently proposed fuel cells are summarized in Table 3.9. The hydrogen–oxygen cell proposed by Bacon appears to be most efficient to date (Fig. 3.11). The chemistry of the Bacon cell has a number of attractive features:

1. Because hydrogen is electrochemically more reactive than carbon, the cell can be operated at relatively low temperatures.
2. This in turn means that aqueous electrolytes or solid-state ion-exchange electrolytes can be utilized.
3. Because the resistance of such electrolytes is low, porous electrodes having large specific areas can be used.

The electrolyte is separated from the gas by a layer of porous metal, and a layer of finely porous metal is used to keep the liquid from flooding the whole electrode. A large area of metal is therefore in contact with a liquid film through which gases can diffuse rapidly. This large surface area results in a minimal amount of chemical polarization.

The electrolyte in any fuel cell plays the vital role of preventing the molecular forms of fuel and oxidant from mixing and transferring their electrons directly. At

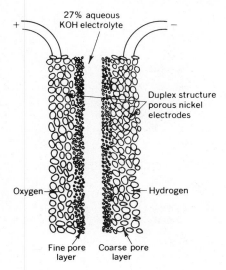

27% aqueous
KOH electrolyte

Duplex structure
porous nickel
electrodes

Oxygen

Hydrogen

Fine pore
layer

Coarse pore
layer

FIGURE 3.11. The Bacon cell (from Grubb and Niedrach[36]; reprinted by permission of the publisher, The Electrochemical Society, Inc.).

the same time it must ensure that electrons pass from the fuel to the oxidizing electrodes only via the external circuit. An ideal electrolyte is also permeable to only one ionic species. A variation of the Bacon cell has been investigated which utilized a heterogeneous ion-exchange membrane (Amberplex) as a solid-state electrolyte replacing the aqueous caustic solution (Fig. 3.12). The conductivity approximates that of a 0.1 N H_2SO_4 solution. The ion-exchange membrane electrolyte allows a very simple construction (minimum of external components and controls) including small unit thickness. This favorable geometry helps compensate for the

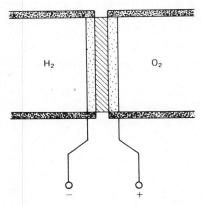

H_2

O_2

$-$ $+$

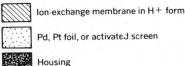

Ion-exchange membrane in H+ form

Pd, Pt foil, or activated screen

Housing

FIGURE 3.12. Schematic diagram of ion-exchange membrane fuel cell (from Grubb and Niedrach[36]; reprinted by permission of the publisher, The Electrochemical Society, Inc.).

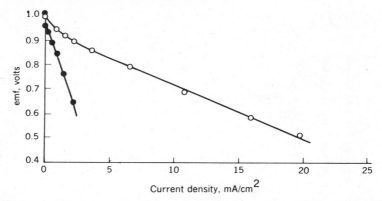

FIGURE 3.13. Effect of free acid on cell operation, leached membrane H^+ form; membrane H^+ form in equilibrium with 6 M H_2SO_4 (from Grubb and Niedrach[36]; reprinted by permission of the publisher, The Electrochemical Society, Inc.).

lower current density of the cell relative to the unmodified Bacon cell. In addition, as much as 67% CO_2 can be present in the H_2 feed with little effect on performance. Furthermore, no dilution occurs from the water formed at the O_2 electrode during cell operation because it is rejected from the saturated electrolyte. Because gas permeability is low, ion-exchange membranes can function as the gas separator as well as the electrolyte.

The efficiency of the cell can be increased by lowering membrane resistance through equilibrating the membrane with 6 M H_2SO_4 (Fig. 3.13). However, this approach is not without its disadvantages since free acid is leachable and introduces formidable corrosion problems.

The fuel cells are operated by admitting hydrogen and oxygen to the two gas chambers and measuring the electric output. To maximize their conductivity, the membranes are kept at 100% relative humidity by bubbling the feed gases through water. Water formed by the cell during operation also helps maintain the membranes in a hydrated condition. The voltage-current-density plots for these fuel cells vary considerably with the nature of the electrodes. All the screen electrodes yield higher currents than the foil types. Among the screen electrodes, efficiency increases with increasing area of contact between electrode and electrolyte. In view of this, the initial Bacon asymmetric nickel electrodes used together with ion-exchange membranes would appear worthy of investigation in conjunction with solid-state electrolytes.

REFERENCES

1. J. Ross, J. Riseman, and K. Kreuger, *Pure Appl. Chem.*, **36,** 473 (1973).
2. M. Mascini, *Anal. Chem. Acta,* **62,** 29 (1972).
3. U. Fiedler and J. Ruzicka, *Anal. Chim. Acta,* **67,** 179 (1973).

4. S. Updike and G. Hicks, *Nature,* **214,** 986 (1967).
5. M. Notin, R. Guillien, and P. Nabet, *Ann. Biol. Chim.,* **30,** 193 (1972).
6. S. Grobler and G. Rechnitz, *Talanta,* **27,** 283 (1980).
7. T. Brock, *Membrane Filtration: A User's Guide and Reference Manual,* Science Technology, Industries, Madison, WI, 1983.
8. P. Chin, *Cellulose Acetate Electrophoresis,* Ann Arbor Science Pub., Ann Arbor, MI, 1971.
9. A. Goetz and N. Tsuneishi, *J. Am. Water. Works Assoc.,* **43,** 943 (1951).
10. M. Young, ASTM STP 673, American Society of Testing Materials, Philadelphia, PA, 1979, pp. 40–51.
11. K. Sladek, R. Suslavich, B. Soln, and F. Dawson, *Appl. Microbiol.,* **30,** 685 (1975).
12. R. Kesting, A. Murray, K. Jackson, and J. Newman, *Pharm. Tech.,* **5**(5), 53 (1981).
13. J. Raistrick, Proceedings of the World Filtration Congress III, London, (1982), p. 310.
14. D. Gillespie and S. Spiegelman, *J. Mol. Biol.,* **12,** 289 (1965).
15. D. Denhardt, *Biochem. Biophys. Res. Commun.,* **23,** 641 (1966).
16. F. Southern, *J. Mol. Biol,* **98,** 503 (1975).
17. W. Brade and T. Davis, *J. Cell. Plastics,* 309 (September–October 1983).
18. W. Good, in *Polymeric Delivery Systems,* R. Kostelnik, Ed., p. 139, MMI monograph Vol. 5, Gordon and Breach, New York, 1978.
19. D. Wise, J. Gregory, P. Newberne, L. Bartholow, and J. Stanbury, in *Polymeric Delivery Systems,* R. Kostelnik, Ed., p. 121, MMI monograph Vol. 5, Gordon and Breach, New York, 1978.
20. D. Wise, A. Schwope, S. Harrigan, D. McCarty, and S. Howes, in *Polymeric Delivery Systems,* R. Kostelnik, Ed., p. 75, MMI monograph Vol. 5, Gordon and Breach, New York, 1978.
21. *Chemical Week,* May 23, 1984, p. 32.
22. R. Langer and J. Folkman, in *Polymeric Delivery Systems,* R. Kostelnik, Ed., p. 175, MMI monograph Vol. 5, Gordon and Breach, New York, 1978.
23. R. Langer, *CHEMTECH,* 98, (February 1982).
24. *Chemical Week,* December 9, 1981.
25. R. Capozza, in *Polymeric Delivery Systems,* R. Kostelnik, Ed., p. 261, MMI monograph Vol. 5, Gordon and Breach, New York, 1978.
26. R. Tutunjian, Proceedings of the Third World Filtration Congress, London, 1983, p. 519.
27. P. Carr and L. Bowers, *Immobilized Enzymes in Analytical and Chemical Chemistry,* Wiley, New York, 1980, p. 183.
28. J. Quinn, Membrane Conference at Bend, Oregon, November 1983.
29. W. Juda and W. McRae, U.S. Patent 2,636,851 (1953).
30. E. Pitzer, U.S. Patent 2,607,809 (1953).
31. C. Morehouse, U.S. Patent 2,771,381 (1956).
32. P. Robinson, U.S. Patent 2,786,088 (1957).
33. W. Grubb, *J. Electrochem. Soc.,* **106**(4), 275 (1959).
34. *Chem. Eng. News,* October 12, 1981, p. 34.
35. R. Watson, *Research,* **7,** 34 (1954).
36. W. Grubb and L. Niedrach, *J. Electrochem. Soc.,* **107,** 131 (1960).

4 MEMBRANE POLYMERS

The term *polymer* is derived from the Greek words *poly* (many) and *meros* (unit). The individual units of which a high-molecular-weight (MW) polymer or *macromolecule* is composed of are called in turn, *monomers*, that is, single units. Although the use of the term monomer was initially restricted to molecules of a single type which could be polymerized, it has now been expanded to include each of the two bifunctional halves of step polymer units. Thus both the diamine and diacid portions of polyamides are designated monomers. In this treatise the term monomer has been further extended to include side-chain modifiers which can react with functional groups on preexisting polymers.

A distinction is made between polymer classes based on their method of synthesis: Chain propagated or addition polymers are formed from monomers with unsaturated carbon–carbon linkages (double bonds). Addition of monomer units to a growing chain is rapid and ceases at chain termination at which time the polymer becomes unreactive. Step or condensation polymers, on the other hand, are the result of the reaction of polyfunctional monomers, oligomers, or polymers with one another at a slow steady rate. Molecular weight (MW) increases throughout polymerization and the resulting polymers retain their ability to grow unless their reactive end groups are capped with a monofunctional moiety.

In this chapter, essential polymer characteristics which relate to processing and function will be treated. This will be followed by a detailed discussion of individual neutral and charge-bearing polymers. Finally, the important topics of water structure and membrane–water interactions, polymer modification, and new polymer design will be covered.

4.1 ESSENTIAL MEMBRANE POLYMER CHARACTERISTICS

The large average size of macromolecules, their size distribution, their architecture, the specific nature of their chemical groups, the arrangement of these groups in the chain, and the state of aggregation of macromolecules, are the fundamental polymer properties which in turn are responsible for all of their derived characteristics. The single feature which distinguishes polymers from any other class of compounds and which make them ideal membrane materials is their fibrillar nature and great size which in turn result in cohesive forces which extend to the macroscopic level.

A convenient way of describing macromolecular size is in terms of degree of polymerization (DP), the number of monomer units contained in the polymer. The influence of DP upon the physical properties of materials is schematically depicted in Figure 4.1.

A polymer is not, of course, a single monodisperse species of a given MW, but rather a polydisperse mixture of species with a distribution of MWs. Both processing and end-use characteristics of membrane polymers are influenced not only by several average MWs but also by the breadth and shape of the MW distribution curve. The latter is best obtained by a technique known as gel permeation chromatography (GPC), which separates molecules by virtue of their size differences.[1-3] A capillary column is filled with a heteroporous, solvent-containing polymer gel which varies in permeability over several orders of magnitude. As a polymer solution permeates the gel, the larger molecules are denied access to all but the largest pores, whereas the smaller molecules diffuse into any suitably sized aperture. The result is that the larger molecules permeate the column more rapidly than do the smaller ones. Calibration with narrow MW fractions determined by some absolute method is utilized to yield the complete molecular-weight distribution curve which includes number average, $\overline{M}_n$, weight average, $\overline{M}_w$, and Z-average, $\overline{M}_Z$, molecular weights. These are defined as follows:

$$\overline{M}_n = \sum N_i M_i = W/N$$

$$\overline{M}_w = \sum N_i M_i^2 / \sum N_i M_i = \sum W_i M_i$$

$$\overline{M}_Z = \sum N_i M_i^3 / \sum N_i M_i^2 = \sum W_i M_i^2 / \sum W_i M_i$$

where N_i = number of molecules of molecular weight M_i
N = total number of molecules,
W_i = weight fraction of molecules of molecular weight M_i, and
W = total weight.

The various molecular-weight averages for a typical polymer are shown on a hypothetical curve (Fig. 4.2). The entire MW distribution is of importance because different properties depend upon different MW averages. Colligative and most mechanical properties depend upon $\overline{M}_n$; melt and solution viscosities upon $\overline{M}_w$ and $\overline{M}_v$; and viscoelastic properties upon $\overline{M}_Z$ (Fig. 4.3). Calibration curves which utilize narrow fractions of known MW determined by an absolute method must be constructed for each polymer. These curves permit the rapid determination of MW

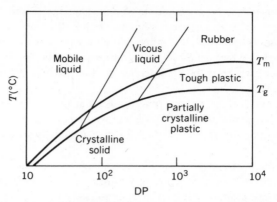

FIGURE 4.1. Approximate relations among DP, T_g, T_m, and polymer properties.

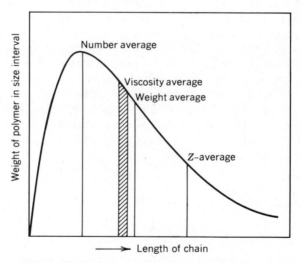

FIGURE 4.2. MW distribution in a typical polymer (from van Krevelen and Hoftyzer[4]).

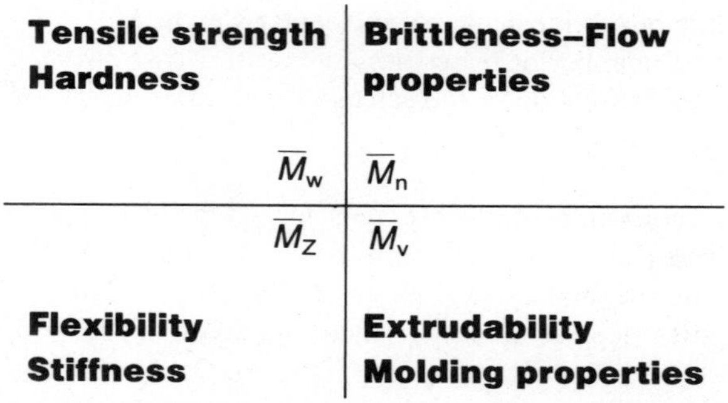

Tensile strength Hardness	Brittleness–Flow properties
$\overline{M}_w$	$\overline{M}_n$
$\overline{M}_z$	$\overline{M}_v$
Flexibility Stiffness	Extrudability Molding properties

FIGURE 4.3. Various MW averages and the properties which they affect (from Waters Associates Inc.[2]).

108

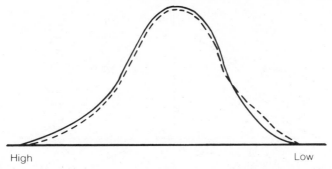

FIGURE 4.4. Schematic MW distribution curves for two lots of membrane polymer (from Waters Associates Inc.[2]).

distribution. A particularly useful means of employing GPC is in the comparison of new lots of polymer with historically accepted standards (Fig. 4.4). With experience, correlations can be made as to probable behavior with respect to processing properties such as melt or solution viscosity, and to end-use properties such as tensile strength, brittleness, flex life, impact strength, burst strength, toughness, elastic modulus, hardness, tear strength, and stress-crack resistance. The MW distribution curve is superior to solution viscosity and other single-point polymer-lot evaluation techniques because it provides more information. The shape of the peaks may be symmetrical or skewed. The former indicates a statistically uniform distribution of macromolecules, whereas the latter is biased to either the high-MW or the low-MW ends of the MW distribution. The properties of the polymer are affected by the amount and direction of the bias (Table 4.1). Shoulders on the curve at the high-MW end may indicate the presence of gel components, whereas those at the low end usually suggest that material has been added to the polymer. It is possible for two lots to have identical viscosities but be composed in one instance of a material with a normal MW distribution and in another of a blend of high- and low-MW material (Fig. 4.5). (The latter could conceivably have been put together in an attempt to satisfy a particular viscosity specification.) However, although the

TABLE 4.1 EFFECTS OF HIGH- AND LOW-MW COMPONENTS ON SOME POLYMER PROPERTIES[a]

Polymer Property	Effect of High-MW Components	Effect of Low-MW Components
Strength	Increases	Decreases
Viscosity	Increases	Decreases
Required processing temp.	Higher	Lower
Chemical resistance	Increases	Decreases

[a]From Waters Associates Inc.[2]

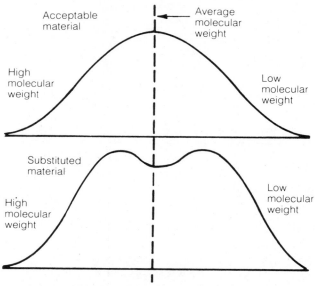

FIGURE 4.5. Schematic MW distribution curves for a single viscosity grade and a blend of two viscosity grades of membrane polymer (from Waters Associates Inc.[2]).

viscosity might be the same in both cases, the physical properties of films prepared from the blend of different viscosity grades is usually significantly inferior (Table 4.2).

Less dramatic than two separate peaks, but similar in nature, is the effect of varying peak widths at a given melt index (MI), a measure of melt viscosity, which

TABLE 4.2 MECHANICAL PROPERTIES OF CELLULOSE NITRATE FILMS[a,b]

Nitrocellulose Viscosity Grades	After 24-h Drying at 60°C			After 60-h Exposure in June under Cloudless Skies		
	Elongation (%)	Double Flexes	Breaking Load (kg/mm)	Elongation (%)	Double Flexes	Breaking Load (kg/mm²)
E 1160 + E 400	6.9	29	7.4		Brittle	
E 1160 + E 510	8.6	34	7.8		Brittle	
E 950 + E 400	9.3	38	8.1	3.8	0	4.9
E 950 + E 510	9.5	41	8.3	4.7	13	5.6
E 730 + E 400	9.7	49	8.6	5.6	21	6.4
E 730 + E 510	10.4	53	8.7	6.2	29	7.2
Standard E 620	15.4	93	8.7	8.1	53	7.8

[a] From Reference 5.
[b] The films were prepared from solutions of equivalent viscosity achieved by blending various viscosity grades as compared to those of a sample viscosity grade.

decreases as MW increases,[6] upon the physical properties of the resultant membranes.[3] The MI is a single bulk point which can be obtained from materials of varying MW distribution. However, for a given MI, the broader the sample peak, the lower will be the tensile strength (Fig. 4.6). This is so because the broader the MW, the greater the amount of low-MW material, and the presence of even a small amount of the latter adversely affects the mechanical properties of a polymer. On the positive side, it is the expectation of significantly improved physical properties for polymers of narrow MW distribution that has prompted the considerable amount of research in this area. In this decade new narrow MW distribution grades of existing polymers rather than entirely new polymers are expected to account for many of the improvements in membrane properties.

Although the subject of MW and MW distribution is of paramount importance to the development of practical synthetic polymeric membranes, it has not been given sufficient attention in the membrane literature. Among the guidelines which pertain to preformed polymers for membrane use which this author advocates are the following:

1. The highest-MW polymer grade should be evaluated first, then the next highest, and so on. This means that, where available, film-forming or fiber-quality grades should be given precedence over extrusion grades. Injection molding grades should, in general, only be employed as minor components in polymer blends. Flexibility, elasticity, and toughness are strongly dependent on $\overline{M}_Z$. Membrane integrity, particularly skin integrity in integrally-skinned membranes, tends to increase with MW. This may be related to increasing probability for chain entanglement with increasing MW. However, it does not follow that the highest MW grade will invariably be the most suitable since optima do exist. Occasionally, for example, MWs are so high that solubility is inadequate. In such cases, grainy, elastic gels rather than free-flowing solutions may result. Another problem is the tendency for nascent membranes of a high-MW polymer to shrink in the transverse direction during the manufacturing process. Although in such cases it is frequently possible to hold the edges to restrain shrinkage, the stress of contraction is then often relieved by the formation of tears. This condition may be alleviated by the utilization of a lower-MW grade. This discussion obviously applies primarily to those poly-

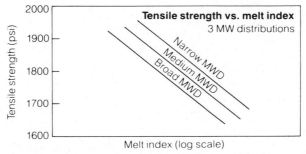

FIGURE 4.6.   The influence of MWD peak width on tensile strength (from Waters Associates Inc.[2]).

mers such as the cellulosics where a multiplicity of viscosity grades are available. Unfortunately, for many polymers only two grades, usually a medium-viscosity extrusion and a low-viscosity injection-molding grade, exist. In such cases, if the extrusion grade is inadequate, synthesis of higher-MW grades may be the only recourse. This is sometimes justifiable insofar as the cost of membrane polymer(s) tends to play a minor role in the overall membrane cost.

2. Membranes made by the dry phase-inversion process generally require higher-MW polymers than do those prepared by the wet process.[7] This is so because dry-process solutions generally contain a high concentration of nonsolvent pore former with the result that the attainment of sufficient viscosity (for processing reasons) often cannot be achieved by simply increasing polymer concentration. In such a situation, the alternative is to increase polymer MW.

3. For reasons cited above, narrow MW distributions are preferable to broad distributions and single viscosity grades are preferable to blends of two viscosity grades. The present author knows of no instance in which blends of two viscosity grades of a given polymer are justifiable for membrane applications. This rule does not apply to blends of two different polymers—for which the minor component must often be of a lower MW to satisfy compatibility requirements.

In addition to MW and MW distribution, macromolecules are also characterized by their molecular architecture. This category is distinct from chemical structure which refers to the molecular group level. Polymer chains may be linear (Fig. 4.7a), branched (Fig. 4.7b), or cross-linked (Fig. 4.7c). If a polymer is only lightly cross-linked, it may retain its solubility. If it exhibits a high cross-link density it becomes an insoluble three-dimensional network. When such highly cross-linked materials

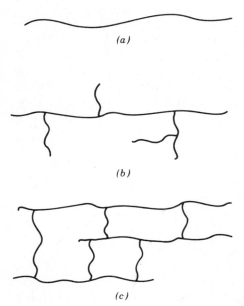

(a)

(b)

(c)

FIGURE 4.7. Schematic representation of (a) linear, (b) branched, and (c) network polymers.

(a)—AABAAABBABABBAAAB—   (b)—ABABABABAB—
(c)—AAAAAAAAABBBBBBBAAA—   (d)—AAAAAAAAAAAA—
                                                                   BBBBBB

FIGURE 4.8. Copolymer types: (a) random, (b) alternating, (c) block, (d) graft.

occur as a result of heat treatment, they are known as *thermoset* resins. Polymers which retain their solubility and melt processability are *thermoplastics*. Hybrid types are also known, as, for example, in the case of a soluble thermoplastic resin containing unsaturated groups which can be cross-linked after membrane formation in a posttreatment step.

A polymer which contains a single monomer or monomer pair is known as a homopolymer. If, on the other hand, the macromolecule contains two or more monomers (chain propagation polymer), or three or more monomers (step polymer), it is termed a copolymer. Among the possible copolymer structures are: random (Fig. 4.8a), alternating (Fig. 4.8b), block (Fig. 4.8c), and graft (Fig. 4.8d).

In addition to macromolecular architecture, the steric position of monomer units relative to one another within a chain segment is a significant factor affecting the state of molecular aggregation and hence processing and end-use properties. *Stereoisomerism* in polyvinyl polymers arises because, in polymers from substituted vinyl groups, every other carbon is asymmetric, that is, has four different groups attached to it. Since asymmetric carbon atoms have two forms (D and L) which are mirror images of one another, they may be joined together in a variety of sequences among which three are recognized as having fundamental significance: atactic, syndiotactic, and isotactic (Fig. 4.9).[4,6,8] In the atactic form (Fig. 4.9c), the D and L forms are joined together in a completely random manner; in the syndiotactic form (Fig. 4.9b) the D and L forms alternate with one another; and in isotactic form (Fig. 4.9a) the sequence is entirely D or L. Because the atactic structure is the least regular, it will exhibit lower T_m and T_g values and greater solubility than the other forms. However, a considerable difference exists even between syndiotactic and isotactic isomers.

Cis–trans isomerism is related to stereoisomerism. If both substituents are on the same side of a structure such as a ring or a double bond which prohibits free rotation, they are cis to one another. If they are on opposite sides, they are trans. Natural rubber is made up of *cis*-1,4-polyisoprene (**I**),

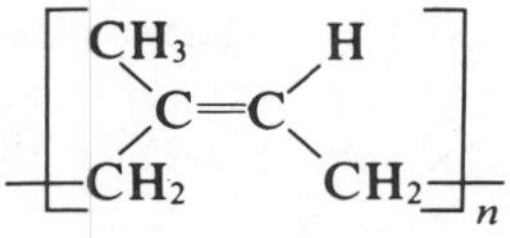

$$\left[\begin{array}{cc} CH_3 & H \\ \diagdown \!\! C\!\!=\!\!C \!\! \diagup \\ CH_2 & CH_2 \end{array} \right]_n$$
I

, whereas gutta percha (**II**),

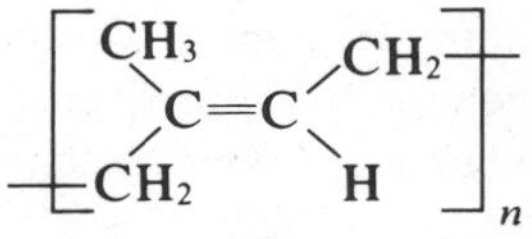

$$\left[\begin{array}{cc} CH_3 & CH_2 \\ \diagdown \!\! C\!\!=\!\!C \!\! \diagup \\ CH_2 & H \end{array} \right]_n$$
II

is the trans isomer.

The former is a rubber and the latter, a stiff, leathery material. Depending upon such factors as overall macromolecular architecture, steric relationships at the

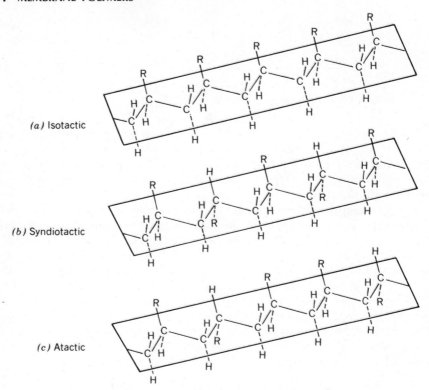

FIGURE 4.9. Tacticity in polyvinyls (from Brydson[6]).

polymer segment level, the strength of intermolecular and intramolecular interactions between polar groups, and temperature, several possible states of molecular aggregation in polymers are possible: *liquid, amorphous (glassy* or *rubbery)*, and *semicrystalline*. More extensive discussions of polymers in the liquid state will be found in Chapter 5 (Polymer Solutions), Chapter 6 (Dense Membranes), and Chapter 9 (Liquid and Dynamically Formed Membranes).

Polymers with sufficient steric irregularities at the molecular or segmental levels as to prohibit more often than permit intermolecular or intramolecular association of polymer chains tend to be amorphous. At low MWs (oligomers) the polymer will generally be solid below the flow temperature and a liquid above this temperature. At high MWs (the region of interest for membranes) a rubbery intermediate zone is usually observed. There are, therefore, two transition temperatures—the glass transition temperature (T_g) at the transition from the rigid solid (glass) to the flexible rubber state and the transition (T_m) from the rubber to the liquid state. In the strictest sense, T_m refers to the crystalline melting temperature, but is also loosely used as the flow temperature for amorphous polymers. In the rubbery state, considerable segmental motion is possible and time- and temperature-dependent interchain oscillations result in transient pores of significant size. Segmental motion is favored because rubbers consist of long, linear chain molecules of high intrinsic flexibility. They also have a paucity of polar groups, an absence of crystallinity, and are only lightly cross-linked (typically, a DP of 300 mobile segments between cross-links).[9] As a result, permeability is generally high, and permselectivity low, when the permselective portion of a membrane is in this state.

In the glassy state motion is restricted to bond vibrations. The glassy state is intermediate in density between the crystalline and the rubber states. It is the state which is encountered in the dense skin layers of highly permselective membranes for gas separations and hyperfiltration. By annealing and/or solvent exchange it is possible to increase the density—by decreasing the average interchain displacement and the attendant free volume within a glassy state—without reaching the density of the crystalline state. Since polymer chains in concentrated polymer solutions are more tightly coiled in good than in poor solvents, this author feels that the casting of membrane polymers from good solvents is essential to obtaining the tightest coils and hence maximum permselectivity for membranes in the glassy state.

Since the glassy state is less dense than the crystalline state, a given mass of the former occupies a larger volume. The difference between the volume occupied by a completely crystalline polymer and that occupied by a polymer in the glassy state is known as the free volume. It is through this free volume that permeation takes place. Actual sorption sites are associated with gaps between chain segments. Free volume in turn is thought[10] to consist of two fractions, namely the difference between the volume occupied by the crystalline state and that occupied by a completely densified glass and any additional free volume, referred to as *unrelaxed volume* (because it is associated with unrelaxed gaps between chain segments). Annealing and/or plasticization effect redistribution of these intersegmental gaps and generally result in densification through a gradual disappearance of the unrelaxed volume. Since plasticization by a penetrant species results in the transient appearance of large interchain gaps, it has an adverse effect on permselectivity. High permselectivity, therefore, is associated with rigid (high T_g and plasticization-resistant) polymer chain backbones with small intersegmental gaps.

For polymers of low crystallinity, which includes many of those providing barrier layers of high permselectivity and permeability, the T_g is an essential property. Both macromolecular and supermacromolecular (macromolecular aggregate) architecture influence the value of T_g. On the macromolecular level, the nature of the groups which comprise the polymer backbone, the nature and size of side-chain groups which increase the energy required for rotation, MW, MW distribution, and the presence of any factor within the macromolecule which can influence intermolecular association whether by steric or cohension forces, all influence the value of T_g. On the supermacromolecular level, the size of side-chain substituents, the presence and positional regularity of polar groups (for secondary valence bonding), the presence of primary valence bonding (cross-linking), and the presence of internal (attached), or external (free-standing) plasticizers are of importance. The presence of aliphatic $-C-C-$, $-C-O-$, or $-OSiO-$ bonds in the polymer chain leads to flexibility (and hence low T_g) because of the ability of these atoms to rotate more or less freely about these bonds. Any factor which hinders free rotation decreases flexibility and increases T_g. The introduction of ring structures, particularly planar aromatic rings, greatly increases chain stiffness. Even methyl groups hinder rotation about the $C\!\!-\!\!C$ bond. Thus polypropylene has a higher ($T_g = -27°C$) than does polyethylene (T_g $-120°C$). Poly(isobutylene), on the other hand, has a lower ($T_g = -65°C$) than polypropylene because the former has two methyl groups attached to the same carbon atom, thereby lowering the dipole moment (polarity). The same is true of poly(vinylidene chloride) ($T_g = -17°C$) relative to poly(vinyl chloride) ($T_g = +80°C$). Hydrogen bonding is among the strongest of the secondary valence forces which act to increase T_g. Poly (ϵ-caprolactone), a polyester with little or not capacity for forming hydrogen bonds, has a T_g of $-70°C$, whereas nylon 6, otherwise closely related but with a strong capacity for hydrogen bonding by virtue of the presence of the $-CONH-$ groups, has a T_g of $\sim 50°C$.

The crystalline state is the most dense state of polymer aggregation. Since a perfect crystallite is virtually devoid of free volume, it is assumed that little permeation occurs through crystallites themselves. The short- and long-range order inherent in crystallite formation requires a high degree of regularity at the chemical group, segmental, and molecular levels. This restricts crystallinity to linear or slightly branched macromolecules. Any randomness in structure at any of these levels acts either to eliminate crystallite formation altogether or to limit it to certain domains within the bulk matrix where the requirements for structural regularity can be met. In practice, therefore, the perfect crystalline state is seldom encountered. What is more often found is the *semicrystalline* state which can be considered as an amorphous matrix in which crystallites are embedded. With few exceptions, the exact nature and size of crystallites in the barrier layer of permselective membranes has not yet been satisfactorily resolved. Indeed it would appear (Chapter 7) that crystallinity is best minimized in this layer and that the presence of crystallites serves primarily to decrease the area available for permeation, and to restrict the swelling of the amorphous matrix in which they are embedded. In the latter role, crystallites function as *virtual* cross-links. However, crystallinity is of importance with respect to mechanical properties. The decreased intermolecular distances which occur in crystallites increase intermolecular cohesion and such properties as stiffness, tensile strength, and T_g. Indeed, were it not for the presence of crystallites, the polyolefins would be rubbery at room temperature. The presence of crystallites in the polyolefins and polyfluorocarbons is fundamental to the development of pores by the stretching and annealing procedures employed in the manufacturing of Celgard®[11] and Gore-Tex®[12] (Chapter 8). Three limiting conformations are possible for solid linear macromolecules: (*a*) the random coil, (*b*) the folded chain, and (*c*) the extended chain. According to van Krevelen and Hoftyzer,[4] the fringed micelle is effectively a mixture of (*a*), (*b*), and (*c*) (Fig. 4.10). Other than the random coil, which is encountered in both rubbery and glassy amorphous states, the fringed micelle, which is believed to be present in semicrystalline polymers with a low degree of crystallinity, and folded chain lamellar crystallites in the Celgard® protomembranes, these macroconformations are of unproven importance to synthetic polymeric membranes.

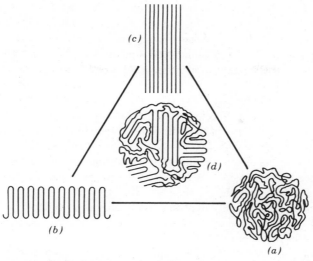

FIGURE 4.10. Limiting macromolecular conformations of linear polymers (after Wunderlich[13]).

Knowledge of crystalline morphology is of importance in understanding the permeability and permselectivity of such polymeric membranes as dense films (including the thin skin layers of asymmetric cellulose acetate membranes), dialysis membranes, and membranes for the separation of gases. In addition to its effects on material-transport phenomena, crystallinity will affect the various chemical-accessibility and mechanical-property parameters, which in turn will influence a membrane's functional behavior in a time-dependent manner.

The fringed-micellar model was once widely employed to describe the mechanical and transport behavior of dense polymeric membranes. This model viewed films of semicrystalline polymers as consisting of crystalline regions embedded in an amorphous matrix. Since X-ray diffraction studies indicated that most crystalline regions extend no more than a few hundred angstrom units, which is less than the completely extended length of polymer chains, it was assumed that the crystallites contain only sections of molecules. It was therefore concluded that the overall semicrystalline structure consists of a composite single phase in which individual polymer chains participate in both crystalline and amorphous regions. More recently, however, the concept of polymer single crystals has been gaining ascendancy. According to this model, polymer crystallites are formed from polymer molecules which fold back on themselves in the shape of spirals. In such a situation, individual molecules need not participate in both crystalline and amorphous regions. In the case of highly crystalline polymers, the amorphous regions may represent only chain ends and minor dislocations within the crystallites.

The most generally applicable technique for the characterization of the crystallinity of polymeric membranes is the X-ray powder method of Debye and Scherrer,[14] and Hull.[14] According to this procedure, a membrane is placed in the path of a monochromatic parallel X-ray beam. Because of the more or less random orientation, a lattice plane will always be available at the correct angle in some of the crystals. The diffraction patterns appear as a number of concentric circles on a flat photographic plate. After the lattice spacings and reflection intensities have been determined, the crystals are classified according to system, class, translation group, and space group by systematic application of crystal-structure theory in conjunction with tables.

An adequate description of the crystalline structure of a polymeric membrane includes the dimensions of the unit cell, percentage crystallinity, crystallite size, and orientation. The volume of the unit cell is established by calculating its angles and axes. The centers of gravity of the constituent atoms are determined by a Fourier analysis of the electronic-charge clouds of the various atoms in the elementary cell. Once the dimensions of the unit cell have been established, it is possible to calculate the density of the purely crystalline polymer.

The percentage crystallinity may be conveniently determined from density measurements by reference to a plot of density versus percent crystallinity. In such a plot, the point of maximum density is deduced from the value calculated from the dimensions of the unit cell. Samples which have been shown by X-ray diffraction to be completely amorphous are utilized to determine the density at the other extreme of the curve. In cases where infrared adsorption bands characteristic of crystalline and amorphous regions exist, it is sometimes possible to substitute infrared measurements for the more tedious density determinations. A direct technique for determining the volume fraction of polymer present in an ordered state involves the width at half or maximum intensity of the radial-intensity curve.

Crystallite size and orientation can be determined by electron-microscopic and X-ray methods. In the latter, increasing orientation causes the diffraction rings to give way to arcs and points. Variations in microcrystalline order can be observed by differential-scanning calorimetry (DSC), even when crystallite size is too small to be detectable by X-ray diffraction.

Highly crystalline polymeric membranes are usually less permeable than amorphous ones. Permeant species are generally considered insoluble in the crystalline regions so that transport occurs in the amorphous phase. Increasing crystallization therefore decreases the volume of amorphous material available to carry these species and increases the tortuosity of the path across the membrane. Lasoski and Cobbs[15] varied the crystallinity of poly(ethylene terephthalate), a polymer which can be quenched from the molten state to a stable, completely amorphous material, be annealing for varying lengths of time at a temperature just above 100°C. They found that the steady-state permeation of water vapors through these membranes decreased as crystallinity increased from 0 to 40%. Reitlinger and Yarko[16] have observed the inverse relationship existing between polymer density and permeability during the isothermal crystallization of natural rubber. The fact that permeability depends upon the microcrystalline structures of polymeric membranes is also responsible for the linear relation which exists betwen moisture regain and the amorphous fraction in cellulose.[17] Furthermore, where hydrolytic or bacterial degradation of polymeric membranes occurs, it is the more accessible amorphous regions which are first affected.[18] This fact has been the basis of an experimental technique for determining accessibility, and hence the magnitude of the amorphous fraction, by kinetic studies of the rates of hydrolysis.[19]

The crystalline/amorphous ratio is of importance to the mechanical properties of polymeric membranes[20] through which it can be causally related to such performance characteristics as the rate of decrease in permeability with time owing to cold flow.[21] The individual crystallites, of course, are highly rigid structures, especially in the direction of their long axis. However, the crystals can be more readily enlarged in the transverse direction by overcoming van der Waals forces between neighboring chains. Because of their rigidity, crystallites inhibit the compaction of polymeric membranes by acting as virtual cross-links. Although the crystallites may be considered as ideal elastic regions which do not flow readily, the intervening amorphous regions may be quite subject to deformation. The extent to which such deformation occurs can vary widely with such environmental factors as temperature [below the glass transition temperature (T_g) amorphous rigidity can be substantial] and the presence or absence of plasticizers (which enhance deformation)[22]. There are also mesomorphic or paracrystalline regions of varying degrees of crystalline order which exhibit resistance to deformation midway between crystalline and amorphous structures.

Crystallinity in polymeric membranes can be attributed in part to the steric regularity of their component macromolecules and in part to the strength of intermolecular forces. In many cases, rapid quenching from a melt yields a highly amorphous film which can be crystallized by annealing above the T_g. However, rapid isolation from solution by evaporation of the solvent can also be made to yield a highly crystalline product if casting is done from a solution containing higher-boiling poor solvents, swelling agents, or nonsolvents.[23]

Of great significance with respect to the state of macromolecular aggregation is the existence of bonds between polymer chains. Known as *cross-links,* they can be attributed to primary valence forces (covalent cross-links) or weaker secondary valence forces such as salt bridges, dipole–dipole interactions, and the presence of crystallites (virtual cross-links). Lightly cross-linked materials behave in the same manner as branched polymers. They may retain their solubility and flexibility and may even participate in crystallite formation. More heavily cross-linked polymers, however, lose their flexibility and solubility and experience an increase in T_g. Covalent cross-linking is of particular importance in the case of ion-exchange membranes where it serves to counteract the substantial swelling forces owing to the presence of fixed ionic charges.[24]

The *mechanical properties* of polymeric membranes received little attention during the early stages of membrane development when more fundamental performance characteristics, such as permeability and selectivity, were necessarily emphasized. The result of this neglect is that the integrity of filter cartridges—particularily those containing microfilters of maximum porosity (and hence minimum strength)—is often less than desired. Mechanical properties are the result of structure at the chemical group, macromolecular, microcrystalline, and colloidal levels. Consider, for example, the implications of structure on one key mechanical property, namely flexibility. Amorphous polymers such as the polycarbonates and the polysulfones have outstanding flexibility in both dense and porous conditions. Highly crystalline and/or highly cross-linked polymers on the other hand, tend to be brittle. Semicrystalline polymers can fit into either category depending on the nature of their secondary valence forces and the manner in which they are handled. Those such as branched low-density polytheylene with weak cohesive forces exhibit adequate flexibility since mobile non-cross-linked amorphous regions act as a form of stress-relieving internal plasticization. On the other hand, semicrystalline polymers which exhibit the potential for hydrogen bonding tend to be friable because intermolecular and intramolecular hydrogen bonds constitute virtual cross-links and friability is proportional to cross-link density. If water-swollen cellulosic or nylon 6,6 membranes are dried directly, capillary forces will promote a high concentration of virtual cross-links with membrane densification and/or embrittlement as the result. If, however, drying is effected via an appropriate (e.g., stepwise isopropanol $\rightarrow$ hexane) solvent-exchange process, cross-link density will be minimized and flexibility will be maintained in the dry state. Both tensile and compressive properties are important. The elastic modulus, stress at yield, stress at break, elongation at yield, and elongation at break are the principal tensile properties of interest (Fig. 4.11). Tensile properties can in turn be empirically related to other less fundamental, but nevertheless, important properties such as flexibility. For example an elongation at break of $\geqslant 10\%$ usually correlates with flexibility which permits convolution without cracking. The elastic modulus is an indication of stiffness, that is, of ability to withstand stress without changing dimension. Stress at yield is related to elasticity, which is the ability to undergo stress without suffering a permanent set. This is also the point at which elastic deformation changes to plastic flow. In plastic flow, the crystallites within the body of the material slip past each other, resulting in permanent deformation. The stress at break is the ultimate ten-

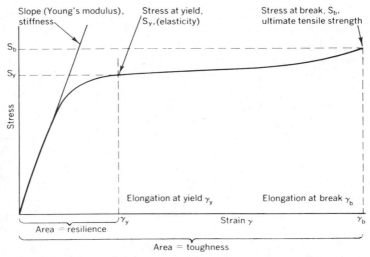

FIGURE 4.11. Mechanical properties based on a typical stress–strain diagram (from Higley[21]).

sile strength of the material and a measure of its ability to carry a dead load. The area under the curve within the elastic limits of the material is a measure of its resilience—the ability to absorb energy without suffering permanent set. The area under the entire curve is a measure of its toughness—the ability to absorb energy and undergo large permanent deformation without rupture.

The mechanical properties of hyperfiltration membranes under compression are of particular interest since they are related to the stresses encountered by the membrane under operating conditions. Compressive-yield studies seek to relate compressive-yield points to permeability decreases during hyperfiltration. They can be determined graphically from the stress–strain curve by drawing a line tangent to the portion of the S-shaped curve of lowest slope, where it is the point at which curve and tangent touch (Fig. 4.12). At a given pressure the slope of the flux–time curve (Fig. 4.13) is a measure of membrane stability. Two slopes are of interest: the initial slope, characteristic of the rapid flux decline which occurs shortly after pressurization, and a second slope after pressurization has been imposed for a period of several hours. The somewhat anomalous existence of positive slopes may be observed both after depressurization and after precompaction (at a pressure higher than the operating pressure), which permits stress relaxation during operation.

Because of their great size and complexity, macromolecules can be considered at various structural niveaux, among which are the macromolecular aggregate, macromolecular, segmental, and chemical group, levels. The finest level with which one is usually concerned in membranology is the chemical group or moiety. The latter is an assemblage of atoms which act in concert. However, the concept of a group or a moiety is somewhat nebulous and depends upon the point of view of the observer. For example, polyethylene (PE) can be seen as an assemblage of methylene ($-CH_2-$) groups. More frequently, it is described as a polymer composed of ethylene ($-CH_2CH_2-$) groups. Under certain circumstances, such as in the

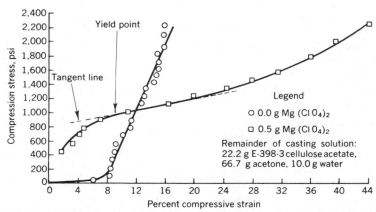

FIGURE 4.12. Compressive stress–strain curves of cellulose acetate hyperfiltration membranes (from Higley[21]).

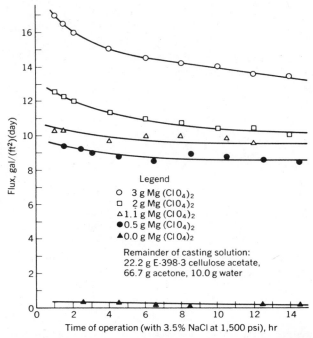

FIGURE 4.13. Flux decline in hyperfiltration membranes versus swelling-agent concentration in casting solutions (from Higley[21]).

case of comparisons with poly(hexamethylene adipamide) or poly (ϵ-caprolactone), PE could be considered as an array of hexamethylene ($-[CH_2]_6-$) moieties. For this reason the listing of a large number of moieties of various sizes and polarities is deemed inappropriate. Instead, a few general principles will be delineated following which the major polymer classes which are of practical interest to membranology will be discussed.

In general, all chemical differences between prospective polymeric materials can be ascribed to group differences in polarity and/or to steric effects. Polarity, which is attributable to unevenness in electron distribution, is quantitatively described in polar group terms as charge density, dipole moment (Table 4.3), and hydrogen-bonding capacity (Table 4.4), as well as by such bulk electrical properties as the dielectric constant and ion-exchange capacity and by surface properties such as the parachor and critical surface tension. Furthermore, since polarity strongly influences solubility and permeability, all of these properties are in turn related to the Hildebrand solubility parameter δ and its various components and refinements (Chapter 5). The dielectric constant ϵ tends to increase as δ increases. Typical values of ϵ and δ are: poly(tetrafluoroethylene), 2.1, 6.35; polyethylene, 2.3, 7.9; polystyrene, 2.6, 8.6; poly(ethylene terephthalate), 2.9, 10.7; cellulose triacetate, 3.2, 9.3; cellulose acetate, 3.6, 9.7; nylon 6,6, 4.0, 13.9. Unfortunately, such correlations are much more qualitative than would appear from this incomplete series

TABLE 4.3 POLAR GROUPS OF IMPORTANCE FOR INTERMOLECULAR ATTRACTION[a]

Polar Group (+ --)	Dipole Moment (Debye units)	Polar Group (+ --)	Dipole Moment (Debye units)
H—F	1.9	>C=NH	2.5
H—Cl	1.1		
H—O\	1.6	N with two =O (O at top and bottom)	3.9
H—S—	0.9	—C≡N	3.8
H—N<	1.6	>C.O.C< (ether)	0.9
>C—Cl	1.7		
H—C	0.4	>C—N=O	1.9
>C=O	2.5	>C=S	3.0
>C—F	1.5		

[a] From Mark.[9] Reprinted with permission from *CHEMTECH*, © 1984, American Chemical Society.

so that their chief utility lies in helping us to understand general trends rather than in the developent of precise predictive functions. The common neutral polymers are dielectrics or insulators, whereas ion-exchange polymers tend, particularly in the presence of water, to be electrical conductors. Ionomers, on the other hand, bridge the gap between the two classes. They tend to be dielectrics at low capacities and conductors at high capacities (≥ 10 mol % ionogenic groups). For high-capacity ionomers, the formation of ionic clusters is favored for low T_g polymers where

TABLE 4.4 HYDROGEN-BOND DONOR AND ACCEPTOR GROUPS[a]

H-Bond Donors		H-Bond Acceptors	
F—H	Strong →	O=C<	Strong
—O—H	Strong →		
>N—H	Strong →	O<(H)(R)	Strong
R>C—H (R,R,R)	Weak if R = C or H →	N<(R)(R)(R)	Strong
R>C—H→ (R,R,R)	Strong if R = F or Cl →	O,O >N—	Strong

[a] From Mark.[9] Reprinted with permission from *CHEMTECH*, © 1984, American Chemical Society.

chain flexibility is pronounced. Cluster formation is less likely in high T_g ionomers. Polymer polarity thus plays an important role in hyperfiltration membranes which require dielectrics and in electrodialysis membranes which require conductors (Chapter 2). In addition, polar groups, especially, ionogenic groups such as $SO_3^- M^+$, strongly influence the mechanical properties of membrane polymers.

Because of weak cohesive forces, nonpolar molecules will generally be more flexible and widely spaced. There are the inevitable exceptions to these rules whenever steric and polarity effects oppose rather than reinforce one another. Thus poly(vinyl acetate), although polar, is (owing to the random positioning of the bulky acetate groups) sterically incapable of undergoing crystallization. Polyethylene, on the other hand, despite its weak cohesive forces, possesses such a simple, highly symmetrical structure that it crystallizes readily.

4.2 NEUTRAL POLYMERS

The *polyethylenes* (PEs), $H{+}CH_2{-}CH_2{+}_n$, are characterized by low cohesive energy ($\delta = 7.9$) and low T_g ($-120°C$) but high crystallinity. There are literally hundreds of grades of polyethylene which differ from one another in degree of branching (and hence in density and crystallinity), in MW and MW distribution and in the presence of impurities. Three specifically distinct types are generally recognized (Table 4.5). All of the PEs, but especially the high-density types, have excellent resistance to solvents with a $\delta > 9^6$.

Environmental stress cracking refers to the phenomenon whereby fracture of a film or a membrane occurs under much lower stress in certain (chemical) environments than in the absence of the environment.[6] It was once believed to be a fatal weakness of the PEs, but is much less of a problem with high-MW grades (MI < 0.4). Considerable variation in properties is possible through the inclusion of small amounts of a second olefin such as propylene or 1-butene. The effect of a second monomer is to produce a controlled amount of short-chain branches with a concomitant decrease in crystallinity and crystallite size. In spite of their appreciable crystallinity, low-density PE films exhibit gas permeabilities in the same range as

TABLE 4.5 TYPES AND PROPERTIES OF POLYETHYLENE

Name	Process	Density	Molecular Architecture (branches/1000 C atoms)	% Crystallinity
Branched	High pressure	Low (0.91–0.94)	Both short and long 30 branches	50–60
(Ziegler)	Low pressure	Medium (0.940–0.955)	Short side chains 5–7 branches	>80
(Phillips)	Low pressure	High (≥ 0.958)	Linear, no branches	>90

that exhibited by rubbers. This is apparently the result of the low cohesive forces in the amorphous regions which permit appreciable segmental mobility. The high degree of crystallinity of the high-density types, on the other hand, results in one-fifth the permeability since there is one-fifth the amorphous (and free) volume. Because of their higher permeability, the low-density polyethylenes are better suited for gas-separation applications where nonporous barrier layers are desired.

For the porous Celgard®[11] membranes prepared by stretching of semicrystalline films (Chapter 8), permeation occurs through submicrometer-sized intercrystalline voids rather than through the free volume in the amorphous phase. Tertiary carbon atoms are located at the branch points in PE because of which it is susceptible to oxidation, although much less so than some competing polymers such as polypropylene. Low-density PE is sparingly soluble in such solvents as carbon tetrachloride and benzene at temperatures below 100°C, but not sufficiently so as to permit practical membrane casting. However, LDPE is sufficiently soluble at 220°C in dihydroxyethyl tallow amine (DHTA)[25] to permit the fabrication of both integrally-skinned and microporous phase-inversion membranes (Chapter 7). On balance, PE appears to be less attractive as a membrane polymer than its closely related cousin, polypropylene (PP). However, because PE contains fewer tertiary carbons than PP, the former is more resistant to sterilization by gamma irradiation,[3] a factor of increasing importance with respect to the utilization of microporous membranes for clinical applications. Polypropylene readily degrades as a result of irradiation ($\overline{M}_n$, $\overline{M}_w$, and $\overline{M}_Z$ all decrease), whereas PE undergoes not only degradation ($\overline{M}_n$ decreases), but also cross-linking ($\overline{M}_w$ and $\overline{M}_Z$ increase).

In addition to homopolymers and copolymers with other olefins, ethylene has been copolymerized with many other monomers. A number of copolymers of ethylene and vinyl acetate (VA) are commercially available. The greater the VA concentration, the less crystalline the copolymer. Those which contain 45% VA are rubbery, whereas those which contain between 10 and 30% VA are waxy. However, polymers which contain only about 3% VA can be considered as modified low-density PEs. Ethylene–vinyl alcohol (EVAL) resins have recently become available and are perhaps of greater interest for membrane applications than the parent copolymers. Copolymers of ethylene and acrylic acid which also contain Na^+ or Zn^{2+} ions were the first commercially available ionomeric polymers.[26] The $-COO^-$ groups in the presence of inorganic counterions cause the ionomers to behave as cross-linked materials which therefore exhibit greater hardness and temperature stability than unmodified PE, while simultaneously retaining thermoplastic behavior at melt temperatures. Certain ionomer grades can even be solvent cast as thin films from toluene–isobutanol solutions. Because they are less crystalline than the PEs, the ionomers are also more permeable to gases.

The first high-volume stereospecific polymer, isotactic PP, is enjoying considerable and increasing success as a membrane material in both Celgard® (Chapter 8) and thermal phase-inversion processes.[25] Like PE, PP is soluble in DHTA at about 200°C. PP has a number of advantages over PE, its chief competitor: an approximately 50°C higher T_g (which means a higher maximum service temperature), greater flexibility, greater resistance to environmental stress cracking, and lower density. On the negative side, it has a higher T_g, a too high brittle point at $\sim 0°C$, and is more susceptible to degradation by oxidation and irradiation. However, the incorporation of up to 3% of ethylene together with propylene results in block copolymers, known as *polyallomers*,[6] which have increased impact strength and a brittle temperature of $-20°C$ (for high-viscosity grades).

Although isotactic PP is a stiff, crystalline polymer with a high T_m and excellent solvent resistance, *atatic* PP, which is formed at the same time, is a useless amorphous polymer which fortunately can be extracted with hexane. Commercial isotactic PP is 90–95% crystalline. The percentage of material which is isotactic is often reported as the isotactic index, a value which is estimated by the percentage of polymer which is insoluble in *n*-heptane. Although both D and L helices occur, they can crystallize together.

The effect of increasing MW on the properties of PP is unusual in some respects. Melt viscosity and impact strength increase, which is normal, but hardness, yield strength, stiffness, and T_g decrease. This has been attributed to the lower crystallization tendency of the high-MW material. In addition, to its use as a membrane material in separations, a bright future can be predicted in controlled-release applications[27] for microporous PP prepared by the thermal phase-inversion process.

The chemical structures, thermal characteristics, available forms, and processing conditions for candidate neutral fluorocarbon membrane polymers are listed in Table 4.6. The important class of perfluorinated ionomers will be discussed separately in a later section of this chapter. Polytetrafluoroethylene (PTFE) is a linear polymer with little or no branching.[28] Because the fluorine atom is too large to fit into the planar zigzag crystalline habit of PE, the macromolecule assumes a twisted configuration with the fluorine atoms spiraling tightly around the carbon backbone. The compact arrangement of the fluorine groups results in stiff molecules and highly crystalline aggregates with a high T_m and thermal stability. However, because of very low cohesive forces ($\delta = 6.2$), no room-temperature solvents exist and tensile strength and creep resistance are not so high as other polymers with comparable T_m's. At elevated temperatures PTFE is soluble in fluorinated kerosenes, that is, in what amounts to its own oligomers. The MW of the PTFEs is unusually high ($400,000 < MW > 9,000,000$). The lower-MW range which consists of dispersion polymers with fibrillar character results in membranes with superior mechanical properties. Small PTFE spheres ($\sim 0.2 \ \mu m$ diam) in a naphtha paste can be ram extruded as a thin film, heated to remove the naphtha and sintered and stretched to develop pores in the intercrystalline regions (Chapter 8). Poly(vinylidene fluoride) (PVDF), the other fluorinated polymer which is found in commercial microfiltration membranes,[29] is cast from dimethyl acetamide (DMAC) or dimethyl formamide (DMF) solutions in a wet phase-inversion process (Chapters 5 and 7). Although not commercially available, excellent wet-process ultrafiltration membranes have been developed employing triethylphosphate (TEP) as a solvent with glycerol as a pore former.[30] Other polymers such as poly(fluoroethylene–propylene)(PFEP) and polyperfluoroalkoxy (PPFA) have properties which are similar to those of PTFE, but are processable as thermoplastics. PPFA is superior to PFEP at temperatures above 150°C. An interesting candidate membrane polymer is

TABLE 4.6 NEUTRAL FLUORINATED MEMBRANE POLYMERS

Polymer P-	Structure (repeat unit)	T_m (°C)	Use Temp. (°C)	Forms Suitable for Membranes	Processing Conditions
—TFE[a]	$-(CF_2CF_2)-$	327	260	Powders, H_2O dispersion	Press (2000–10,000 psi) → preform, sinter 360–380° Ram extrusion
—FEP	$-(CF_2CF_2 co CF_2CF)-$ CF_3	290	200	Naphtha paste Powders, H_2O dispersion	Extrusion
—PFA	$-(CF_2CF_2CFCF_2CF_2)-$ O R_f where $R_f = -C_nF_{2n+1}$		260	Powders	Extrusion
—ECTFE	$-(CH_2CH_2CF_2CF)-$ $C\ell$	245	180	Powders	Extrusion
—ETFE	$-(CH_2CH_2CF_2CF_2)-$	270	180	Powders	Extrusion
—VDF[a]	$-CH_2-CF_2-$	170	150	Pellet + powder	Wet-process phase-inversion DMF, DMAC, TEP solvents

[a]Commercially available in membrane form.

poly(ethylene–chlorotrifluoroethylene) (PECTFE), which has significantly greater strength, wear resistance, and creep resistance than the other fluorinated polymers. Its mechanical properties are similar to those of nylon 6. The bulky chlorine atoms provide steric resistance to slippage of polymer chains. However, PECTFE is swollen by hot chlorinated solvents and attacked by hot amines.

The fluorinated membrane polymers and copolymers are more stable in a greater variety of chemical environments than any other class of membrane polymers. In addition, they are hydrophobic and stable at elevated temperatures. However, their high cost restricts their use to those applications where membranes of other less costly materials will not suffice. One such area is the microfiltration of chemicals for use in the electronics industry.

The acrylic polymers, $CH_2=CR-X$, where R = H, CH_3, or CH_2CH_2OH and X = COOH, $COOCH_3$, or CN, have generated a significant amount of interest as membrane materials. One reason for this interest is the stability of the $-C-C-$backbone. Any degradation usually occurs on the polar side groups which are themselves quite resistant to attack. The acrylic polymers also cover the entire range from water-soluble materials such as poly(acrylic acid), through extremely hydrophilic but insoluble poly(hydroxyethyl methacrylate)(HEMA) hydrogels, to polar but hydrophobic resins such as poly(methyl methacrylate)(PMMA) and poly(acrylonitrile)(PAN). Copolymers are also encountered—both to control swelling as in the case of tetraethyleneglycol dimethacrylate which is utilized to cross-link HEMA hydrogels—and to disrupt order and/or increase hydrophilicity in the otherwise too crystalline and hydrophobic PANs. Finally, certain acrylic monomers have been added as side groups to preexisting polymers as, for example, in the formation of cellulose acetate methacrylate (CAM) from cellulose acetate and methacroyl chloride.

Poly(acrylic acid) and the slightly weaker poly(methacrylic acid) are weak-acid cation exchangers. Because a separate section on ion-exchange polymers follows, it need only be stated here that they have been investigated both as hydrophilic comonomers, as grafts on hydrophobic membrane substrates,[31] and as hydrophilic intermediates to prevent penetration into the cellulose nitrate microporous support layer of dilute cellulose triacetate solutions which were utilized to cast ultrathin films.[32] None of these applications has reached the commercial stage. However, they are being utilized commercially in series with dynamically formed membranes of zirconium oxide.[33] An important application of weak-acid cation-exchange polymers to membranes is to "protect" strong-acid cation exchangers by acting in series with the latter. Because sulfonic acid groups have a strong negative charge they tend to bind irreversibly to polycations, thereby becoming permanently fouled. This tendency to foul irreversibly is diminished if the incoming polycation first encounters a weak acid with which it complexes less strongly. During cleaning cycles such relatively weak complexes can be broken thereby increasing membrane life.[34]

Because of their low chemical reactivity, high strength, and high permeability, the hydrogels (hydrophilic polymers prepared from monomers such as HEMA) have attracted considerable attention for medical applications.[35, 36] The hydrogels

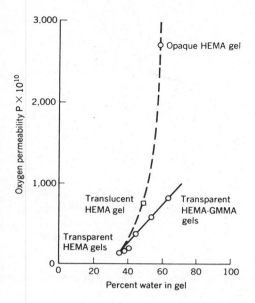

FIGURE 4.14. Effect of hydrogel water content on oxygen permeability (from Yasuda et al.[36]; © 1966).

are typical of those polymeric species which, while soluble at low conversion, become infusible and insoluble at higher conversions. The equilibrium water content of the hydrogel membranes determines their permeability (Fig. 4.14). The water content of the polymerization mixture has a decisive effect on the swelling and deswelling behavior of the hydrogels subsequent to their formation (Fig. 4.15). For less hydrophilic hydrogels there is a critical initial dilution to produce a gel which will not change its volume after it has formed. With increasing gel hydrophilicity, this critical value increases. There is a point, however, beyond which hydrophilicity becomes so pronounced that the gels will swell. The water content of the hydrogels

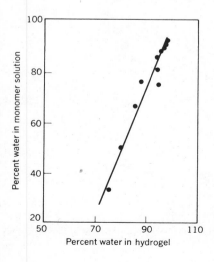

FIGURE 4.15. Effects of the amount of water in the monomer solution upon the amount of water in glyceryl methacrylate hydrogels (from Refojo[35]; © 1965).

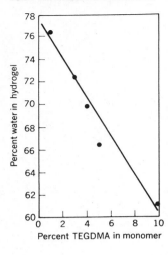

FIGURE 4.16. Effect of cross-linking on the water content in glyceryl methacrylate hydrogels (from Refojo[35]; © 1965).

decreases with increasing concentration of the cross-linking agent in the monomer solution (Fig. 4.16) and increases with increasing concentration of initiator (Fig. 4.17).

Although permeability is approximately proportional to water content in transparent (homogeneous) HEMA gels, it is much greater in translucent and opaque gels with comparable degrees of swelling. This abrupt increase in permeability is an indication that the dense (homogeneous) membrane structure has been replaced

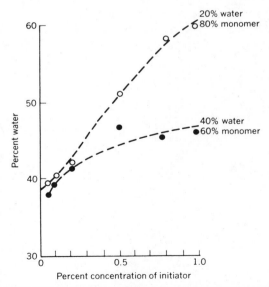

FIGURE 4.17. Effect of the initiator concentration on the water content of HEMA hydrogels (from Yasuda et al.[36]; © 1966).

by heterogeneous structures composed of low-resistance voids within aggregates of microgel particles (see Chapter 7).

Poly(acrylonitrile) and copolymers of AN which contain a few mole percent of another monomer to increase tractibility during processing, and flexibility in end use, have been solvent cast in the wet phase-inversion process to yield tubular[37] and hollow-fiber[38] UF membranes, respectively. Poly(acrylonitrile) membranes are resistant to both hydrolysis and oxidation. The CN group is among the most polar with the result that the Hildebrand parameter is extremely high ($\delta = 12.3$). However, although it is a common misconception that polarity invariably denotes hydrophilicity, PAN is in reality somewhat hydrophobic. Typical solvents are DMF ($\delta = 12$) and 95% HNO_3. Acrylonitrile has also been copolymerized with a number of ionogenic monomers (Chapter 5), among which sodium methallylsulfonate (SMAS) is worthy of special mention because hemodialysis membranes prepared from the AN/SMAS copolymer are commercially available.[39]

Atactic poly(methyl methacrylate) is the epitomy of the amorphous glassy state in polymers. It is hard, rigid, and transparent. Because of the steric protection afforded by the presence of the methyl group in the alpha position, PMMA is very resistant to hydrolysis. By way of contrast, poly(methyl acrylate)(PMA) lacks this protection and can be readily hydrolyzed. Atactic PMMA is soluble in toluene ($\delta = 8.8$), ethylacetate ($\delta = 9.1$), and chloroform ($\delta = 9.3$). It has not been extensively investigated as a membrane material because of marginal toughness—greater than that of polystyrene but less than that of the cellulosics. Recently, however, blends of syndioactic and isotactic PMMA have been utilized to prepare hollow-fiber membranes for dialysis and other blood-handling applications.[40] Increased incompatibility in solution due to structural order *and* the use of blends has increased polymer–polymer interaction and made possible phase inversion and the preparation of microporous membranes. Previous attempts to do so with atactic PMMA failed because the latter's lack of structural regularity resulted in strong polymer–solvent interaction and in a strong tendency to densify during desolvation.

Silicones, or *polyorganosiloxanes*, are thermally stable, hydrophobic organometallic polymers with the lowest P–P interaction ($5 < \delta > 6$) of all commercialy available polymers. This fact coupled with the flexibility of the

$$\left[\begin{array}{c} CH_3 \\ | \\ -O-Si-O- \\ | \\ CH_3 \end{array} \right]$$

backbone results in a low T_g ($-80°C$) and an amorphous rubbery structure for the high-MW poly(dimethyl siloxanes). Silicone rubber membranes are considerably more permeable to gases than membranes of any other polymer (Fig. 4.18). However, silicone elastomers exhibit poor mechanical properties and are, in thin-film form, so subject to pinholes as to require the use of multiple layers to guarantee integrity. As a result, the silicone homopolymer elastomers are rather poor membrane materials. However, two techniques have been developed to circumvent this

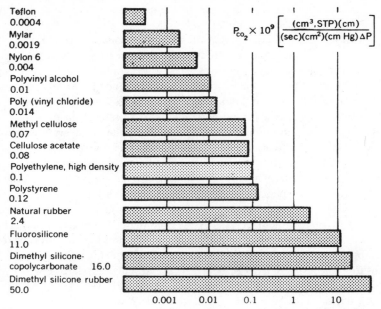

FIGURE 4.18. Carbon dioxide permeabilities in various polymers (General Electric[41]).

difficulty and thereby take advantage of the inherently high gas permeability of silicone. The first is the preparation of block copolymers incorporating hard blocks of a second polymer such as bisphenol-A polycarbonate[42] or polysulfone,[43] which themselves exhibit reasonably high permeability to gases and which simultaneously confer film-forming properties on the copolymers which are greatly superior to those of the silicone homopolymer. The commercially available polydimethylsiloxane-*co*-polycarbonates exhibit excellent gas permeabilities and can be solvent cast to produce thin pinhole-free membranes. Although the permeability of the copolymer is only about one-third that of the silicone homopolymer, this can be overcome by utilizing a thinner copolymer membrane. A range of resins is available in which permeability increases and strength decreases with increasing silicone concentration. Membrane properties can also be varied by changing the nature of the casting solvent. If hexane ($\delta = 7.2$) is utilized as a solvent, the silicone blocks will act as a continuous matrix and the resultant film will be more elastomeric than if methylene chloride ($\delta = 9.7$), a solvent for the polycarbonate blocks, is utilized.[43]

The second technique is to utilize silicone monomers and/or oligomers and form *in situ* an integral silicone membrane as part of a composite in series with a supporting membrane. The latter may itself function as the principal permselective barrier in which case the silicone membrane serves primarily to ensure integrity, that is, the absence of larger and therefore less selective permeability sites.[44] On the other hand, the support membrane may be simply that—a sturdy microporous support for an otherwise fragile but permeable silicone membrane.

Cellulose is the virtually omnipresent naturally occurring high polymer which

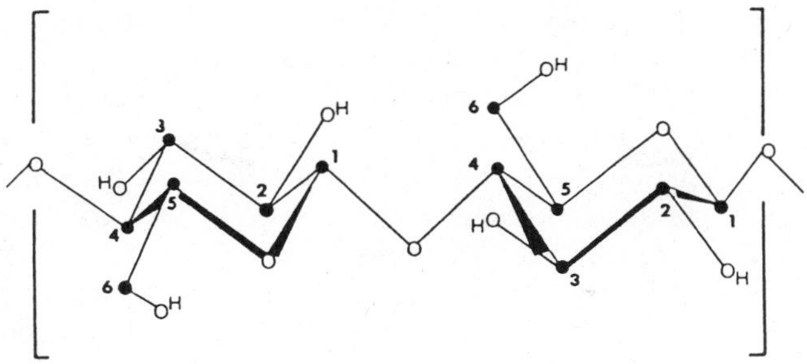

FIGURE 4.19. Chemical structure of cellulose.

consists of anhydroglucose units joined to one another by β-1,4-glucosidic linkages (Fig. 4.19). The anhydroglucose moiety is a cyclic structure in the chair conformation. All of its pendant groups, the single primary and the two secondary hydroxyl, and the β-glucosidic oxygen, are in the equatorial position. As a result, cellulose and its derivatives are linear, rodlike (at low MWs), and rather inflexible molecules.[45] Cellulose itself is an important membrane polymer used primarily in dialysis, particularly, hemodialysis, membranes. The best source of cellulose for membrane applications is cotton linters.[46] These cottonseed hairs possess the highest concentration of alpha cellulose (the high-purity, high-MW fraction). The many hydroxyl groups of cellulose render it extremely hydrophilic so that water-swollen cellulose membranes exhibit a hydrogel character. However, because of its structural regularity and the intermolecular hydrogen-bonding capacity of its hydroxyl groups, cellulose is water insoluble. Cellulose contains crystallites which act as virtual cross-links. Although a few exotic organic solvents for cellulose exist, for example, a mixture of DMSO and paraformaldehyde,[47] N-methylmorpholine N-oxide,[48] and 5–8% LiCl in DMAC,[49] the principal solvent for the manufacture of cellulose membranes is an aqueous solution of cuprammonium hydroxide (Cuoxam). Cellulose is an extremely versatile polymer which can be modified by reactions between suitable "monomers" and the active hydrogens of its three side-chain hydroxyl groups. The products of such reactions are known as cellulose derivatives or cellulosics. Since there are three potential sites for monomer addition, not all of which need be occupied, a range of derivatives with varying degrees of substitution (DS) is possible. In a product with a DS of 3 all three hydroxyl groups have been modified, whereas cellulose itself has a DS of zero. Crystallinity is most pronounced when structural regularity is at its greatest which is the case at both extremes of DS. At DSs < 3, crystallinity diminshes until in some cases at 0.6 < DS > 0.8 crystalline order is totally lacking and water solubility results. The usual DS range for membrane applications lies between 2.3 and 2.8. At a DS ≥ 2.75, the polymer is considered trisubstituted. Polymers with a true DS of 3 are prepared with difficulty and are intractible. They are not items of commerce. The most important derivatives of cellulose for membrane applications are the inorganic (nitrate) and organic (acetate and acetate mixed), esters. Organosoluble cellulose ethers such as ethyl cellulose are also of interest. Water-soluble cellulose ethers such as hydroxyethyl, and hydroxypropyl cellulose are sometimes utilized as wetting agents for membranes of cellulosic and noncellulosic polymers.

Cellulose nitrate (CN), or nitrocellulose, was the first synthetic polymer and also the first polymer utilized to produce a synthetic membrane.[45] Once widely utilized together with camphor as a general plastic known as celluloid, CN has surrendered most of these applications to cheaper and less flammable materials. It survives as an item of commerce today largely because of its widespread use in smokeless powder and lacquer coatings. In the latter, its toughness, adhesive nature, and transparency permit it to remain competitive. Cellulose nitrate was for a long time the principal polymer utilized in filtration membranes. In fact it may still be, although its leadership role is facing a stern challenge from other polymers such as the nylons. For use in membranes, CN polymers known as Type RS or Type E (11.8–12.3 % N, DS ~ 2.5) are utilized. Such polymers are Lewis acids; have a δ of ~ 10.5; and are soluble in a wide range of cheap organic solvents such as acetone and methyl acetate, but only slightly soluble in alcohols. A large number of viscosity grades of CN is available. Microfiltration membranes of CN are more friable than analogous membranes from nylon, polysulfone, or certain acrylic copolymers. To circumvent this problem, CN membranes of increased flexibility are being developed. Although the excellent solubility of CN suggests that the use of membranes of this material would be restricted to aqueous solutions, this is not always the case. For example, CN membranes can be utilized to filter solutions containing alcohols. They are also quite resistant to chlorinated hydrocarbons which are solvents or swelling agents for polymers such as polysulfone, polycarbonate, and poly(vinylidene fluoride). Surprisingly, CN microfiltration membranes are more resistant to shrinkage during autoclaving than are cellulose triacetate or cellulose acetate membranes.

The acetylation of cellulose is carried to completion—actually to a DS of ~ 2.8$^+$ (43% acetyl)—to produce the *primary* or *tri* acetate. Cellulose triacetate (CTA) is a membrane polymer of importance which is utilized in microfiltration, electrophoresis, and hyperfiltration membranes. In the last mentioned it is utilized by itself in hollow-fiber form and in blends with 2.5-DS (~ 40% acetyl) cellulose acetate. CTA has a high degree of structural regularity and a potential for crystallization when cast as a dense film from a single solvent and annealed. However, it is believed to be largely amorphous after it has gone through the micellization inherent in the formation of phase-inversion membranes.

Cellulose triacetate is a Lewis base with a δ = 9.5. It is, therefore, soluble in acidic solvents such as methylene chloride (δ = 9.7) and chloroform (δ = 9.3). It is also soluble in trifluorethanol, an acidic alcohol, and in certain strong aprotic solvents such as dimethylformamide (DMF) and N-methylpyrollidone (NMP). Strong nonsolvents for CTA include water, glycerol, and ethylene glycol. The utility of CTA as a hyperfiltration membrane polymer is due to the increasing permeability ratio of water to salt with increasing acetyl content (Fig. 4.20). The higher permselectivity of CTA-, relative to CA-hyperfiltration membranes, may be at least in part due to the fact that molecules of the former are more tightly coiled in solution than are those of the latter.[51] This tendency would be expected to be carried over

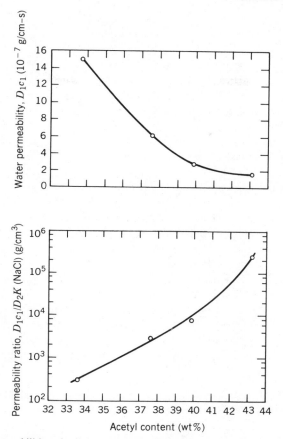

FIGURE 4.20. Permeabilities of cellulose acetate to water and sodium chloride versus acetyl content at 25°C (from Lonsdale et al.[50]; © 1965).

into the solid state with the result that the average distance between neighboring chains, in other words, the average pore size, would be smaller, the higher the acetyl content. However, water permeability decreases with acetyl content (Fig. 4.20) because as the latter increases, the concentration of hydrophilic hydroxyl groups decreases. This is the reason for the success of Cannon's widely utilized CA/CTA blend desalination membrane, which is a compromise between permeability and permselectivity.[52]

Because CTA is difficult to dissolve in cheap solvents, it has been less widely utilized as a membrane polymer than has the polymer known as secondary cellulose acetate, cellulose diacetate, or simply cellulose acetate (CA) which is the result of the partial deacetylation of CTA. Cellulose acetate has a DS of ~ 2.5. It has both basic (acetyl) and acidic (hydroxyl) character; a $\delta = 10$; and is soluble in acetone, methyl formate, propylene oxide, dioxolane, dioxane, and methylene chloride/methanol (4/1), as well as in many other solvents, including formamide. It is available in a wide range of viscosity grades and in a modest range (38–40%) of acetyl contents. It is to be found in microfiltration membranes by itself and in blends with CN.

Ultrafiltration, hyperfiltration, and electrophoresis membranes are frequently prepared from CA. In addition to use of CA as a membrane polymer per se, hollow fibers of CA also serve as protomembranes which are converted by deacetylation to their final cellulose form for use in dialysis. Cellulose acetate was the most permselective polymer discovered by Reid and Breton[53] in their survey of candidate hyperfiltration membrane polymers, and the polymer utilized by Loeb and Sourirajan[54] in their classic development of integrally-skinned membranes which inaugurated the modern membrane era.

The hydroxyl groups of CA are hydrophilic sites which form hydrogen bonds with water and are therefore responsible for the high water permeability of its membranes in hyperfiltration. The acetyl groups, on the other hand, are hydrophobic sites which serve as virtual cross-links to restrict the swelling about the intervening hydroxyl groups so that, to a first approximation, only water and not hydrated ions can permeate the glassy polymer matrix. The balance of properties, which led to the development of 2.5-DS CA by Miles[55] in the first place, is also related to its utility as a hyperfiltration membrane polymer (Table 4.7).

Cellulose acetate is a versatile membrane polymer. Its low cost, availability in a wide variety of viscosity grades, outstanding tractibility, and reasonable resistance to oxidation by chlorine, guarantee its continued utility for the foreseeable future. Its weaknesses are a borderline T_g of 68.6°C (which limits its utility at

TABLE 4.7 RELATIONSHIP BETWEEN ORIGINAL FILM AND SUBSEQUENT MEMBRANE CHARACTERISTICS OF CELLULOSE (2.5 DS) ACETATE

Original Film Property Sought	Method by Which Achieved	Membrane Hyperfiltration Characteristic
Transparency	Uniformity, random substitution, achieved by the homogeneous deacetylation of CTA in solution	High permselectivity because no high local concentration of hydroxyl groups are present
Good solubility coupled with dimensional stability in humid environments	Deacetylation to ~ 2.5 DS; lower-DS products are prone to swell; higher-DS products are poorly soluble	Nearly optimum hydrophilic–hydrophobic balance maximizes permeability while retaining adequate permselectivity
Toughness and flexibility	Optimized in cellulose derivatives by the retention of a low but definite concentration of hydroxyl groups	Ability to be handled and withstand pressure without creeping. Integrally-skinned CTA membranes are more prone to collapse under pressure than are CA membranes.

elevated temperatures and pressures), a tendency to undergo hydrolysis in alkaline media, and a lack of resistance to attack by microorganisms. Techniques to at least partially circumvent these shortcomings have been found. Maintaining pH between 4 and 6 and temperature $\sim 25\,°C$ for CA membranes permits their continuous operation in hyperfiltration for three to four years. Operation at pressures < 30 atm have held permeability declines due to compaction within acceptable limits. The feasibility[56] of increasing CA's bulk modulus (and hence the resistance of its membranes to compaction) by the incorporation of unsaturated monomers and the cross-linking of these moieties *in situ* subsequent to membrane formation has been demonstrated. This approach may become economically viable in the near future if plans (currently being evaluated) to commercialize a polymerizable CA are carried through. Grafting of high-modulus materials such as polystyrene[57] onto CA has also been shown to lessen compaction. However, here also a commercial source of the graft copolymer would be required. Biological attack is successfully averted during storage by: (1) the addition of formaldehyde to wet (ultragel) membranes; (2) the development of techniques for drying wet membranes to permit their storage in the dry condition,[58] and (3) the development of dry (microgel) membranes which are wet–dry reversible.[59] Biological attack during operation is controlled by the chlorination of the feed and by the utilization of more biostable CA membranes such as those from CA/CTA blends.[50] More complete inhibition of biodegradation can be achieved by the modification of the CA polymer with monomers containing quaternary ammonium groups.[60]

The CA mixed esters of commerce such as CA propionate (CAP) and CA butyrate (CAB) are not particularly good membrane polymers. This is so because the ratios of other acyl to acetyl groups are too high. Acyl groups which are larger than the acetyl act as internal plasticizers to increase chain flexibility and decrease the T_g. Therefore, the concentration of the second acyl component must be low to retain sufficient strength and rigidity in the CA mixed ester. The best way of circumventing this difficulty is to utilize commercially available (2.5 DS) CAs as starting materials for the preparation of CA mixed esters. A virtually infinite variety of interesting membrane polymers is accessible by this approach. A 2.45-DS CA can be readily further acylated to any desired combined DS up to about 2.8. (It is seldom possible to achieve a total DS of 3 and solubility usually diminishes when the combined DS > 2.75.) This means that, for practical purposes, a 2.45-DS CA has approximately 0.3 DS, and a 2.3-DS CA approximately 0.45 DS, of —OH groups which are available for substitution. Since meaningful modifications can be made with the addition of 0.1 DS or even less, it is apparent that a wide range of membrane polymers is available by this approach. If we denote CA as CAOH (to stress the potential of free —OH groups in CA to undergo reactions involving active hydrogens), the formation of CA mixed esters (III) can be schematically depicted as

$$CAOH + RCOCl \xrightarrow{R_3'N} CAOCOR + R_3'N \cdot HCl \tag{1}$$
$$\textbf{III}$$

where R can be virtually any group.

Long linear saturated paraffinic moieties serve to increase both solubility and (through increasing side-chain entanglements) structural integrity.[61] Monomers which contain double bonds such as methacroyl (CH_2=C (CH_3)—CO—), crotonyl (CH_3—CH=CH—CO—), and undecenoyl (CH_2=CH(CH_2)$_8$CO—) chlorides result in cross-linkable CA mixed esters. Reactions of CA with anhydrides are also feasible.[45]

$$CAOH + (CH_3CO)_2O \longrightarrow CAOCOCH_3 \qquad (2)$$
$$\textbf{IV}$$

$$(3)$$

Reaction of CA with isocyanates to yield carbamate esters (**VI**) has been carried out.[45]

$$CAOH + RNCO \xrightarrow{R_3N} CAOCONHR \qquad (4)$$
$$\textbf{VI}$$

Here R can be a saturated or unsaturated hydrocarbon or contain ionogenic groups.

If dry CA membranes are reacted with the proper concentration of diisocyanates in inert aprotic solvents which are nonsolvents for CA, cross-linked CAs (**VII**) will result,[62]

$$2CAOH + OCN(CH_2)_6NCO \longrightarrow CAOCONH(CH_2)_6NHOCOCA \qquad (5)$$
$$\textbf{VII}$$

This reaction can result in membranes which are not only insoluble in, but even wet–dry reversible in, acetone.

An interesting approach is the utilization of blocked isocyanates (**VIII**):

$$RNCO + BH \longrightarrow RNHCOB \qquad (6)$$
$$\textbf{VIII}$$

where BH is a compound such as acetone oxime or imidazole which contains a labile H. Different compounds split at different temperatures. When VIII is heated to its splitting temperature, the reaction will reverse itself, thereby freeing the RNCO groups.

If the blocked isocyanate is included in the solution from which a CA film or membrane is subsequently cast, it will be encapsulated within the resultant CA matrix. Heating such a structure in the dry condition above the splitting temperature will unblock the isocyanate group which may become available for reaction with the free hydroxyl groups of CA:

$$\text{CAOH} + \text{RNHCOB} \xrightarrow{\Delta} \text{CAOCONHR} + \text{BH} \qquad (7)$$
$$\text{VI}$$

This provides a facile means for the *in situ* preparation of CA carbamate esters[62, 63] and constitutes an economic alternative to the separate preparation of preformed polymers.

A means of reducing the cost of membranes from preformed CA mixed esters is the preparation of a polymer with a higher DS of substituent groups than is required in the final product. This high-DS polymer may then be included in a casting solution together with unmodified CA starting material to yield a blend membrane with a desired average DS of substituent groups.[61] This approach has been particularly successful in those cases in which the CA mixed ester has contained C_{10}–C_{12} chains as second acyl components because such polymers have proven not only to be very soluble when utilized by themselves, but also to exhibit excellent compatibility with unmodified CA polymers. A further advantage of the blending approach is increased reproducibility since the optimum DS is more readily achieved in the final membrane by varying the ratio of CA to CA mixed ester than by the synthesis of a CA mixed ester with the DS of the second acyl compound controlled within narrow limits.

Organosoluble cellulose ethers such as ethyl cellulose (EC) and ethyl hydroxyethyl cellulose (EHEC) are stable in high-pH environments, although they are more subject to hydrolysis at low pH than are the cellulose esters. Membranes of the organosoluble cellulose homoethers are less permselective and tend to be more permeable to gases than are those of the cellulose homoesters. This is due to the greater flexibility of the $-\text{C}-\text{O}-\text{R}$ compared to the $-\text{COOR}$ groups. Of course mixed esters such as CAB in which intermolecular association is minimized as a result of side-chain irregularity are also highly permeable to gases. Porous cellulose ether membranes are difficult to prepare by dry casting because they are soluble in the alcohols which serve as pore formers for this process. They are, however, accessible by wet casting procedures.[64]

The polyvinyl polymers which are of greatest interest as membrane materials are poly(vinyl chloride)(PVC), $\text{+CH}_2\text{—CHCl+}_n$, and its copolymers. Because of the presence of the chlorine atom, rotation about the C–C backbone in PVC is severely limited. Also the C–Cl dipole increases the strength of cohesive forces so that the $\delta = 9.5$ compared to a $\delta = 8.0$ for polyethylene. Polyvinyl chloride is a Lewis acid and is soluble in basic solvents, particularly, tetrahydrofuran (THF), $\delta = 9.5$, and cyclohexanone, $\delta = 9.9$. It is a stiff polymer with a relatively high T_g of 80°C. Commercial PVC is produced by free-radical initiation at about 50°C. It is primarily atatic and amorphous with a significant amount of branching, but does possess about 5% of crystalline syndiotactic segments.[4] Polymerization at lower temperatures favors the formation of unbranched syndiotactic polymers which are higher melting and less soluble. Postchlorination of PVC results in chlorinated PVC, a product which is in effect a copolymer of vinyl chloride and 1,2-dichloroethane. Chlorinated PVC is more dense and exhibits greater thermal stability than the parent PVC. Poly(vinyl chloride) homopolymers are tough, but inflexible unless plasticized, and exhibit excellent stability in the presence of water and many chemicals.

The presence of a small amount of a second monomer such as vinyl acetate, vinylidene chloride, ethylene, or propylene has the effect of an internal plasticizer and greatly increases the flexibility and tractibility of the copolymers relative to the homopolymer PVC. In PVC–VA copolymers containing 10% vinyl acetate, the impact strength is double that of the PVC homopolymers. The same would be the case for plasticized PVC which contains a certain level of external plasticizer. However, at plasticizer concentrations below this level, the effect is actually that of an *antiplasticizer* and the impact strength is less than that of the unplasticized polymer. Propylene–vinyl chloride copolymers containing between 2 and 10% propylene are resistant to most acids, alkalis, alcohols, and aliphatic hydrocarbons and exhibit good stress cracking properties with chlorinated solvents. Because of its stiffness, low cost, and solvent resistance, microporous PVC has been investigated as a support for thin-film composites.[65] Films of PVC are also widely utilized as membranes for ion-selective electrodes.[66]

Polysulfone (PS) is the polymeric product of the reaction between the disodium salt of bisphenol-A and di-*p*-dichlorodiphenyl sulfone:

PS

Among the properties which qualify PS as an outstanding membrane polymer are a T_g of 195°C, an amorphous glassy state, thermal and oxidative stability, excellent strength and flexibility, resistance to extremes of pH, and low creep even at elevated temperatures. The closely related polyether sulfone (PES),

PES

is totally devoid of aliphatic hydrocarbon groups because of which it exhibits even higher thermal stability. It has a T_g of 230°C, and other properties similar to those of PS. Neither of these polymers is particularly solvent resistant. Both PS and PES are Lewis bases (sulfone, aromatic, and ether groups) and are soluble in acidic solvents such as hexafluoroisopropanol (HFIP) and chlorinated hydrocarbons such as chloroform ($\delta = 9.3$) and methylene chloride ($\delta = 9.7$). They are also soluble in the polar ($\delta = 12$) solvents: dimethylformamide (DMF), dimethylacetamide

(DMAC), and dimethylsulfoxide (DMSO) which are employed in wet-process casting solutions that are utilized in membrane preparation.

The combination of phenyl rings attached to sulfone groups results in a high degree of resonance stabilization.[19] The sulfone group acts as a sink for the electrons in the aromatic groups and confers both thermal and oxidative resistance. Two additional stabilizing effects of resonance are increased bond strength and a planar configuration which contributes to rigidity even at elevated temperatures. The ether groups in the backbone provide some flexibility which results in inherent toughness. Because of their aromatic moieties, membranes from both polymers are resistant to high-energy irradiation. Another desirable property of the aromatic groups is that they may be sulfonated, thereby providing an excellent means for the introduction both of strong cation-exchange groups and potential cross-linking sites. Membranes from unmodified PS and PES are utilized in both flat-sheet and hollow-fiber form in hyperfiltration (HF), ultrafiltration (UF), microfiltration (MF), and gas-separation membranes, both by themselves (UF and MF), and in series with other membranes (HF and gas separation). They are widely utilized as porous supports in thin-film composite membranes for HF[67] and as the principal (albeit imperfect) barrier layer in hollow-fiber gas-separation membranes which are utilized in series with permeable barriers of silicone and other elastomers.[44] Sulfonated PS has been widely investigated as the barrier layer in both integrally-skinned (asymmetric)[39] and thin-film composite HF membranes. Sulfonated PS with a capacity of sulfonic acid moieties sufficient to render it water soluble is currently undergoing evaluation as an HF barrier layer (after cross-linking) in series with microporous supports of unmodified PS.[68] The advantage of this combination is that both support and barrier layers are chlorine resistant, the one quality lacking in all other HF membranes today.

Attempts have been made to increase the permeability of polysulfone gas-separation membranes by the introduction of methyl groups into the aromatic rings of the bisphenol A and/or sulfone monomer. The methyl groups exert steric and inductive (decreased basicity of methylated relative to unmethylated aromatic moieties) effects, which tend to decrease polymer–polymer interaction and hence increase average interchain displacement. The placement of methyl groups is critical. Thus, whereas the utilization of tetramethyl bisphenol-A increases permeability with no loss in permselectivity, the incorporation of bisphenol-L not only results in a selectivity loss but also in a catastrophic decline in physical properties.

The relatively low melting points and the poor hydrolytic stability of the aliphatic polyesters makes them excellent candidate membrane polymers for controlled-release applications where polymer degradation or erosion is desired (Chapter 3). The best of these polyesters is poly(ϵ-caprolactone) (PCL), a semicrystalline polymer with a melting point of about 60°C and a M_w (PCL-700) of about 40,000.[69]

$$RO \left[\begin{matrix} O \\ \| \\ C \end{matrix} - (CH_2)_5 - O \right] H$$

PCL

Poly(ε-caprolactone) has an extended zigzag conformation quite similar to that of polyethylene. Its a and b dimensions are almost identical to those of polyethylene, 'but its c dimension (fiber axis) is larger. Despite the fact that PCL is more polar than polyethylene, it has a greater degree of rotational freedom about the chain backbone and hence a lower melting point ($\sim 60°C$) than the latter ($136°C$). Poly(ε-caprolactone) is basic and, therfore, soluble in such acidic solvents as methylene chloride and chloroform. It is also soluble in benzene, toluene, carbon tetrachloride, tetrahydrofuran, cyclohexane, and dioxolane. Although soluble in warm acetone, methyl ethyl ketone, dimethylformamide, and ethyl acetate, its solutions tend to separate from these solvents after a week at room temperature. The strong polarity of PCL results in a volume resistivity which is about four orders of magnitude lower than that of most polymers. It is comparable in these respect to cellulose nitrate and phenol–formaldehyde polymers. Perhaps the most important characteristic which distinguishes PCL from other polymers with similar properties is its unique ability to form compatible (that is, single T_g) blends with a variety of membrane polymers, such as poly(vinyl chloride), cellulose nitrate, and cellulose acetate butyrate. It also forms mechanically compatible blends with poly(vinyl acetate), polystyrene, poly(methyl methacrylate), polysulfones, polycarbonates, and poly(vinyl butyral) and poly(vinyl alkyl) ethers. This ability to form blends with polymers, coupled with the related charcteristic of chain flexibility, permits PCL to be utilized as a polymeric plasticizer for PVC.

Linear thermoplastic aromatic polyesters are considered polymers in which the diacid monomer is aromatic and the dialcohol is a saturated aliphatic type. Although a considerable variety of monomers is available, terephthalic acid,

is the only diacid encountered in commercial resins for fiber and film use. This acid is esterified with ethylene glycol to yield polyethylene terephthalate (PET),

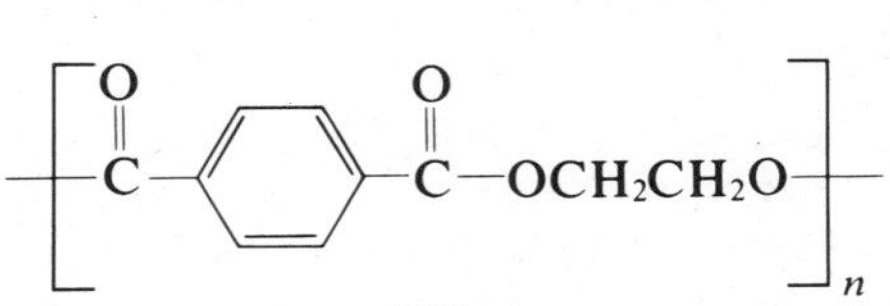

PET

with 1,4-butanediol to yield polybutylene terephthalate (PBT),

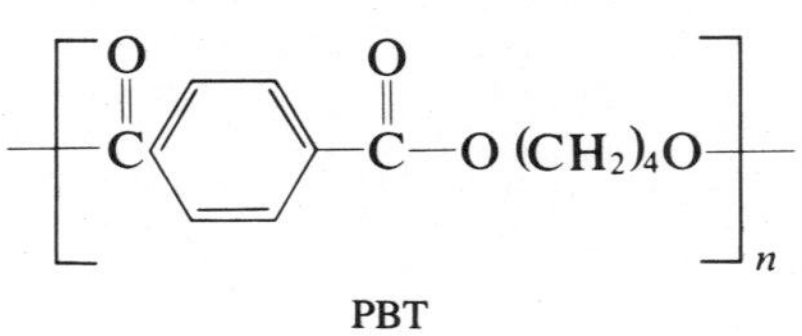

PBT

or with cyclohexanedimethanol to yield poly(1,4-cyclohexylene dimethylene tere-phthalate) (PCHDMT),

PCHDMT

The densest and most crystalline of these three candidate membrane polymers is PET. The terephthaloyl,

moiety is highly planar because of resonance stabilization and because the carbonyl group is able to rotate to permit dense packing. Nevertheless, because its T_g is 80°C it is amorphous as supplied and must be dried and crystallized before it is melt processed into film or fiber form. Membranes of PET are strong (8000 psi tensile) and tough (300% elongation at break). They have low water absorption and good chemical resistance. Although PET has a δ of 10.7, its basic nature requires the utilization of acidic solvents such as hexafluoroisopropanol (HFIP) or o-chlorophenol. Even in these, its solubility is borderline from the standpoint of phase-inversion processing into membrane form. However, thin ($\sim 10~\mu$m) dense films of PET are utilized as one substrate in the preparation of radiation track membranes.[70] Because the flexibility of its chain is greater and close packing less likely, the solubility and amorphous character of PBT is greater than that of PET. Solutions of the former in HFIP are clear, whereas those of the latter are turbid.[71] However, PBT membranes are still resistant to most solvents and withstand contact with aliphatic and aromatic hydrocarbons, alcohols, ketones, esters, and even slightly acidic species such as the chlorinated hydrocarbons. Hollow-fiber membranes of PET were once utilized in gas separations. They met with limited commercial success presumably because of uneconomically low permeability, a result of several factors such as crystallinity and the fact that dense (high resistance throughout the entire thickness of the membrane), rather than skinned, (high resistance only in dense thin skin layer) membranes were employed. Despite the fact that PCHDMT is more hydrolytically and thermally stable than the other linear aromatic polyesters, its somewhat poorer physical properties have thus far precluded its use in membranes.

The utilization of isophthaloyl,

rather than terephthaloyl moieties significantly lowers the crystallinity and flow temperature and increases the solubility of the resulting (PEI) polyesters. The polyester from tetrachloro BPA (TCBPA) and isophthaloyl chloride, for example, is soluble in methylene chloride.[72] Nonwoven polyester fabrics contain PET/PEI (80/20) blends.[73] They are passed between heated rollers at a temperature above the flow temperature of PEI which has the effect of strongly binding the mass together without losing the desirable characteristics of the PET fibers. Because of their excellent strength, solvent resistance, and dimensional stability, highly calendered nonwoven polyester fabrics of this type constitute an ideal membrane support for separations at high differential pressures.[74]

Polyester block copolyethers,

$$\left[\!\!\left[\begin{array}{c} O \\ \| \\ C \end{array}\!\!-\!\!\left\langle\bigcirc\right\rangle\!\!-\!\!\begin{array}{c} O \\ \| \\ COCH_2CH_2OC \end{array}\!\!-\!\!\left\langle\bigcirc\right\rangle\!\!-\!\!\begin{array}{c} O \\ \| \\ C \end{array}\!\!-\!\!(OCH_2CH_2)_n\!\!-\!\!O\right]_n\right.$$

with a hydrogel character (strong affinity for water) were investigated by Lyman et al.[75] for hemodialysis applications. This system represents an attempt to increase the hydrophilicity of PET while simultaneously retaining its other properties. A linear relationship exists between the dialysis rates for urea and the polyoxyethylene glycol content (Fig. 4.21). A comparison of the dialysis rates of various compounds

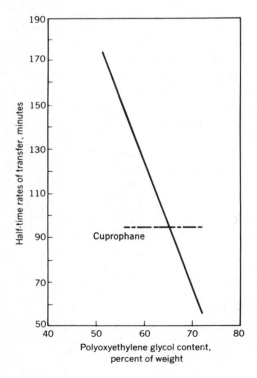

FIGURE 4.21. Effect of the weight percent of polyoxyethylene glycol in the copolymer on the half-time rates of transfer of urea (from Lyman et al.[75]; reprinted with permission from *Biochemistry*, © 1964 American Chemical Society).

TABLE 4.8 RELATIVE DIALYSIS RATES OF VARIOUS COMPOUNDS[a]

| | | Half-Time Escape Rates (min) | |
Compound	Molecular Weight	Copolyester[b] Membrane	Cellulose[c] Membrane
Urea	60.8	58	68
Creatinine	113.1	117	123
Uric acid	168.1	217	320
Ascorbic acid	176.1	135	178
Glucose	180	268	223
Thiamine chloride	337.3	150	160
Sucrose	342	397	270
Raffinose	504.5	930	450

[a] From Lyman et al.[75] Reprinted with permission from *Biochemistry,* © 1964 American Chemical Society.
[b] Containing 71% Carbowax 1540.
[c] Cuprophane.

through cellulose (Cuprophane) and the copolyester membranes indicates that dialysis rates differ somewhat from species to species and membrane to membrane (Table 4.8). Although they and the closely related polycarbonate block copolyether,[76] and polyurethane block copolyether,[77] hydrogels resulted in membranes with interesting dialysis properties, none has achieved commercial success.

The commercial aromatic polycarbonates (PCs) are a class of polyesters derived from carbonic acid and bisphenol-A:

PC

In spite of structural regularity on the molecular level, the PCs are glassy amorphous polymers which do not crystallize under ordinary conditions. The long and inflexible but nonplanar bisphenol-A groups and the polar carbonyl of the ester linkage yield extremely tough polymers (and membranes) with outstanding impact strength. Although the polymer segments are brought into motion with difficulty thereby resulting in the high T_g of 140°C, nevertheless their nonplanarity results in a large average interchain displacement with the result that gas permeability is much higher through PC films than through those of the corresponding planar and close-packed PETs. The presence of aromatic and ester moieties confers a basic character on PC so that it is readily soluble in acidic solvents such as methylene chloride and chloroform. However, the large average interchain displacement and resultant accessibility of its groups renders PC soluble as well in a wide variety of solvents over the δ range (9.5–12.0). Solubility diminishes markedly with increasing MW and the highest-MW grades (not available from domestic sources) form elastic gels rather than true solutions in methylene chloride. Thin ($\sim 10~\mu M$) dense films of high-MW PC are the most common substrate for the preparation of microporous radiation track membranes.[70]

Because of its outstanding physical properties and solubility, PC has often been investigated as the *hard* block in the preparation of block copolymers with more flexible *soft* blocks such as the polyethylene glycols (PEGs)[70] and the silicones.[41,43] The flexibility of the block copolymer chains is considerably greater than that of the corresponding PC homopolymer chains. As a result, the former are more soluble and remain so at higher MWs than do the latter. The addition of only 5 wt% of PEG 4000 to BPA resulted in the ready synthesis of a high-MW PC-PEO copolymer, which was required for the casting of microporous PC membranes by the dry phase-inversion process.[7] Such membranes were commercially available for a number of years. The addition of between 20 and 40 wt % of PEGs 6000 or 20,000 to BPA, on the other hand, resulted in hydrophilic polymers suitable for use in hemodialysis membranes.[76] Silicone block copolycarbonates with varying ratios of hard and soft blocks remain commercially available.[41] They combine the excellent film-forming characteristics of PC with gas permeabilities which are second only to those of silicone homopolymers. Both flat-sheet and hollow-fiber membranes of the silicone polycarbonates have been investigated for use in membrane blood oxygenators.

Polyamides are the step reaction products of primary or secondary diamines and dibasic acids. The most important structural feature of polyamides (and the large number of related polymers) is the $-$CONH$-$ group, the amide link, which has a strong capacity for hydrogen bonding. The formation of hydrogen bonds significantly enhances intermolecular associations and where sterically possible, crystallite formation, with the result that such polymers tend to exhibit the related characteristics of high T_g's and poor solubility. Polyamides of three classes are commercially available: fully aliphatic (both the diacid and the diamine monomers are aliphatic); aromatic (the diacid monomer is aromatic and the diamine is aliphatic); and fully aromatic (both the diacid and the diamine are aromatic). Every class of the polyamides, including those prepared from monomers both of which are linear aliphatic compounds, has representatives which are suitable for use in membranes. In fact the variety of such macromolecules and their resultant properties is so great that the polyamides and the closely related materials are rightly considered to be the most versatile class of membrane polymers.[78] An excellent review of this subject is that of Blais[79]. This section will emphasize the structural principles which govern the utility of this class of membrane materials and then consider those species which are either already encountered in commercially available membranes or whose properties are such as to render likely such an eventuality. Among these are the linear aliphatic fiber-forming polyamides or nylons 4/6, 6, 6/6, 6/10, 11, and 12; the alcohol-soluble binder nylons (methoxymethylated nylon 6/6 and the nylon multipolymer (6/6, 6/10, 6); the aromatic polyamides poly(piperazine phthalamides) and poly(piperazine thiofurazanamides); the fully aromatic polyamides (Aramids); poly(benzimidazoles) (PBIs); poly-(benzimidazalone) (PBIL); a number of polyimides; and interfacial polyconden-

sates formed *in situ*, such as the NS-100 polyureas, the NS-101 aromatic polyamide, and the FT-30 fully aromatic polyamide.

The liner aliphatic nylons 4/6, 6/6, and 6 are structurally regular chains consisting of a series of methylene groups joined together by the strongly hydrogen-
bonding amide linkage.[78] The hydrogen bonding results in strong intermolecular
associations which have the related effects of increasing thermal stability and decreasing solubility relative to those of the analogous aliphatic polyesters such as
poly (ϵ-caprolactone). However, the weak van der Waals forces exerted by the methylene group sequences which separate the amide groups results in flexible segments, the combination of which, together with rigidity due to hydrogen bonding,
confers overall toughness on these materials (Table 4.9). Nylon 4/6 exhibits a higher
melting point and is stiffer, tougher, and more abrasion resistant than the other
aliphatic nylons. Should it become commercially available, it will very likely prove
to be an excellent membrane polymer. A unique feature of these polymers is their
hydrophilicity owing to the presence of amide groups in amorphous regions, which
to the extent that they are not involved in intermolecular hydrogen bonding, are
available for interaction with water. For this reason, nylon 6/6 and nylon 6 are the
only commercially available polymers whose microporous membranes do not require the presence of a wetting agent to promote *instantaneous* wetting.[80] Nylon
4/6 will undoubtedly exhibit this characteristic as well. Only acidic solvents such
as formic acid (FA) (77.5–100% FA) and trifluoroethanol (TFE) are capable of
serving as practical casting solvents for the nylons. In the case of FA, hydrolysis
occurs in all but the 100% solvent[81] so that the shelf life of casting solutions even
at 25°C is limited to about 3 weeks. These polymers are also unique among membrane polymers in that solutions which consist only of polymer and solvent without
any explicit pore formers will form high-void-volume membranes by the dry casting technique.[81] Ordinarily, utilization of this process with such a solution would,
owing to the combined effects of plasticization by the solvent and gravity, result in
a film of low porosity. In this case, however, intermolecular (P–P interaction) forces
result in crystalline gels which are strong enough to resist densification.[83] As the
ratio of methylene to amide groups is increased, hydrophobicity increases with the
result that membranes of nylons 11 and 12 will not wet spontaneously.[84]

TABLE 4.9   PROPERTIES OF ALIPHATIC NYLONS

Nylon	mp (°C)	Tension Modulus ($\times 10^{-5}$) (lb/in.^{2})
4-6	300	5
6-6	260	4.3
6	220	4
6-10	215	3
11	185	2
12	175	2
6/6,6/10,6 (40:30:30)	160	2

The desirable combination of properties offered by the nylons, namely strength and flexibility, thermal stability, solvent resistance, wettability, and resistance to basic hydrolysis, is not found in any other family of membrane polymer. It therefore appears likely that nylons will assume a leading position in microfiltration membranes for process filtration before the end of this decade.

Methoxymethylated nylon 6/6 (referred to as type 8 nylons) and nylon (6/6, 6/10, 6) are linear polymers whose structural regularity has been so diminished that intermolecular hydrogen bonding between amide groups is significantly less likely. The result is that these polymers are in the amorphous state (with properties somewhat between those of glass and rubber) and are alcohol soluble. Type 8 nylon may be cross-linked by the inclusion of citric acid in its solutions and heating the finished membrane.[85] Both polymers exhibit considerable potential for use in membranes and in ancillary areas. Indeed, the excellent strength and flexibility of amorphous nylon 6/6, 6/10, 6 multipolymer, together with its compatibility with crystalline high-MW nylon 6/6 homopolymer in FA solutions, has resulted in its inclusion in high-strength nylon-blend membranes.[83] The MW of the nylon 6/6 homopolymer and the ratio of homopolymer P_1 to multipolymer P_2 proved to be decisive factors insofar as the preparation of integral, flexible, and strong membranes was concerned. Whereas low-MW P_1 resulted in brittle nonintegral structures and high-MW P_1 in integral, but skinned and friable membranes, strength, flexibility, and filtration properties increased with increasing concentration of P_2 up to a P_1/P_2 ratio of approximately 3. Although porosity was maintained at P_1/P_2 ratios as low as 1, at values of $P_1/P_2 < 1$, dense films resulted. At the optimum P_1/P_2 ratios of approximately 4, porosity and bubble point are controlled by varying the total $(P_1 + P_2)$ concentrations of polymer in the casting solution (Table 4.10). In addition to its end-use function as a polymeric plasticizer, the nylon multipolymer serves in the processing phase as a gelation promoter to eliminate skinning and to increase porosity beyond that which is attainable with the crystalline polymer by itself. As a further processing subtlety, the MW of P_2 is also a factor. At MWs in excess of a critical value, compatibility between P_1 and P_2 decreases with deleterious effects upon uniformity.

Although the flow temperatures of nylon 6 and 6/6 and 4/6 are low enough to permit their polymerization in the melt, this is not true of their aromatic and fully aromatic counterparts. These must be synthesized by more costly interfacial and low-temperature solution polycondensation techniques.[86] The poly(piperazinamides) are aromatic polyamides with the general formula

Most frequently, R is a phthalic acid moiety:

TABLE 4.10 EFFECT OF CONCENTRATION OF P_1 + P_2 AT P_1/P_2 = 4 UPON THE PROPERTIES OF DRY-PROCESS NYLON-BLEND MEMBRANES[a]

Type	Membrane Casting Solution Concentration (g/100 mL solvent)	Thickness (in.) ($\times 10^3$)	Water Bubble Point (psi)	Nominal Pore Size (μm)	Air Flow (L/min cm^2 at Δp = 10 psi)	Burst Strength (psi)
Blend nylon	17	6.2	28	0.45	4.00	32
Blend nylon	18.5	6.3	35	0.45	2.61	44
Blend nylon	19	6.2	55	0.2	1–1.5	60

[a] From Kesting.[83]
[b] P_1 = Vydyne 66B (Monsanto).
[c] P_2 = Elvamide 8061 (du Pont).

149

(ortho); (iso); or (tere).

However, Credali et al. have extensively reported upon the unusual polymer, poly(*trans*-3,5-dimethyl piperazinthiofurazan -3,4-dicarboxamide)[87]

Because the piperazines are disecondary amines, the poly(piperazinamides) and poly(substituted piperazinamides) which result, lack the NH groups which render the polyamides as a class so susceptible to oxidative degradation by chlorine and several other widely utilized sterilizing agents. Since membranes from these polymers are intended for use in hyperfiltration applications where both long lifetime (3–4 yr) and chlorine resistance are desired, this is a most attractive feature and work continues to produce both preformed polymers for phase-inversion membranes and *in-situ* formed thin films for composite membranes. Unsubstituted piperazine is somewhat too hydrophilic, so that the methyl-substituted monomers have received the most attention. The utilization of 2-methylpiperazine minimizes structural regularity and hence increases polymer solubility. However, the best balance of processing and end-use properties is achieved when *trans*-3,5-dimethylpiperazine is employed. In the case of the poly(piperazine phthalamides), solubility decreases in the series ortho (O) > iso (I) > tere (T). The T-phthalamides require the strongest solvents, such as formic acid and *N*-methyl pyrollidone (NMP), whereas the O-phthalamides are reported to be soluble in weakly acidic solvents such as methylene chloride and chloroform.

The fully aromatic (Aramid) polyamides

are membrane polymers with exceptional strength, bulk modulus, thermal and hydrolytic stability, and permselectivity. Richter and Hoehn[88] optimized such polymers for application to integrally-skinned hyperfiltration membranes. These Aramids have a T_g of ~280°C. The optimum balance between processing and end-use properties was achieved by randomizing the molecular structure through the use of T and I diacids and *o*- and *m*-phenylenediamines. The result is a completely amorphous skin layer in the glassy state. Even so the preformed polymer is rather intractible. It is cast from DMAC solutions which also contain LiCl to enhance solubility and increase permeability in the final membrane. More recent Aramid membranes are prepared from all *m*-substituted monomers, approximately 10% of whose aromatic rings have been sulfonated. Such polymers are soluble in DMSO. The weakness of Aramid membranes—lack of chlorine resistance—is not as was first thought, due to the presence of —NH— groups in the polymer backbone. Instead the aromatic rings are halogenated by electrophilic substitution.[89] The bulkiness of halogen substituents in turn results in changes in hydrogen-bonding modes from intermolecular to intramolecular. This leads to chain deformation and alteration of the average interchain displacement within the membrane skin structure. The end result is increased permeability and decreased selectivity.

Polybenzimidazoles (PBIs) are a class of membrane polymer which like the Aramids were originally developed for use as textile fibers. The PBI which is being utilized in hyperfiltration membranes is poly(2,2'-(*m*-phenylene)-5,5'-bibenzimidazole)[90]

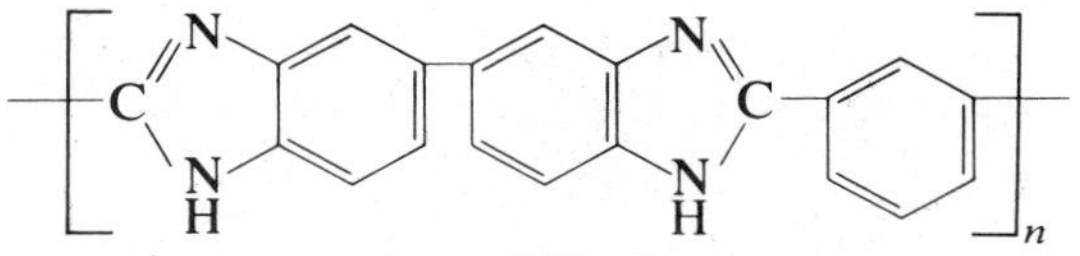

PBI

The tractibility of PBI is sufficient that it may be isolated as a preformed polymer and redissolved to form a casting solution. However, this does not mean that it will dissolve in acetone at room temperature as will cellulose acetate. The actual procedure involves the addition of 16% PBI to DMAC containing 2% LiCl at 240°C under nitrogen. The solution is held and cast at 100°C, partially dried under low humidity at 90°C with an air flow rate of 1200 ft/min, and gelled in cold water. The primary membrane gel is converted to the final secondary gel by annealing in ethylene glycol at 180°C for 10 min. The higher T_g ($140°C < T_g > 180°C$) of PBI compared to that (68.6°C) of CA is responsible for the capacity of PBI to withstand higher operating temperatures.

Poly(benzimidazolone) (PBIL) has the structural formula[91]

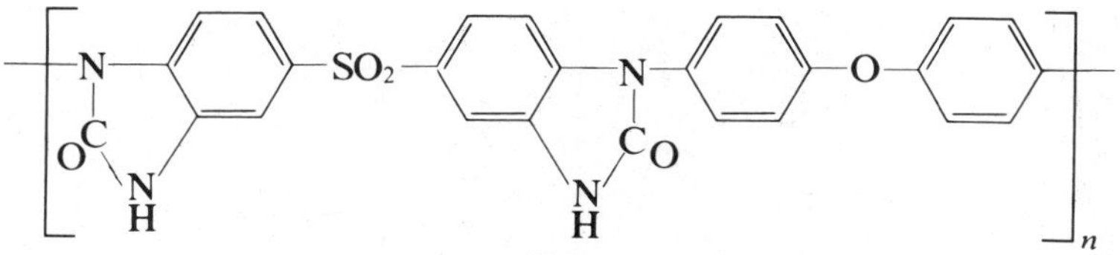

PBIL

Integrally skinned membranes of PBIL can be made by casting NMP solutions containing ~15% polymer and varing concentrations of membrane salts such as LiCl or LiNO$_3$. Evaporation of the cast film for ~10 min at 130°C is followed by immersion in water and annealing for 10 min at 80°C. The fact that PBIL membranes may be annealed under conditions comparable to those employed for CA membranes, would appear to suggest that its T_g if similar to that of CA and that permeability decline at higher than ambient temperatures and pressures should be appreciable. However, water permeability through PBIL membranes is reported to be linear up to 60°C (compared to about 45°C for CA membranes). One possible explanation for this anomalous behavior could be that PBIL, because of its concentration of aromatic rings, exhibits a higher bulk modulus and hence greater resistance to compaction than does CA. Flux stability and resistance to extremes of pH is greater than PBIL than for CA. The chlorine resistance of PBIL was initially reported to be excellent. This was somewhat unexpected in view of the presence of the N–H group. However, subsequent studies failed to confirm the long-term resistance of PBIL to low doses of chlorine. More than once the supposition of long-term resistance to chlorine based on the extrapolation of short-term data at higher exposure levels has proven too sanguine.

The reaction of diamines with dianhydrides yields first polyamic acids (IX) and then polyimides (X):

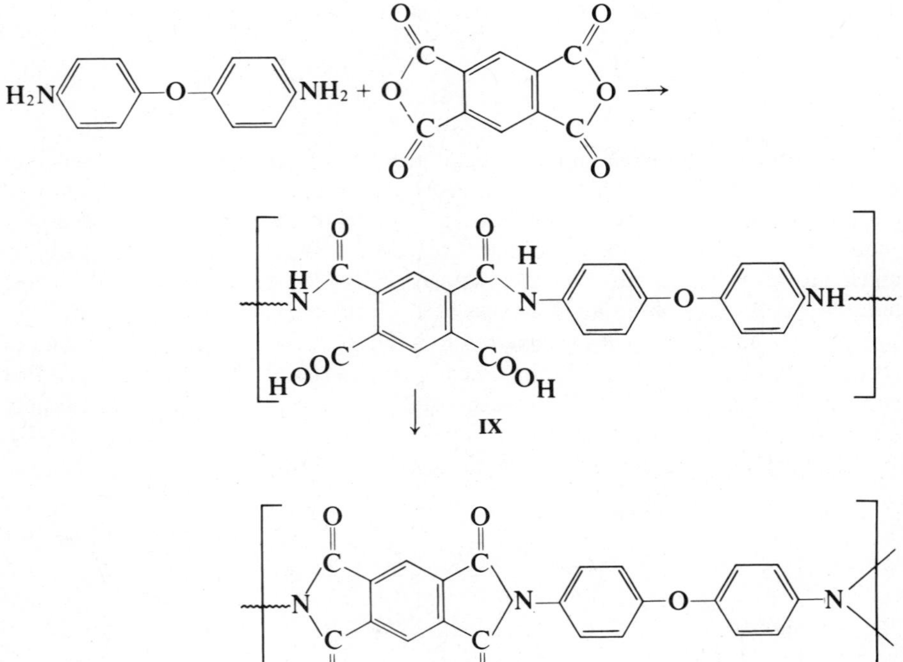

The polyimides have excellent thermal ($T_g \sim 350$°C) and oxidative stability and constitute rigid stable polymers.[92] The original aromatic polyimides (Kapton®) were so intractible that they were processed in the polyamic acid form and only fully condensed in their final physical (flat-sheet) configurations. Because of the supposed absence of $-NH-$ groups in their backbones, it was initially felt that polyimides would be resistant to oxidation by chlorine. For this reason polyimides have been intensively investigated as candidate hyperfiltration membrane polymers. However, the chlorine resistance of the polyimides, possibly because conversion of the polyamic acid to polyimide, is not quantitative and, possibly because of ring halogenation,[89] has proven inadequate and interest in polyimide hyperfiltration membranes has waned. Present activity in polyimide membrane polymers centers about their use in gas-separation membranes. Hoehn[92] has increased the solubility of polyimides by the incorporation of hexafluoroisopropylidene groups which act to decrease cohesive forces. Interchain displacement, and hence permeability, has been given an additional boost by the utilization of a naphthyldiamine which forms a *kink* in the otherwise linear polymer chain. The resultant polyimide (XI) is a sophisticated example of tailoring a polymer with both processing and end-use considerations in mind:

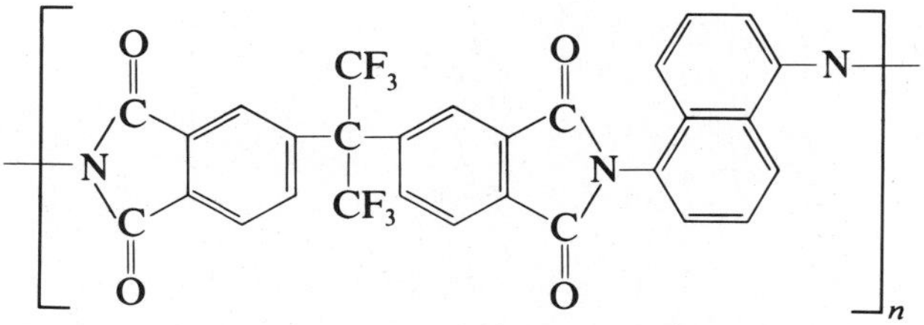

XI

The fact that the high melting temperatures of aromatic and fully aromatic polyamides and related polymers requires their synthesis by interfacial or low-temperature solution polycondensation techniques does not mean that such procedures are restricted to the preparation of *preformed* polymers for formulation and casting into phase-inversion membranes. Indeed, the interfacial polycondensation approach[86] may also be employed to produce the thin permselective barrier layers *in situ,* that is, at the surface of their microporous supports. The latter approach has been pioneered and developed by Cadotte[67] who had earlier first developed techniques for the transfer and subsequently for the direct application of thin films of preformed polymers to microporous supports. Interfacial polycondensation is unusual in that neither monomer purity nor exact stoichiometry of reagents is required.[86] Three examples will be utilized to illustrate the application of this technique to polyamide and related membrane polymers: (1) NS-100, a polyurea; (2) NS-101 (= PA300), an aromatic polyamide; and (3) FT-30, a fully aromatic polyamide. In each case a microporous PS membrane with surface pores (~ 0.1 μm in diameter was first wetted with a dilute aqueous solution of the water-soluble component: polyethyleneimine (PEI), or *m*-phenylenediamine. After removing the excess solution, the polysulfone (PS) membranes were coated with hexane solutions of tolylenediisocyanate (TDI), isophthaloyl chloride (IPC), or trimesoylchloride (TMC),

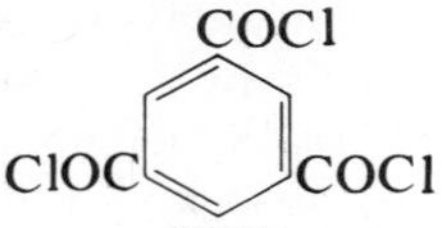

TMC

and annealed in air. Reaction of the branched PEI with TDI or IPC resulted in the formation of highly cross-linked, and hence brittle, polyurea and polyamide thin films. Both types of composite resulted in excellent hyperfiltration membranes from the standpoint of flux and rejection characteristics. However, in both cases the skin layers were so thin and fragile that production yields were low. Furthermore, because of the presence of NH groups in the cross-linked polymers, chlorine resistance was negligible. Thus, although a modified PA 300 continues to be an item of commerce, the polyurea has given way to the fully aromatic and less highly cross-linked polyamide. FT-30,[98] whose approximate structure may be depicted as

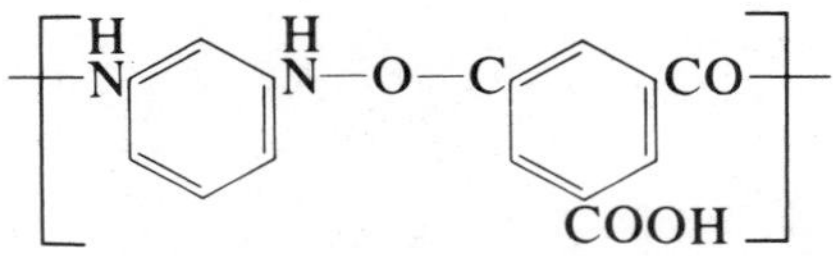

FT-30

Although almost all of the TMC moieties in FT-30 contain one free carboxyl group, a small but significant number is present in which all three carboxyl groups have reacted. Also significant is the greater thickness (0.25 μm) of the skin of FT-30 compared to those ($\sim$0.1 μm) of its predecessors. Increased skin thickness has yielded increased resistance to abrasion and a resultant ease of handling comparable to that of integrally-skinned CA membranes. Unfortunately, as was the case with PBIL membranes, expectations of long-term chlorine resistance (based upon short-term exposure to a high bolus) have not been realized.

The extreme sensitivity to oxidation in a nitrogeneous polymer is exemplified by the "cross-linked polyether" PC membrane which is formed by the condensation of one part of 1,3,5-trishydroxyethyl isocyanurate (XII) with two parts of furfuryl alcohol (XIII) with sulfuric acid as the catalyst.[94]

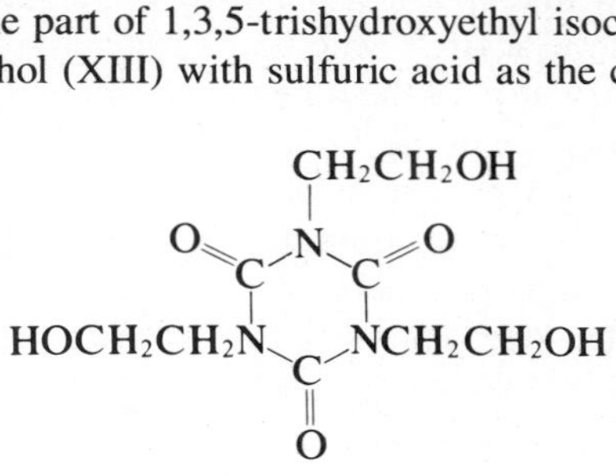

XII 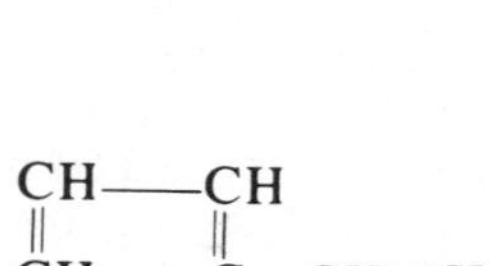   XIII

As with the related NS-200 sulfonated polyfuran membrane, a low degree of sulfonation appears to accompany the formation of the modified furan resin. Furans are believed to cross-link as follows[6]:

$$\text{—CH}_2\text{—} \quad \longrightarrow \quad \text{—CH}_2\text{—}$$

The cross-link density is high and the segments between the cross-links are quite rigid. This results in black, brittle thin films whose physical integrity must be safeguarded by an overlay of a flexible and permeable cross-linked poly(vinyl alcohol) film. The extremely high (99.8%) rejection on Persian Gulf ($\sim 5\%$ NaCl) seawater is phenomenal and is probably attributable to high cross-link density and rigidity separating the large number of hydrophilic sites. The sensitivity of all furan resins to oxidation is shared by both the NS-200 and the PC membranes. The latter in fact is oxidized even by dissolved molecular oxygen and must be kept in a reducing environment by the continuous addition of $NaHSO_3$ to the feed.[95]

Another type of interfacial condensation of thin nitrogeneous films *in situ* is that effected by *plasma* polymerization.[96] Plasma is an ionized gas produced by electric discharge. It is composed of electrons, ions, gas atoms, and various molecules in ground and excited states. The reactions in a plasma are similar to those encountered in high-energy irradiation except that the former are restricted to surfaces. As is the case with high-energy irradiation, monomers are converted into reactive forms which are usually unrelated to the forms which obtain under ordinary, for example, vinyl, polymerizations. As a result, hydrophobic monomers can become hydrophilic thin films and vice versa. The different but active monomers polymerize at any surface, including the surfaces of microporous support membranes. Under the proper conditions plasma films can become strongly bonded and highly cross-linked. As with other high-energy irradiation reactions, the results are interesting if somewhat upredictable. Plasma-produced films from 4-picoline and 4-ethylpyridine, neither of which are considered to be monomers in nonplasma systems, are superior to those from vinyl monomers. At present the chemical and physical structures of plasma-produced films and the relationship between such structures and their transport characteristics are under investigation.

4.3 ION-EXCHANGE POLYMERS AND IONOMERS

The terms *ion exchange* and *ionomer* are both used to qualify membrane polymers which bear fixed acidic and/or basic groups or their salts. Although they are sometimes used interchangeably, there is a difference in degree, with ion exchange used to refer to those polymers with higher capacities, whose strong tendency to hydrate and swell is held in check by a high density of covalent cross-links. The latter are usually introduced in the form of polyfunctional monomers, such as divinyl benzene (DVB), during polymerization. Ionomers, on the other hand, usually contain between 2 and 15 mol % of monomers with ionic groups present in side chains. Owing to their lower capacities, they often do not require cross-linking to inhibit swelling. Another difference between traditional ion-exchange membranes and the newer ionomeric types is the generally superior film-forming characteristics of the latter. The presence of ionic groups in ionomers, particularly of the SO_3^- M^+ variety, acts to increase T_g and the modulus, even in water-swollen membranes. Chain stiffness is inversely proporitonal and polymer solubility is directly proportional to the size of the M^+ counterion. Quaternary ammonium groups, on the other hand, do not exert so great an influence on physical and solubility properties as do sulfonate groups.

Ion-exchange membranes frequently begin as a monomer mixture which is cast in the final membrane configuration on glass plates with dammed edges. Because they are relatively thick (≈ 2 mm), ion-exchange membranes are properly classified as sheet rather than as film and are utilized exclusively in stacks, that is, the plate and frame configuration. Ionomers, on the other hand, can be cast or spun either by themselves or as blends with nonionogenic (neutral) polymers into strong and flexible films and hollow fibers. Of course, as the ion-exchange capacity (IEC) of an ionomer is increased, so will its tendency to hydrate and swell with the result that its utility as a stable water-insoluble film former diminishes. Ultimately, this may result in the formation of water-soluble polymers known as *polyelectrolytes*. In the remainder of this section, a general treatment of ion-exchange phenomena will be followed by a discussion of polystyrene and styrene–divinylbenzene copolymers, the most common neutral starting material for ion-exchange membranes, and then a discussion of the ion-exchange membranes themselves. Finally, several ionomeric polymers, including the extremely important class of perfluorinated ionomers, will be considered.

The characteristic which distinguishes ion-exchange and ionomer membranes from other types is the presence of charges or ionic groups in their component polymer molecules. Mobile ions bearing a charge opposite to that born by the fixed ions are known as counterions; those bearing the same charge ar co-ions. Polymers containing positively charged groups are polycations. Because of the electroneutrality requirement they will have a stoichiometric amount of exchangeable anions in association with their fixed cations. Since such anions are mobile and can be exchanged for others in an external solution, polycations are known as anion exchangers. For similar reasons polyanions are termed cation exchangers. There are also amophoteric types, which are capable of exchanging both cations and anions, and redox types for oxidation and reduction.

The properties of ion-exchange polymers ar exemplified by a cation-exchange membrane (containing no sorbed electrolyte) which is placed in a dilute solution of a strong electrolyte (Fig. 4.22). The cation concentration is higher in the membrane (because the cations are attracted to its fixed negatively charged groups). The mobile-anion concentration, on the other hand, is higher in the solution. These concentration differences cannot be leveled out by diffusion because electroneutrality

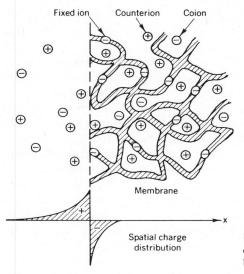

FIGURE 4.22.   Schematic representation of the distribution of ions in the vicinity of the membrane–solution interface (from Schlögl[97]).

would then be disturbed, owing to the electric-charge-bearing properties of the mobile ions. The flux of cations into the solution and of anions into the membrane results in an accumulation of positive charge in the former and negative in the latter. The first diffusing ions therefore result in a potential difference between the two phases. This *Donnan potential* effectively draws cations back into the membrane and anions back into the solution, with the eventual establishment of equilibrium between the electric field and the tendency of the ions to eliminate concentration differences by diffusion.[98, 99] For this reason counterion concentration in the membrane is higher, and co-ion concentration lower, than in the external solution. Since co-ions are repelled from the membrane, the electrolyte itself is too, because of the electroneutrality requirement. This exclusion of the elctrolyte from the membrane is known as *Donnan exclusion*.

Electrolyte exclusion increases with the magnitude of the Donnan potential. Therefore, this potential and its effects upon ions are of prime importance to understanding the sorption and transport of ions by and through ion-exchange and ionomer membranes. Electrolyte sorption and transport depend chiefly upon the distribution of the co-ion since electrolyte and co-ion uptakes are stoichiometrically equivalent. Among the parameters which influence the magnitude of the Donnan potential are the capacity of the dry resin, the degree of swelling or its inverse, the cross-linking density, solution concentration, and ion charge density.

Capacity most often refers to the counterion content (in milliequivalents per gram) of anhydrous resin before any electrolyte sorption has occurred. This definition is a characteristic of the membrane material and is independent of the experimental conditions. The concentration of fixed ionic groups is subject to change with such variables as degree of swelling and electrolyte solution concentration.

The volume concentration of fixed ionic groups determines the charge density of the membrane and hence its Donnan potential. Obviously, since swelling decreases this concentration, it must be minimized if a maximum Donnan potential is to be attained. The means most commonly employed to this end are the physical restriction of interchain displacement by covalent cross-links. Donnan potential, and hence electrolyte sorption, is inversely proportional to the degree of swelling and directly proportional to the cross-link density. Since Donnan equilibria are electric field phenomena which depend on the overall charge of both fixed and mobile groups, these values affect the magnitude of the Donnan potential. The decreasing efficiency of electrolyte exclusion with increasing solution concentration is due to the increasing tendency of the ions to eliminate concentration differences by diffusion at a constant electric field. (The electric field is constant because the concentration of fixed charges in the membrane is constant). The equilibrium between these opposing tendencies is therefore displaced, with the result that the Donnan potential and electrolyte exclusion decrease. Counterions of high charge density (small size and/or high valence) and co-ions of low charge density minimize electrolyte exclusion. This effect is due to maximum attraction and minimum repulsion by the membrane's fixed ionic groups of counterions and co-ions, respectively. In addition to the effects upon interaction with, and repulsion by, the membrane, high counterion and low co-ion charge density inhibit the formation of co-ion pairs between mobile ions so that external forces, such as the electric field exerted by the membrane's fixed charges, have a correspondingly larger effect than would obtain if strong associations between the electrolyte's component ions were in effect. When, on the other hand, ion pairs or complexes are formed by counterions and co-ions, the combination can act as a unit with an effective charge density corresponding to the relative proportions of positive and negative charges. In such a case co-ion exclusion, and hence Donnan efficiency, may be diminished owing to the entry of the former as a disguised portion of a counterion complex.

Likewise where association between fixed ionic groups and counterions results in ion-pair formation, the effective charge density of the former, and with it the magnitude of the Donnan potential, is reduced. In extreme cases where multivalent counterions associate with the fixed ionic groups, the membrane can reverse its "fixed" charge. Helfferich[24] cites the example of thorium ions (Th^{2+} or ThO^{2+}), which are strongly associated with the negative groups of cation-exchange membranes:

$$-R^- + ThO^{2+} \rightarrow RThO^+$$

The excess of positive charges confers anion-exchange character upon the membrane, which is then compensated by an influx of anions.

An important characteristic of ion-exchange membranes is selectivity, that is, the sorption of one counterion in preference to another. Selectivity is influenced by the magnitude of membrane–solute and solute–solute interactions. Thus a membrane tends to prefer counterions of high charge density (for electrostatic reasons) and small size (for steric reasons). It also tends to prefer those counterions which are least encumbered by interactions with other mobile ions. Selectivity sequences for some of the more common cations are[100–105]

$$Ba^{2+} > Pb^{2+} > Sr^{2+} > Ca^{2+} > Ni^{2+} > Cd^{2+}$$
$$> Co^{2+} > Zn^{2+} > Mg^{2+} > UO_2^{2+}$$
$$Tl^+ > Ag^+ > Cs^+ > Rb^+ > K^+ > NH_4^+ > Na^+ > Li^+$$

The corresponding anion-exchange sequence is

$$citrate > SO_4^{2-} > oxalate > I^- > NO_3^- > CrO_4^{2-}$$
$$> Br^- > SCN^- > Cl^- > formate > acetate > F^-$$

The position of H^+ varies with the acid strength of the membrane; for strong acids it usually lies between Na^+ and Li^+, whereas for weak acids it depends on the strength of the fixed anions. The position of OH^- also varies; for strong bases it usually lies between acetate and fluoride, and for weak bases it depends on the strength of the fixed cations.

In addition to the various fixed-ion–mobile-ion associations which occur for largely electrostatic reasons, ion-exchange membranes can also sorb weak electrolytes and nonelectrolytes from solution. Sorption of such solutes is generally, but not always, reversible and is governed by the same factors which govern solute sorption by nonionic membranes. Owing to the close association, or *binding*, of solvent molecules with fixed ionic groups, not all the solvent with ion-exchange membranes is available for solvation of solutes. Bound solvent molecules have diminished mobility and hence are incapable of rapidly reorienting themselves about solute particles. For this reason only free (nonbonded) solvent is usually available for interaction with the solute. Salting out of solute can therefore occur in the same manner as when a soluble salt is added to one liquid phase in a liquid–liquid distribution equilibrium. Salting-out effects are most pronounced when fixed ions and counterions are highly hydrated and solvent uptake is limited by a highly cross-linked network. Both conditions tend to minimize the concentration of free water in the membrane. In the case of acetic acid sorption by sulfonated polystyrene resins, the free water available in the membrane amounts to the total water content less four molecules of water bounded to each $-SO_3H$ group.[106]

Salting-in is a less frequently observed phenomenon but does occur, for example, when acids are added to certain membrane–alcohol systems such that an increase rather than a decrease in the solubility of the aliphatic alcohols is noted. This phenomenon is probably attributable to a lowering of the dielectric constant of the bound water, making it more alcohollike. In this connection it should be remembered that the solvent power of solvents for nonelectrolytes, unlike solvents for electrolytes, need not be directly related to the dielectric constant.

In ion exchange as well as in nonionic membranes, specific interactions such as dipole–dipole interactions between the polar groups of the solute and those of the membrane, enhance solute sorption. The nonpolar portion of organic solutes confers surface activity upon the solute because of the conflicting tendency for the polar portion to remain in solution. As a result, solute sorption usually increases with increasing solute hydrophobic/hydrophilic ratio, except where sieving effects predominate. Since the average mesh width of fully swollen ion-exchange membranes is from 6 to 30 Å,[24] the common sulfonated polystyrene types with 8–12% divinylbenzene (DVB) cross-linking sorb simple phenyl and naphthyl derivatives and glucose without much steric hindrance. Larger molecules are accommodated only in less highly cross-linked resins.

Solute sorption increases with increasing solution concentration, but the solute distribution coefficient usually decreases owing to saturation of the membrane at low concentrations. Increasing the temperature decreases the extent to which solvent molecules are bonded by fixed ionic groups and hence increases the percentage of free solvent. However, temperature has similar effects upon solute–solvent interactions, so that its net effect upon sorption and other membrane phenomena is rather unpredictable.

Neutral polystyrene and poly(styrene–divinyl benzene) copolymers are generally not suitable for film and membrane applications per se. Atactic polystyrene is a completely amorphous clear glassy polymer with a softening temperature of approximately 90°C.

$$\left[\!\!\begin{array}{c} CH_2-CH \\ \\ \end{array}\!\!\right]_n$$

By way of contrast, the less commonly encountered isotactic polystyrene is a crystalline, opaque material with a T_g of 230°C. The C–C backbone confers hydrolytic and thermal stability, while the aromatic ring confers stiffness and relatively strong polar forces. With a δ of 9.1, polystyrene is soluble in aromatic hydrocarbons such as benzene and toluene. Furthermore, because of its basic (aromatic) groups it is also soluble in such acidic solvents as methylene chloride and chloroform. The utility of poly(styrene–divinyl benzene) copolymers as starting materials for ion-exchange polymers and membranes lies in the reactivity of the aromatic rings of both monomers for such pertinent reactions as sulfonation, nitration, and chloromethylation which can lead to strong acids and strong and weak bases. Commercial divinyl benzene (DVB), actually a mixture which contains 40% ethyl vinyl benzene, is the most commonly utilized cross-linker. In a standard copolymerization, styrene is mixed with a given percentage (4–30) of DVB, which amount is referred to as the nominal percent cross-linking. The typical value is 8%.[34] Polymerization of styrene–DVB where [DVB] $\leqslant$ 10% results in matrices which after modification are known as *homogeneous* membranes. These are comparatively dense and isotropic structures with voids in the 5–30 Å range, because of which the membranes are transparent. However, if a linear polymer such as polystyrene is incorporated within the monomer phase during polymerization (for postpolymerization leaching),[24,34] or if polymer is precipitated within a high DVB content gel before polymerization is complete, then opaque *macroporous* or *macroreticular* membranes are the result. These two techniques can be applied to the polymerization of other monomers, for example, HEMA, as well and represent two general methods for the preparation of microporous membrances during polymerization (Chapter 8).

Strong-acid cation-exchange membranes are the result of the sulfonation of the poly(styrene–DVB) copolymer with sulfuric or chlorosulfonic acids.

On the styrene moieties sulfonation is primarily at the para position. After careful displacement of the excess acid with successively lower concentraitons of acid, the membranes in the acid form are converted to the sodium form by neutralization with a slight excess of alkali.

Weak-acid cation-exchange membranes are made by copolymerizing acrylic or methacrylic acids with DVB,[34]

where R = H or CH_3, respectively.

Strong-base anionic-exchange membranes can be made by the chloromethylation with methyl chloromethyl ether of the same (styrene–DVB) copolymers as are utilized for the strong-acid cation-exchange membranes mentioned earlier, followed by amination with an appropriate tertiary amine. However, because the formation of methylene bridges is a side reaction of chloromethylation, they can also be prepared by the chloromethylation of linear polystyrene which does not contain any DVB[34]:

Weak-base anion-exchange membranes can be made by the reaction of primary, RNH_2, or secondary, R_2NH, amines with chloromethylated polystyrene:

$$\left[CH_2-CH\underline{\hspace{4cm}}CH_2CH\underline{\hspace{3cm}}\right]_n$$

with pendant groups: $-CH_2\underset{H}{N}R$ and $-CH_2NR_2$, CH_2, and

$$\left[CH-CH_2\right]_n$$

Redox membranes are solid oxidation and reduction agents which are usually classified as ion-exchange resins in spite of the absence of charged groups in the polymer matrix. They incorporate components such as quinone–hydroquinone, which can be oxidized and reduced. Electron-exchange resins can be prepared by condensation and by addition polymerization. Addition polymerization of esterified hydroquinone, styrene, and DVB to yield stable but hydrophobic resins has been accomplished.[107-110] The hydroquinone must be esterified prior to polymerization and only subsequently hydrolyzed, because the unesterified compound otherwise acts to inhibit polymerization.[111, 112] Sulfonation[113, 114] increases hydrophilicity without interfering with the redox properties of the electron exchanger. The redox capacity of redox ion exchangers, however, is removed by such a procedure.

Redox ion exchangers are conventional ion-exchange resins containing reversible redox couples such as Fe^{3+} and methylene blue–leucomethylene blue. These couples are held in the resin either as counterions or as the result of both specific and nonspecific sorption. Anion exchangers containing cupric ions have been developed for removal of dissolved oxygen from water. Both electron exchangers and redox ion exchangers are characterized by their redox capacity (the redox equivalent of ion-exchange capacity), the redox potential (analogous to the membrane potential), and their rates of reaction. Reaction rates are generally slower than for analogous ion-exchange resins.

Whereas the cross-linked ion-exchange resins and the water-soluble polyelectrolytes are situated in the high-capacity range of ion-containing polymers, the ionomers tend to occupy the low-capacity range. The first representatives of this class were addition copolymers (ethylene/acrylic acid) in which the minority component was ionic.[26] Today, however, the term encompasses condensation polymers as well. Because of the tendency, once a threshold concentration has been surpassed, of the ions in ionomers to form domains in which the ions strongly associate with one another in clusters, the bulk properties of ionomers differ considerably from those of their nonionogenic counterparts. This holds true even in the presence of ion-solvating substances such as water. The ionic groups act as virtual

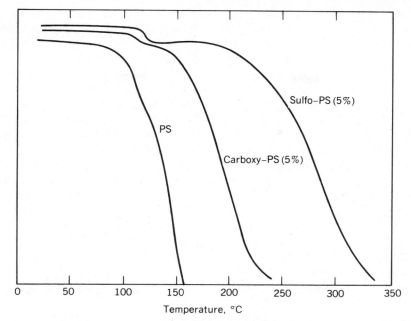

FIGURE 4.23. A comparison of the softening behaviors via TMA of PS, C-PS, and S-PS (load = 10 g; heating rate = 10°C/min) (from Lundberg and Makowski[115]; reprinted with permission from ACS, © 1980 American Chemical Society).

cross-links to stiffen the polymer and their ability to do so varies directly with both concentration and the magnitude of their charges. Thus with respect to softening temperatures (Fig. 4.23):

polystyrene < carboxylated polystyrene < sulfonated polystyrene

This holds true for related properties such as modulus, tensile strength, elongation, melt and solution viscosities, and so on. The nature of the cationic counterion also has a considerable influence upon physical properties. In comparison of the properties of sulfonated EPDM (ethylene–propylene–diene monomer terpolymer) with various associated cations, only Pb and Zn increased melt flow and improved physical properties.[116] Cations with excessively stiffening effects included Hg, Mg, Ca, Co, Li, Ba, and Na. These results are consistent with a more pronounced covalent character (lower charge density) for the Zn and Pb ions. The water absorption of these materials was also strongly influenced by the nature of the associated cation. The influence of quarternary ammonium groups on the bulk physical properties of ionomers is much less than that of sulfonate groups. This is probably related to the greater organic character of the former.

First developed during the mid-1960s, the importance of cation-exchange membranes consisting of *perfluorinated ionomers* containing sulfonate and/or carboxylate groups has been growing steadily, until they now are more significant than all other ion-exchange and ionomeric membranes combined. The reason for their spectacular growth lies in their use in membrane cells used in chlor-alkali production where their strength and chemical stability are required. Perfluorinated ionomer membranes compete favorably with the older diaphragm cell technology which employs an asbestos diaphragm on the cathode to separate chlorine from caustic soda and hydrogen. Since the diaphragm is permeable to both water and brine, it yields a dilute (8–10%) caustic solution which contains 15% salt. The concentration of caustic to 50% and the removal of salt require a considerable amount of energy. The greater permselectivity of the perfluorinated ionomer membrane, on the other hand, results in the rejection of chloride ions and hence in the production of higher-strength (15–40%) caustic with less (0.02%) salt. Therefore, less energy is required for concentration of the caustic and a more nearly salt-free (0.05%) product is obtained.[117, 118]

The first membranes of this type, Nafion 900®, contained only $-SO_3^-$ Na^+ anions. They were put into sheet form, suitably reinforced by an open-weave Teflon fabric, by extrusion from the melt of a thermoplastic nonionogenic precursor, known as XR resin:

$$\text{+ CF}_2\text{CF}_2\text{ +}_n\text{CFO (CF}_2-\text{CFO +}_m\text{CF}_2\text{CF}_2\text{SO}_2\text{F}$$
$$\begin{array}{cc} | & | \\ \text{CF}_2 & \text{CF}_3 \\ | & \end{array}$$

XR resin

Treatment of the membranes in the sulfonyl fluoride form with caustic converts them to the sulfonated ionomeric form:

$$\text{XR resin + NaOH} \longrightarrow \text{+CF}_2\text{CF}_2\text{+}_n\text{CFO+CF}_2-\text{CFO+}_m\text{CF}_2\text{CF}_2\text{SO}_3^-\text{ Na}^+$$
$$\begin{array}{cc} | & | \\ \text{CF}_2 & \text{CF}_3 \\ | & | \end{array}$$

Nafion

where $m = 1$ and $5 < n > 11$. This yields an equivalent weight of from 1000 to 1500 g of dry hydrogen ion from polymer per mole of exchange sites.

Compromises have to be made between membrane resistance and permselectivity. The higher the capacity of $-SO_3^-$ groups, the lower the resistance. This is ideal insofar as energy consumption is concerned. On the other hand, sulfonate groups hydrate readily so that if they are too close to one another, the membrane will swell excessively and produce channels which are large enough to permit the back-migration of the hydroxyl ions into the anolyte, thereby decreasing the concentration of the caustic which is produced. Several ways around this dilemma have been investigated:

1. Decreasing the capacity by increasing equivalent weight. However, this increases membrane resistance excessively.

2. Surface treatment with ethylene diamine. This converts surface $-O-CF_2CF_2SO_2F$ groups in the XR resin precursor to $-OCF_2CF_2SO_2NHCH_2CH_2NH_2$. The N-β-aminoethylsulfonamide group is a weaker, and therefore less hydrated, acid group than is the sulfonate. However, the lifetimes are too short in practice.

3. Surface oxidation of sulfonyl to carboxylate groups by conversion of SO_2F to SO_2Cl, oxidation of the alcohol wet surface, and hydrolysis.

4. Formation of a separate carboxylate resin and lamination of this resin to XR resin membranes followed by hydrolysis. This yields a composite membrane with the more permselective carboxylate layer facing the cathode and shielding the more conductive, but too highly swollen, sulfonate layer. A typical carboxylate resin is

$$
\begin{array}{c}
+ (CF_2CF_2)_n CF-CF_2 +_x \\
\quad\quad\quad\quad | \\
\quad\quad\quad\quad O \\
\quad\quad\quad\quad | \\
\quad\quad\quad\quad CF_2 \\
\quad\quad\quad\quad | \\
\quad\quad\quad\quad CF-CF_3 \\
\quad\quad\quad\quad | \\
\quad\quad\quad\quad OCF_2CF_2CF_2CO_2^- Na^+
\end{array}
$$

The perfluorinated ionomers contain a certain small amount of crystalline domains which serve as virtual cross-links to restrict swelling. Two distinct amorphous phases are present as well: a hydrophobic fluorocarbon phase and hydrophilic ionic domains. Hydration and swelling can be enhanced by boiling the membranes in water. Water content increases by up to 50% over that found in membranes which were in equilibrium at room temperature. This increased water uptake is not reversible upon cooling to room temperature. Such membranes will also absorb appreciable concentrations of alcohols and, if the equivalent weight of the ionomer is less than 970, will dissolve in them. Higher (1100 and 1200) equivalent weight polymers may be dissolved in alcohol–water mixtures by superheating them in an autoclave.[119]

As was the case for the covalently cross-linked ion-exchange membranes, copolymerization with ethylenically unsaturated monomers containing ionogenic groups is an excellent technique for the preparation of ionomeric membrane polymers. The first example of this approach was the copolymerization of acrylic acid (AA) with ethylene (E) to yield EAA ionomers.[120] In this case inorganic ions were also incorporated into the melt so that salt-bridge cross-links formed in the resultant extruded film after cooling. Because cross-links are relatively weak, they are revsersible on heating, so that the polymer retains its thermoplastic behavior at elevated temperatures while exhibiting the advantages of a cross-linked polymer at its use temperatures. Ethylenically unsaturated monomers which contain sulfonate moieties have been successfully copolymerized with acrylonitrile to yield tractible ionomers.[39] The same is true of monomers which contain quaternary ammonium or weakly basic moieties.[121] It is, of course, also possible to form copolymers containing halogen groups and to convert them to basic moieties by post polymerization, or indeed even by post membrane formation, reactions with appropriate amines.[122]

Although less common, the polymerization of ionogenic, or potentially iono-genic, monomers can also be utilized to form ionomeric step reaction polymers. When the charges are located on atoms which make up the polymer backbone, these materials are known as *ionene* polymers. Thus —NCO terminated oligo-meric polyurethanes have been reacted with N-methyldiethanolamine to yield po-lyurethanes (**XIV**) which contain tertiary amine groups in the polymer backbone[123]:

$$\text{OCN} \sim\sim\sim \text{NCO} + \text{HOCH}_2\text{CH}_2\overset{\overset{\displaystyle CH_3}{|}}{\text{N}}\text{CH}_2\text{CH}_2\text{CH}_2\text{OH} \longrightarrow$$

$$\begin{array}{c} \overset{\displaystyle CH_3}{|} \\ \text{+NHCOOCH}_2\text{CH}_2\text{NCH}_2\text{CH}_2\text{CH}_2\text{OOCNH}\text{+}_n \end{array} \quad (8)$$

XIV

Reaction of **XIV** with CH_3I results in a polyurethane (**XV**) with corresponding quaternary ammonium groups:

$$\textbf{XIV} + \text{CH}_3\text{I} \longrightarrow \begin{array}{c} \overset{\displaystyle CH_3 \, I^-}{|} \\ \text{+NHCOOCH}_2\text{CH}_2\overset{+}{\text{N}}\text{CH}_2\text{CH}_2\text{OOCNH]}\text{+}_n \\ \overset{|}{\displaystyle CH_3} \end{array} \quad (9)$$

XV

Likewise, the reaction of 1,6-dibromohexane with a suitable ditertiary amine will yield a polymer (**XVI**) with quaternary ammonium groups in the backbone:

$$\text{Br(CH}_2)_6\text{Br} + \text{N} \underset{}{\bigcirc} \text{N} \longrightarrow \left[(CH_2)_6 - \overset{+}{N}\underset{}{\bigcirc}\overset{+}{N} - (CH_2)_6 \right]_n \quad (10)$$

$$ Br^- Br^-$$

XVI

Neutral preformed step reaction polymers can also be converted into ionomers. If the latter are hydrolytically stable and contain aromatic groups, then sulfonation can be effected under conditions which are comparable to those employed with the poly(styrene–DVB) copolymers, that is, utilizing sulfuric or chlorosulfonic acids as the sulfonating agents. In the case of the ionomers, however, sulfonation is usually carried out on a linear polymer prior to membrane formation. Examples of such polymers are polysulfone,[124] polyether sulfone,[125] poly(dimethylphenylene ox-ide),[126] fully aromatic polyamides,[89] and polyimides.[127] Sometimes such polymers can be sulfonated to an extent sufficient to confer water solubility which can then be reversed by heating to convert a fraction off the sulfonic acid groups to sulfone cross-links.[68] Obviously, chloromethylation and reaction with tertiary amines to yield strongly basic quaternary ammonium groups or with secondary and primary amines is also possible. The side reaction of chloromethylation, namely the introduction of methylene bridges, is controllable and reproducible, and can be terminated before the formation of an insoluble gel network. Thus chloromethylation can be utilized as a means for increasing the viscosity of an aromatic membrane polymer while simultaneously retaining its solubility.

Ionomers may also be prepared by the grafting of ionogenic or potentially ionogenic materials onto neutral polymers or even onto already formed films or membranes. Insofar as true grafting, that is, the covalent bonding of grafted materials to the film substrates, occurs, this method results in stable structures. It has long been recognized, however, that an appreciable amount of homopolymerization generally accompanies the desired graft–copolymerization reaction, so that a significant amount of material is present as a physically included, rather than as a chemically attached, species. Therefore the potentials of grafted membranes tend to decrease somewhat with time.

Although grafting of both condensation and addition polymers is known, the latter, because of their greater chemical and thermal stability, have received greater emphasis. The reaction whereby free radicals or ions are produced upon the film substrate for subsequent addition of vinyl monomers can be accomplished by chemical means involving peroxides[128] or redox catalysts[129] and by means of high-energy (bombardment by electrons or gamma rays)[130] or low-energy (ultraviolet)[131] irradiation. Grafting of neutral but potentially ionogenic materials is preferred, since the presence of ionic monomers often interferes with the propagation of polymerization by a free-radical mechanism. The grafting of a polyacrylate ester onto cellophane followed by hydrolysis to polyacrylic acid has been utilized to prepare a weak-acid cation-exchange membrane.[132] Styrene has been grafted onto polyethylene for subsequent sulfonation to form a strong-acid cation-exchange membrane.[133]

Considerable variation in the structure of the grafted membrane is possible. For a given amount of grafted material, for example, the length of the grafted chains can be controlled. Everything else being equal, membrane homogeneity can be expected to increase with decreasing length of the grafted chains, to the point where the chain length is that of a single monomer unit. Chain length decreases with increasing dose rate in the mutual-irradiation technique (monomer present during radiation). The MW of grafted chains is generally greater in the preirradiation technique (monomer added after irradiation).[134] Chain length can be decreased by increasing the concentration of initiator or by the inclusion of chain transfer agents. However, control of the length of the grafted chains is not the only means for varying the homogeneity of grafted membranes. The degree of inhomogeneity in depth can be modified by a procedure in which only one surface of the film substrate is allowed to come into contact with the monomer solution. In such a case grafting may be a diffusion-controlled phenomenon in which grafting occurs from the exposed surface inward.[135]

Less hydrolytically stable polymers such as cellulose acetate are converted into ionomers by taking advantage of the reactivity of residual free hydroxyl groups$_{45, 61, 136}$ [Eqs. (11)–(14)]:

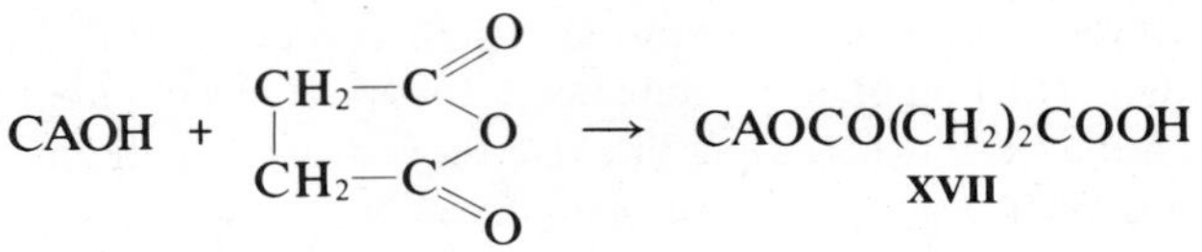

$$CAOH + \begin{matrix} CH_2-C\overset{O}{\underset{O}{\diagup}} \\ | \\ CH_2-C\overset{\diagup O}{\underset{O}{}} \end{matrix} \longrightarrow CAOCO(CH_2)_2COOH \qquad (11)$$

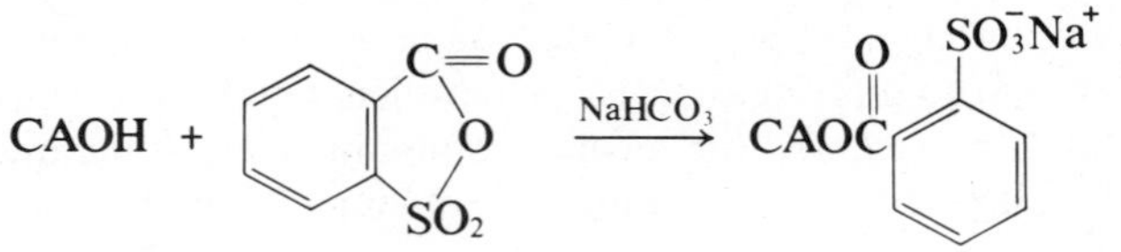

XVII

$$CAOH + \underset{SO_2}{\overset{C=O}{\underset{O}{\diagdown}}} \xrightarrow{NaHCO_3} CAO\overset{O}{\overset{\|}{C}} \overset{SO_3^-Na^+}{\diagup} \qquad (12)$$

XVIII

$$CAOH + ClOC(CH_2)_{10}Br + (CH_3)_3N \longrightarrow CAOCO(CH_2)_{10}N^+(CH_3)_3 \quad Br^-$$

XIX $\qquad$ (13)

Polymers **XVII** and **XVIII** are CA mixed-ester ionomers which contain weak and strong cation-exchange groups, respectively. Polymer **XIX,** the trimethylammonium salt of CA 11-bromoundecanoate, is also accessible by another route, namely the reaction between the trifluoroacetyltrimethyl ammonium salt of 11-bromoundecanoic mixed anhydride (**XX**) and CA:

$$CAOH + (CF_3\overset{O}{\overset{\|}{C}}\!\!+\!O\!+\!\overset{O}{\overset{\|}{C}}(CH_2)_{10}N^+(CH_3)_3 \quad Br^-) \longrightarrow XIX \qquad (14)$$

XX

The mixed anhydride **XX** is readily prepared by the reaction of the trimethyl ammonium salt of 11-bromoundecanoic acid with trifluoroacetic anhydride (TFAA):

$$TFAA + HOOC(CH_2)_{10}N^+(CH_3)_3 \quad Br^- \longrightarrow XX \qquad (15)$$

Although **XVII** is soluble in acetone, **XVIII** and **XIX** require the addition of water or methanol to solvate the strong-acid and strong-base salts, respectively, before complete dispersion as a clear solution in acetone is possible. In the absence of any high dielectric constant solvent to shield the ions, the ionomers will either not dissolve at all (if their capacity is high) or will form grainy solutions (if their capacity is low).

Ionomers can either be utilized as the sole membrane polymer or they can be utilized in blends with other, usually nonionogenic polymers to form *blend*[61] or *interpolymer*[137] membranes. Where solution compatibility between two polymers is sufficient, there are advantages to utilizing blends of ionomers and neutral pol-

TABLE 4.11 HF PERFORMANCE CHARACTERISTICS[a] OF DRY-RO MEMBRANES OF E-394-60 CA, OF QCTE[b] AND OF BLENDS OF QCTE WITH E-394-60 CA[c]

Polymer(s)	IEC (meq/g)	Polymer Cost ($/kg)	Product Flux (gal/ft^2 day)	NaCl Rejection (%)	SRF[d] = Feed Conc. / Product Conc.
E-394-60 CA	—	3	6.0	91.4	12
QCTE (I)	0.21	100	9.0	93.4	15
QCTE (II)	0.36	—	very high	—	—
QCTE II + CA	0.14		6.4	92.2	13
	0.22	68	10.6	91.6	12
	0.29		very high	—	—
QCTE III	0.56		very high	—	—
QCTE III + CA	0.19	36	7.4	96.5	29

[a] 0.5% NaCl feed at 400 psig and 25°C.
[b] QCTE = TEA salt of E-394-60 CA 11-bromoundecanoate (I.E.C.s as indicated).
[c] From Kesting et al.[61]
[d] SRF is the salt reduction factor.

ymers compared to utilizing an ionomer by itself. One advantage is lower polymer cost since a blend of expensive ionomer with inexpensive neutral polymer costs less than an equal amount of expensive ionomer. A second advantage is increased reproducibility. It is easier to vary blend ratios to obtain a final membrane with a given capacity than to synthesize an ionomer with a capacity defined within narrow limits.

The feasibility of blending ionomers with nonionogenic polymers is demonstrated by the equivalence between HF membranes prepared exclusively from ionomeric homopolymers of a given ion exchange capacity (IEC) and membranes with the same overall IEC but prepared from blends of higher-capacity ionomers and a nonionogenic polymer (Table 4.11). Here the ionomeric (QCTE) homopolymers are similar to **XIX** above, but prepared with triethylamine rather than trimethylamine. The utilization of neat QCTE (I) IEC = 0.21 meq/g resulted in a membrane with a flux of 9.0 gal/ft^{2}day and 93.4% rejection, which is roughly equivalent to that (10.6 gal/ft^{2}day at 91.6% rejection) from a blend of QCTE (II) IEC = 0.56 meq/g with CA in such a ratio that the overall IEC of blend membrane was 0.22 meq/g. Because of the close equivalence between the HF performance characteristics of the two membranes, it is assumed that compatibility extends to the molecular level. This is also suggested by the clarity of the blend solutions (once methanol is added), and the glossiness of the skins of the resultant membranes. More definite evidence for a high degree of compatibility between the polymers in the blend and equivalency of HF performance characteristics from membranes with equivalent IECs is found in Tables 4.12 and 4.13. Here two different neutral CA polymers were utilized to prepare ionomers with different capacities and yet equivalent IEC blend membranes yielded virtually identical results.

TABLE 4.12 HF PERFORMANCE CHARACTERISTICS[a] OF DRY-RO MEMBRANES OF JLF-68 CA, AND OF BLENDS OF JLF-68 DERIVED-QCTE[b] WITH JLF-68 CA[c]

Polymer(s)	IEC (meq/g)	Polymer Cost ($/kg)	Product Flux (gal/ft² day)	NaCl Rejection (%)	SRF = Feed Conc. / Product Conc.
JLF-68 CA	—	3	4.2	94.6	18
QCTE + JLF-68 CA	0.07		1.5	98.7	77
	0.11		1.7	98.7	77
	0.15		2.6	99.0	100
	0.17		3.3	99.1	111
	0.19	36	5.3	98.5	67
	0.20		5.5	97.5	40
	0.22		6.4	74.3	4
QCTE	0.56	100	very high		—

[a] 0.5% NaCl feed at 400 psig and 25°C.
[b] QCTE = TEA salt of JLF-68 CA 11-bromoundecanoate (0.56 meq/g).
[c] From Kesting et al.[61]

TABLE 4.13 HF PERFORMANCE CHARACTERISTICS[a] OF DRY-RO MEMBRANES OF JLF-68 CA, AND OF BLENDS OF E-383-40-DERIVED QCTE[b] WITH JLF-68 CA[c]

Polymer(s)	IEC (meq/g)	Polymer Cost ($/kg)	Product Flux (gal/ft^2 day)	NaCl Rejection (%)	SRF = Feed Conc. Product Conc.
JLF-68 CA	—	3	4.2	94.6	18
QCTE + JLF-68 CA	0.14		2.7	98.55	69
	0.16		3.1	98.85	87
	0.18		4.8	98.72	78
	0.21	31	5.3	98.72	78

[a] 0.5% NaCl feed at 400 psig and 25°C.
[b] QCTE = TEA salt of E-383-40 CA 11-bromoundecanoate (0.72 meq/g).
[c] From Kesting et al.[61]

4.4 WATER STRUCTURE AND MEMBRANE–WATER INTERACTIONS

Because of the importance of membrane separations in aqueous systems, the properties of water and water–solute interactions will be used to present the general case of solvent–solvent and solvent–solute interactions. The unique properties of water and aqueous solutions have been the object of intensive research and have been extensively reviewed.[138, 139] In spite of this, there are widely differing opinions of what the structure of water really is. According to the flickering-cluster model of Frank and Wen,[140] the formation of hydrogen bonds in liquid water is a cooperative phenomenon in which a number of hydrogen bonds are formed and broken simultaneously between a group of water molecules (Fig. 4.24). The slight covalency of the hydrogen bond leads to a certain amount of charge separation and therefore formation of new bonds by molecules that are already bonded. The water molecules will group together so that the number of hydrogen bonds in the cluster is a maximum (about 100 molecules per cluster at 25°C), resulting in spheres in which the stabilization per hydrogen bond for molecules in the interior is greater than those at the surface. The formation and dissolution of these clusters is governed by local high-energy fluctuations. Their lifetime, although short (10^{-11}–10^{-10} s), is nevertheless two or three orders of magnitude greater than the period of molecular vibration.[142]

Water is open structured and full of cavities similar to those present in ice. The cavities are believed to contain a dense liquid composed of non-hydrogen-bonded molecules. Clustered and non-hydrogen-bonded water are in equilibrium with one another. The sensitive order–disorder equilibrium within liquid water is responsible for its unusual properties, such as its ability to increase in order both by interaction

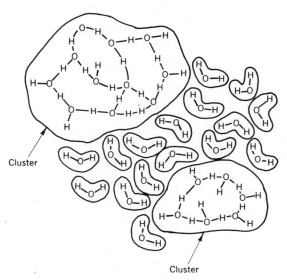

Cluster

Cluster

FIGURE 4.24. Schematic representation of liquid water, showing hydrogen-bonded clusters and unbonded molecules (from Nemethy and Scheraga[141]).

with active species such as cations of high charge density and by its reaction in the presence of inert solutes with which it has no direct or specific interaction. In the presence of the latter, for example, aliphatic hydrocarbons, water undergoes a hydrophobic hydration, that is, a cage or iceberg formation which may be due to repulsion by the solute in such a manner that dense unassociated water molecules turn back on themselves, as it were, and enhance overall structural order by forming more ordered and/or additional flickering clusters (Fig. 4.25). Frank and Evans[143] have ascribed this unexpected phenomenon to an "oil on toubled waters" reaction, in which the feeble polarizability of the inert solute leads to an inability to transmit cluster-disruption influences, with the result that they have a local calming effect and offer "boundary protection" to icelike clusters adjacent to them. That a variety of opposing chemical effects can lead to synergistically enhanced order, which in turn can be further enhanced by a sterically regular matrix comprising, for example, interstices between regularly spaced crystallites of equal size, is consistent with a long-range ordering and immobilization of water within a membrane and at the membrane–solution interface.

When solutes are added to water, water structure will be enhanced or destructured depending upon the nature of the interactions. In structure enhancement (or structure making) the presence of solute shifts the water structure to larger or more ordered clusters at the expense of the dense monomeric species. In structure breaking, on the other hand, the presence of solute shifts the structure equilibrium toward smaller and/or less ordered clusters and increasing concentrations of dense monomeric water. Considering first ionic solutes, Frank and Wen[140] and Frank and Evans[143] have postulated three concentric regions surrounding an ion:

1. An innermost structure-forming region of polarized, immobilized, and electrostricted water molecules.

2. An intermediate destructured region in which the water is less icelike, that is, more random in organization than ordinary water.

3. An outer region containing water having the normal liquid structure.

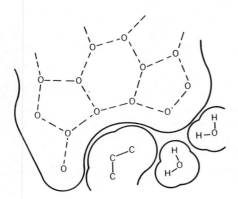

FIGURE 4.25. Schematic cross section of a hydrogen-bonded cluster near a hydrocarbon solute molecule, indicating the formation of a partial cage around the solute (from Nemethy and Scheraga[141]).

In the case of ions with a high charge density, there is a lineup of water dipoles within the hydration layer or layers of the various ions present. These water molecules tend to form a tightly bound, oriented hydration layer around ions of either charge. For cations it is reasonable to assume that the oxygen atoms of water are oriented into the ion, whereas for anions the hydrogen atoms are so oriented. For some ions of high charge density (small crystal radius and/or high charge), the electric field beyond this layer is sufficiently strong for several molecular diameters to orient the water dipoles radially and thus prevent or inhibit the formation of the usual tetrahedral configuration of water.

The net structure-making ions include relatively small or multivalent ions such as Li^+, Na^+, H_3O^+, Ca^{2+}, Al^{3+}, OH^+, and F^-. Water molecules about the ion are less mobile than those in the bulk, a phenomenon which has been called positive hydration by Samoilov.[144, 145] Such ions increase the viscosity of water and in the case of neutral (uncharged) membranes permeate less readily than structure-breaking ions.[146]

Large monovalent ions such as K^+, NH_4^+, Cs^+, Cl^-, Br^-, I^-, NO_3^-, BrO_3^-, IO_3^-, and ClO_4^- generally exert a net structure-breaking effect. Their field is capable at most of orienting water molecules in the layer immediately adjoining the ion, while beyond this layer a structure-breaking effect is noted. Water molecules other than those in the layer adjacent to the ion have a higher mobility in the presence of such ions than in the bulk, that is, negative hydration,[144, 145] so that solution viscosity decreases and membrane permeability increases. Asymmetrical ions such as IO_3^-, SO_4^{2-}, $PtCl_6^{2-}$, and OH^- can have effects which differ at various positions on the surface of the ion; the net effect, however, is to strongly bond the first water layer without any association between the first and second layers. Relatively large ions such as R_4N^+, $Pt(NH_3)_4(HC_2)Cl^{2+}$, $Pt(NH_3)_6^{4+}$, and $Fe(CN)_6^{4-}$ have such a diffuse charge that the ions are essentially unhydrated. They do, however, exert a structure-making effect closely akin to that of neutral solutes so that structure in the bulk solution is increased. In the case of symmetrical negative ions of very low charge density, such as I^- or ClO_4^-, the structure-breaking effect extends all the way into the ion, so that the activation energy in moving a water molecule away from its neighbors is higher than in moving a water molecule away from an I^- ion. The difference between large positive ions such as R_4N^+ (which enhance the structure of water in the bulk solution) and large negative ions such as I^- and ClO_4^- (which destructure water) is somewhat obscure but may be related to the previously mentioned differences between the inward orientation of oxygen or hydrogen atoms depending upon the charge classification (positive or negative) of the ion in question.

The effect of soluble nonelectrolytes upon water structure is of interest both for its own sake and for what can be learned by a consideration of these species as model systems for studying the interaction between polymeric membranes and water. Considering first the structure of simple liquid alcohols, it is apparent that their hydrogen-bonding capacity is less than that of water. They possess only one hydrogen capable of hydrogen-bond formation, and the organic group sterically hinders cooperative association. This does not mean, however, that the formation of dimers is more likely than of higher mers; indeed the reverse is true, and the tetramer of methanol, for example, is much more stable than the dimer.[147] Since alcohols form linear rather than three-dimensional polymers, they do not have the open-structure characteristics of water.

When traces of methanol, ethanol, or n-propyl alcohol are added to water,[148] the interfacial tension between hexane and water (at least at temperatures below 35°C) is increased. This fact is consistent with an increase in structural order within the alcohol–water system. Furthermore, the addition of a little water to t-butanol or to isobutanol causes a decrease in the dielectric constant ϵ, additional evidence for strong association between alcoholic and aqueous OH groups.[142] It may be that at low concentrations alcohols do not disturb the flickering clusters of water so as to lead to a loss of openness. Instead, they may fit into the interstices between water clusters and enhance the structure of dense monomeric water. At higher concentrations, however, a net structural loss and densification are found owing to a loss of the overall density of hydrogen bonds. It is noteworthy that, subject to steric limitations, the proton-accepting ability of the alcoholic OH is greater than that of the water OH and increases in the series MeOH < EtOH < i-PrOH < t-BuOH.[149]

Urea, however, in spite of its hydrogen-bonding capacity, exerts a net structure-breaking effect upon water. This is due to disruption of water clusters because of the steric inability of urea to fit into the tetrahedral water structure.

The behavior of water within membranes and at their surface has been the subject of broad speculation. There is a dichotomy of opinion, with the phenomenologists appearing reluctant to accept the idea of any form of long-range ordering near aqueous interfaces and the structurists tending to credit the existence of considerable amounts of bound water. The existence of bound water is indicated whenever the physical properties of a macromolecular suspension cannot be readily accounted for by the properties of the suspending medium and those of the macromolecules. This concept can also be extended to membrane gels, that is, three-dimensional arrays of macromolecules which have been insolubilized by one means or another.

Bound water relaxes at frequencies between those characteristic for ice and those characteristic for normal water.[150] Therefore, from a structural point of view, bound water stands between normal water and ice. A broader spectrum of time constant is involved in the relaxation of bound water than in that of normal water, which is indicative of variation in the characteristics of the former associated with different activation energies. In other words, not only can water be bound, but it can also be bound to various extents. Very probably a continuum exists, with the more highly (primarily) bound water molecules in the inner hydration shell and less highly (secondarily) bound water molecules in successive hydration shells.[146] The lower ϵ of bound as compared to normal water indicates lower mobility and hence lower capacity for hydration of ionic solutes.

By analogy to the known effects of solutes on the structure of water, Berendson and Migchelsen[151] predicted the following polymer–water influences:

1. Polar side chains should hydrate individually but have a structure-breaking influence beyond the first layer. (Actually, not all polar groups are capable of hydration, so that polarity does not always indicate hydrophilicity.)

2. Nonpolar side chains should induce order of the cage type similar to the effects of nonpolar solutes.

3. Backbone structures with no available hydrogen-bond donors or acceptors will act as nonpolar solutes.

4. Backbone structures able to form hydrogen bonds to water will have structure-breaking or structure-promoting effects, depending on the geometry of hydrogen-bonding sites. If the geometry is such that the sites to which water may be bound form an array fitting into an ice I structure, a structure-promoting influence is to be expected. With hydrophobic backbones, similar effects might occur if short polar side chains repeat in a pattern fitting into a regular water lattice. The effects will be stronger for rigid backbones or side chains.

Understanding the fundamentals of water structure and membrane–water interaction greatly simplifies the task of screening and tailor making new polymeric candidates for hyperfiltration membranes. In the case of cellulose acetate hyperfiltration membranes, for example, it appears that effects of hydration of hydrophilic hydroxyl groups, structure-enhancing formation of water cages about hydrophobic acetate groups, and the rigidity of the cellulosic molecules act synergistically to permit the water to obtain the high viscosity (and hence low ϵ) necessary to exclude salt almost completely.

Nuclear magnetic resonance (NMR) evidence for the presence of collagen hydrates is consistent with the adhesion of water molecules to each other and to collagen molecules to form chains in the fiber direction. The lifetime of a water molecule in a particular chain was found to be 10^{-6} s, that is, 10^5-10^6 times longer than in a flickering cluster and $\sim 10^8-10^9$ times longer than that required for molecular vibration. By comparison to ordinary water, therefore, water bound to collagen may be considered immobile. Ling[152] has cited water-vapor sorption data as evidence that water on collagen is polarized and oriented in multilayers. He also found, significantly, that the equilibrium concentration of the Na^+ ion in bound water is only about 0.1 that in the external medium.

Bound water is believed to be of importance in many separations of aqueous solutions. It is one of the reasons for the selectivity of ion-exchange membranes, and its presence in the skin layer of hyperfiltration desalination membranes is believed to constitute the basis for their salt-retention capacity.

Lack of membrane–water interactions, as in the case of hydrophobic materials such as polyethylene, leads to higher interfacial tension. In such an environment the water molecules tend to form clusters with others of their own kind rather than to wet the membrane surface. Since the passage of large clusters of water molecules quite obviously meets with greater resistance than the passage of small clusters or individual molecules, permeability decreases with decreasing membrane-solvent interaction.

The effect of a number of important structural groups on water sorption as a function of relative humidity has been reviewed by Barrie.[153]

4.5 POLYMER MODIFICATION AND NEW POLYMER DESIGN

Although the essential characteristics which govern the choice of polymer(s) for various membrane applications have already been considered in general terms, it is hoped that the consideration of a few specific examples of problem solving will serve to acquaint the reader with the manner in which these principles are applied in practice.

Microfiltration membranes of CN or CN/CA, although suitable for use as flat filter disks in the plate and flame configuration, exhibit borderline strength and flexibility insofar as their utilization in pleated cartridges is concerned. It is, of course, possible to incorporate low-MW plasticizers such as glycerol to permit the convolution of such membranes, but such a treatment tends to be fugitive, particularly after autoclaving. Therefore, the utilization of an insoluble high MW and, therefore, nonextractible plasticizer represents an attractive alternative. In view of our imperfect understanding of the factors which determine the compatibility of polymer blends, the choice of the polymeric plasticizer, ethyl cellulose (EC), to replace CA in a CN/CA blend was made on a largely empirical basis. Compatibility in this case may be rationalized on the basis of acid–base interaction since CN is a Lewis acid and EC a Lewis base. The blend was included in a standard casting solution which also contained acetone as a solvent, and 1-butanol, as a nonsolvent pore former. The resultant CN/EC membrane was strong and flexible, but totally impermeable. The porosity of the membranes was determined gravimetrically and found to be virtually identical to that of the corresponding CN/CA membrane. Water wettability was poor—despite the fact that glycerol was present—but wettability in isopropanol was excellent. Permeability to aqueous solutions was normal if the membranes were first wet with isopropanol. These facts were consistent with the alcohol solubility of EC. As the acetone evaporated, phase inversion occurred and micelles of alcohol-insoluble CN appeared and formed the membrane gel. However, the EC remained in solution until the alcohol evaporated and then formed thin films which covered the pores in the CN framework—hence the water impermeability of the final blend membrane. Immersion in alcohol removed the thin films of EC and restored permeability to normal. The problem was resolved[154] by the utilization of alcohol-insoluble cyanoethyl cellulose (CEC), an ether and, therefore, a Lewis base which was compatible with CN. When phase inversion occurs in the CN/CEC blend solution, both polymers participate in micelle formation and no polymer remains in alcohol to subsequently glaze the fenestrations in the final membrane.

Insufficient wettability, in the absence of wetting agents, is a deficiency of CN/CA membranes. However, since extractibles should always be minimized, there is an increasing desire to avoid, whenever possible, the use of frank surfactants.

Sometimes the use of an innocuous substance such as glycerol will suffice.[82] A more certain, albeit more costly, solution to the problem is to covalently attach nonionic surfactants to the CA and to blend what amounts to a CA graft copolyether together with CN.[155] The reaction of hexamethylene diisocyanate (HDI) or tolylene diisocyanate (TDI) with a poly(oxyethylene nonylphenol) with 100 moles of ethylene oxide and the subsequent reaction of this adduct with an equal weight of 2.3-DS CA resulted in a polyoxyethylene-grafted CA containing 20% PEG. The low grafting efficiency is due to the fact that the adduct is a mixture of products and to the general kinetic difficulty inherent in grafting one polymer to another. Nevertheless, the desired effect was achieved.

Hyperfiltration membranes of CA when utilized for the demineralization of unchlorinated tap water are subject to attack and degradation by a number of microorganisms. This problem can be overcome by the derivitization of some of the remaining free hydroxyl groups of CA with an Ω haloacyl chloride "monomer" in the presence of a tertiary amine to yield a CA mixed ester bearing quarternary ammonium groups (Table 4.10). The first such polymer made was the pyridinium salt of CA chloroacetate. However, the strong inductive effect of the quaternary ammonium group weakened the ester linkage with the result that hydrolysis occurred. The mixed ester then rapidly reverted to the biodegradable CA starting material. The inductive effect was lessened and hydrolytic stability increased, by increasing the number of carbon atoms which separated the ester from the quaternary ammonium groups.[60] However, at this juncture the membranes tended to swell, that is, permeability increased and permselectivity decreased, with time. The solution was to increase the chain length further to 11 carbons (thereby decreasing hydrophilicity) and to substitute trimethylamine, a nonsolvent for CA, for pyridine, a solvent for CA. This increased P–P interaction and favored the formation of virtual cross-links.[61] An interesting sidelight to this development is that permselectivity was considerably improved by the utilization of the long C_{11} chains. It was also noted that the casting solutions were clearer and exhibited greater tolerance for nonsolvents than those from CA mixed esters containing shorter chains. Finally, the skins of the membranes containing the long-chain monomer were glossier and more uniform than those containing the short-chain monomer. These effects suggested that the long-chain side groups increased the solubility of the mixed ester. It was also postulated that entanglement of side chains took place which effectively reduced the number of defects in the skin layer. To test the effect of chain length as distinct from chain length coupled with a positive charge, CA laurate (CAL) was synthesized and cast into membranes. Solubility, solution clarity, nonsolvent tolerance, membrane glossiness, and permselectivity were all enhanced as had previously been the case, but permeability was one-tenth that of membranes containing the long chain with the ionogenic group. It was concluded, therefore, that the long chain conferred high permselectivity, and the quaternary ammonium group, high permeability, on the resultant membranes. It was further speculated that the obviously superior solubility and permselectivity characteristics were indicative of the formation of tighter coils in the casting solutions and the formation of a denser and more integral glassy state in the skin layer of the membrane. If this hypothesis and its many ramifications prove to be correct, it should contribute to the development of more permselective integrally-skinned membranes for hyperfiltration and gas separations.

Several approaches have been investigated to minimize the compaction and resultant loss or permeability which accompany the operation of hyperfiltration membranes at elevated pressures. The first was to incorporate unsaturated monomers such as methacroyl chloride into the membrane polymer and to polymerize these moieties *in situ* in a post-membrane-formation step.[44] Cellulose acetate methacrylate (CAM) with an acetyl DS of 2.3 and a methacroyl DS of 0.2–0.4 was cast into membranes and cured to 90% acetone insolubility by annealing in the presence of a redox system consisting of a reducing agent ($NaHSO_3$ or Na_2SO_3), an oxidant (H_2O_2 or $K_2S_2O_8$), and an inorganic promoter ($FeSO_4$ or $CuSO_4$). Not only was the flux decline slope m lowered by approximately an order of magnitude—which permitted the membrane to retain 90% of its original permeability after a year or so of operation at 800 psi, compared to 65% for the untreated membrane—but the flux-rejection properties tended to improve as well.[56] Another approach was to blend polystyrene (PS) with CA. However, because CA and PS proved to be incompatible in solution, CA–PS graft copolymers were prepared.[57] The subsequent line of investigation showed that tensile-creep behavior of dense films of CA could be modified by the grafting of styrene onto CA. Although grafts of long chains of PS onto long chains of CA proved to be the best compatibilizers for CA and PS,[156] grafts of short chains of PS onto CA had a greater effect on tensile creep behavior.[57] The solubility behavior of the CA–PS graft copolymers is treated separately in Chapter 5.

As was noted earlier, insufficient wettability of microfiltration membranes can be a problem. The reverse can sometimes also be true. Blood oxygenation membranes must be permeable to oxygen and carbon dioxide but impermeable to blood plasma. The intrusion of plasma into either a skinned or microporous membrane will inhibit the flow of oxygen. Plasma intrusion into EC films was, therefore, controlled—and oxygen permeability increased—by the reaction of EC with perfluorobutyryl chloride to yield EC perfluorobutyrate, a much more hydrophobic polymer than the EC starting material.[157] Surface treatment with silicones can also convert even intrinsically wettable membranes such as those of nylon 66 or nylon 6 into hydrophobic barriers.

Because of conflicting requirements for the maintenance of viscosity while simultaneously exhibiting tolerance for a high concentration of nonsolvent pore formers, polymers for use in the dry process must usually be of higher MW than those which are utilized in the other variations of phase inversion. However, since such polymers are not always commercially available, they must sometimes by synthesized. In the case of bisphenol-A polycarbonates, although a source of high-MW homopolymer was available, the nonsolvent tolerance of solutions of this material was inadequate to the task of obtaining membranes of high porosity. The problem was resolved by the synthesis of a polycarbonate block *co*-polyether (5% PEO) whose chain flexibility (and resultant increased solubility and tolerance for nonsolvents) yielded the desired results.[7] For some applications this copolymer exhibited a too great tensile creep. This was significantly decreased by substituting tetra-chlorobisphenol-A for bisphenol-A.

The block copolymer approach did not prove as applicable to polysulfone which is much more difficult to synthesize than is polycarbonate. The introduction of methylene bridges by the chloromethylation of commercial resin was utilized to provide materials whose solutions exhibited the required viscosity at concentrations low enough to ensure adequate nonsolvent tolerance. The solubility of polyesters, at least of those such as poly (pivalolactone) and the polyester from terephthalic acid and *trans*-2,2,4,4-tetramethyl-1,3-cyclobutanediol which contain gem dimethyl groups, can be increased by chlorination.[158]

Polysulfone was modified to minimize fouling of hollow fibers during recovery of positively charged electrolytic particles by ultrafiltration of an effluent stream. Grafting of vinyl monomers containing quaternary ammonium groups onto the surface of the hollow fibers provided the charge repulsion necessary to significantly reduce fouling and thereby increase the time between wash cycles.[159]

Several different techniques for increasing the chlorine resistance of thin-film composite membranes have been found. The first utilizes the piperazinamides, a class of polyamide which, because its diamine monomer is a di-*secondary* amine, results in polymer backbones lacking NH groups.[87] The second utilizes a more inherently chlorine-resistant polymer, polysulfone, which is sulfonated until it is water soluble, and applied as a thin film from aqueous solutions before being cured to water insolubility by the introduction of sulfone cross-links.[68]

A final example is that of swelling control in the sulfonated perfluorionomers. Two approaches are currently being followed and, as of this writing, it is by no means clear which will dominate. Both utilize a weak-acid carboxylated per-fluoroionomer in series with the strong-acid sulfonated polymer. However, in one case surface treatment is utilized which results in a carboxylate layer which is 2–10 μm thick, whereas in the second case a carboxylate layer $\sim$70 μm thick is laminated to a perfluorosulfonate membrane which is 100–300 μm thick.[118] Energy efficiency favors the utilization of the thinner high-resistance carboxylate layer, while membrane durability and lifetime favor the thicker laminate.

REFERENCES

1. J. More, *J. Polym. Sci.*, **A2**, 835 (1964).

2. *Know More About Your Polymer*, Waters Associates Inc., Milford, MA.

3. *Theory of Gel Permeation Chromatography* TN-86, Perkin Elmer, Norwalk, CT.

4. D. van Krevelen and P. Hoftyzer, *Properties of Polymers*, 2nd Ed., Elsevier, Amsterdam, 1976.

5. *Nitrocellulose*, Wolff Walsrode, Bomlitz, Germany.

6. J. Brydson, *Plastics Materials*, 2nd Ed., Van Nostrand Reinhold, New York, 1970.

7. R. Kesting, U.S. Patent 3,945,926, 1976.

8. F. Billmeyer, Jr., *Textbook of Polymer Science*, 2nd Ed., Wiley, New York, 1971.

9. H. Mark, *CHEMTECH*, **14**(4), 220 (1984).

10. R. Chern, K. Koros, E. Sanders, S. Chen, and H. Hopfenberg, in *Industrial Gas Separations*, T. White, Jr., C. Yon, and E. Wagener, Eds., Chap. 3, ACS Symposium Ser. No. 223, American Chemical Society, Washington, D.C., 1983.

11. *Celgard® Microporous Polypropylene Film*, Celanese Fibers Marketing Co., Charlotte, N.C.

12. Gore-Tex® Membrane Products, M.L. Gore and Associates, Inc. Elkton, MD.

13. B. Wunderlich, *Adv. High Polymer Res*. **5**(4), 568 (1967/1968).

14. P. Debye and D. Scherrer, *Z. Phys.*, **17**, 227 (1916); A. Hull, *Phys. Rev.*, **10**, 661 (1914).

15. S. Lasoski and W. Cobbs, *J. Polym. Sci.*, **36**, 21 (1959).

16. S. Reitlinger and I. Yarko, *Colloid J. (USSR)* (English trans.), **17**, 369 (1955).

17. E. Ott, H. Spurlin, and M. Grafflin, Eds., *Cellulose and Cellulose Derivatives*, Interscience, New York, 1955, p. 1063.

18. R. Siu, *Microbial Degradation of Cellulose*, Reinhold, New York 1965.

19. R. Nickerson, *Ind. Eng. Chem.*, **33**, 1022 (1941); **34**, 85, 1480 (1942).

20. G. Jones and F. Miles, *J. Soc. Chem. Ind.*, **52**, T251 (1933).

21. W. Higley, in OSW R&D Report 154 (November 1965).

22. D. Brubaker and K. Kammermeyer, *Ind. Eng. Chem.*, **44**, 1465 (1952).

23. H. Schnell, *Chemistry and Physics of Polycarbonates*, Interscience, New York, 1964, p. 131.

24. F. Hellferich, *Ion Exchange*, McGraw-Hill, New York, 1962.

25. A. Castro, U.S. Patent 4,247,498 (1981).

26. *Modern Plastics Encyclopedia*, vol. 53, no. 10A, McGraw-Hill, New York, 1976–1977, p. 32.

27. W. Worthy, *Chem. Eng. News*, **56**, 23 (December 11, 1978).

28. *Modern Plastics Encyclopedia*, vol. 53, no. 10A, McGraw-Hill, New York, 1976–1977,

29. Milllipore Filter Corporation, New Bedford, MA.

30. W. Benzinger, U.S.Patent 4,384,047 (1982).

31. V. Stannett and H. Yasuda, "Permeability," in *Crystalline Olefin Polymers*, Pt. 2, Interscience, New York, 1965.

32. H. Lonsdale et al., OSW R&D Report 577 (March 1970).

33. H. Spencer, Paper presented at the 187th National Meeting, American Chemical Society, St. Louis, April 1984.

34. J. Abrams and L. Benezra, *Encyclopedia of Polymer Science and Technology*, vol 7, Ion Exchange Polymers, Wiley, New York, 1967.

35. M. Refojo, *J. Appl. Polym. Sci.*, **9**, 3161 (1965).

36. H. Yasuda et al., *J. Polym. Sci.*, **A1**(4), 2913 (1966).

37. Wafilin Co., Entscheede, The Netherlands.

38. Y. Hashino, M. Yoshino, H. Sawaby, and S. Kawashima, U.S. Patent 3,933,653 (1976).

39. Rhone-Poulenc Industries, Lyon, France.

40. Y. Sakai, H. Tsukamoto, Y. Fryii, and H. Tanzawa, in *Ultrafiltration Membranes and Applications*, A. Cooper, Ed., Plenum, New York, 1980.

41. General Electric Company, Pittsfield, MA.

42. A. Noshay, M. Matzner, and T. Williams, *Ind. Eng. Chem. Prod. Res. Dev.*, **12**, 268 (1973).

43. A. Noshay and M. McGrath, *Block Copolymers: Overview and Critical Survey*, Academic, New York, 1977.

44. J. Henis and M. Tripodi, U.S. Patent, 4,230,463 (1980).

45. R. Kesting, *Synthetic Polymeric Membranes*, McGraw-Hill, New York, 1971.

46. R. Kesting, "Asymmetric Cellulose Acetate Membranes" in *Reverse Osmosis and Synthetic Membranes*, S. Sourirajan, Ed., NRCC Publ. No. 15627, Ottawa, Canada, 1977.

47. R. Seymour and E. Johnson, *Coatings Plastics Prepr.* **36**(2), 668 (1976).

48. D. Johnson, U.S. Patent 3,447,939 (1969).
49. C. McCormick and T. Shen, cited on p. 62, *Chem Eng. News*, (September 7, 1981).
50. H. Lonsdale, U. Merten, and R. Riley, *J. Appl. Polym. Sci.*, **9**, 1341 (1965).
51. W. Moore, "Concentrated Solutions," in *Cellulose and Cellulose Derivatives*, N. Bikales and L. Segal, Eds. Chap. 5, Vol. V, Pt. IV, Wiley-Interscience, New York, 1970.
52. C. Cannon, U.S. Patent 3,497,072 (1970).
53. C. Reid and E. Breton, *J. Appl. Polym. Sci.*, **1**, 133 (1959).
54. S. Loeb and S. Sourirajan, U.S. Patent 3,233,132 (1965).
55. G. Miles, French Patent 358,079.
56. D. Hoernschemeyer, R. Lawrence, C. Saltonstall, and O. Schaeffler, "Stabilization of Cellulosic Membranes by Crosslinking," in *Reverse Osmosis Membrane Research*, H. Lonsdale and H. Podall, Eds., Plenum, New York, 1972.
57. F. Kimura Yeh, H. Hopfenberg, and V. Stannett, "The Preparation and Properties of Styrene Grafted Cellulose Acetate Membranes," in *Reverse Osmosis Membrane Research*, H. Lonsdale and H. Podall, Eds., Plenum, New York, 1972.
58. U. Merten, H. Lonsdale, R. Riley, and K. Vos, OSW R&D Report 265(1967).
59. R. Kesting, U.S. Patent 3,884,801 (1975).
60. R. Kesting, U.S. Patent 4,035,459 (1977).
61. R. Kesting, J. Newman, K. Nam, and J. Ditter, *Desalination*, **46**, 343 (1983).
62. S. Peterson, *Justus Liebigs Ann. Chem.*, **526**, 205 (1949).
63. R. Kesting, J. Ditter, K. Jackson, A. Murray, and J. Newman, *J. Appl. Polym. Sci.*, **24**, 1439 (1979).
64. E. Klein and J. Smith, "The Use of Solubility Parameters for Reverse Osmosis Membrane Research," in *Reverse Osmosis Membrane Research*, H. Lonsdale and H. Podall, Eds., Plenum, New York, 1972.
65. B. Baum, R. White, and W. Holley, Jr. "Porous Tubulets for Desalination Barriers," in *Reverse Osmosis Membrane Research*, H. Lonsdale and H. Podall, Eds., Plenum, New York, 1972.
66. V. Fiedler and J. Ruzicka, *Anal. Chim. Acta*, **67**, 179 (1973).
67. J. Cadotte, U.S. Patent 3,926,798 (1976).
68. J. Quentin, U.S. Patent 3,709,841 (1975).
69. J. Coleske, "Blends Containing Poly(ϵ-Caprolactone) and Related Polymers," in *Polymer Blends*, D. Paul and S. Newman, Eds., Vol. 2, Academic, New York, 1978.
70. Nuclepore Corp. Pleasanton, CA.
71. R. Kesting, U.S. Patent 3,957,651 (1976).
72. H. Hoehn and J. Richter, U.S. Patent Reissue 30,351 (1980).
73. Reemay Products, E. I. DuPont de Nemours, Wilmington, DE.
74. Eaton Dikeman, Division of Knowlton Brothers, Mt. Holly Springs, PA.
75. D. Lyman, B. Loo, and R. Crawford, *Biochemistry*, **3**, 985 (1964).
76. B. Fisher and P. Cantor, *Modified Polycarbonate Membranes for Dialysis*, Report PB 213-150 to NIAMD (January 1972)
77. D. Lyman, *Ann. N.Y., Acad. Sci.*, **146**(1), 113 (1968).
78. M. Kohan, Ed., *Nylon Plastics*, Wiley, New York, 1973.
79. P. Blais, "Polyamide Membranes," in *Reverse Osmosis and Synthetic Membranes*, S. Sourirajan, Ed., NRCC Publ. No. 15627, Ottawa, Canada, 1977.
80. Pall Filter Corp., Glen Cove, NY.
81. D. Pall, U.S. Patent 4,340,479 (1982).

82. R. Kesting, L. Cunningham, M. Morrison, and J. Ditter, *J. Parenteral Sci.*, **37**(3), 97 (1983).

83. R. Kesting, U.S. Patent 4,450,126 (1984).

84. D. Pall and F. Model, U.S. Patent 4,340,480 (1982).

85. BCI 800 Series Nylon Resins, Technical Bulletin VIIIA, Belding Chemical Industries, New York.

86. P. Morgan, *Condensation Polymers: By Interfacial and Solution Methods*, Interscience, New York, 1965.

87. L. Credali, A. Chiolle, and P. Parrini, *Desalination*, **14**, 137 (1974).

88. J. Richter and H. Hoehn, U.S. Patent 3,567,732 (1971).

89. J. Glater and M. Zachariah, Paper presented at 189th National Meeting, American Chemical Society, Philadelphia, PA, August 1984.

90. F. Model, H. Davis, and J. Poist, "PBI Membranes for Reverse Osmosis," in *Reverse Osmosis and Synthetic Membranes*, S. Sourirajan, Ed., NRCC Publ. No. 15627,Ottawa, Canada, 1977.

91. S. Hara, K. Mori, Y. Taketani, and M. Seno, *Proc. Fifth Int. Symp. Fresh Water from the Sea*, **4**, 53 (1976).

92. H. Hoehn, Paper presented at 187th National Meeting American Chemical Society, St. Louis, MO, April 1984.

93. J. Cadotte U.S. Patent 4,277,344 (1981).

94. M. Kurihara, U.S. Patent 4,366,062 (1982).

95. M. Kurihara, T. Nakagawa, H. Takeuchi, and T. Tonamura, *Desalination*, **46**, 101 (1983).

96. H. Yasuda, "Composite Reverse Osmosis Membranes Prepared by Plasma Polymerization," in *Reverse Osmosis Membrane Research*, H. Lonsdale and H. Podall, Eds., Plenum, New York, 1972.

97. R. Schlögl, *Stofftransport durch Membranen*, Steinkopf Verlag, Darmstadt, 1964.

98. F. Donnan, *Z. Elektrochem.*, **17**, 572 (1911).

99. F. Donnan, *Z. Phys. Chem.*, **A162**, 346 (1932).

100. O. Bonner, *J. Phys. Chem.*, **58**, 318 (1954).

101. O. Bonner, *J. Phys. Chem.*, **59**, 719 (1955).

102. O. Bonner and L. Smith, *J.Phys. Chem.*, **61**, 326 (1957).

103. J. Salmon and D. Hale, *Ion Exchange: A Laboratory Manual*, Academic, New York, 1959.

104. *Dowex Ion Exchange*, The Dow Chemical Co., Midland, MI, 1959.

105. *Duolite Ion Exchange Manual*, Chemical Process Co., Redwood City, CA, 1960.

106. D. Reichenberg and W. Wall, *J. Chem. Soc.*, 3364 (1956).

107. G. Manecke, *Z. Elektrochem.*, **57**, 189 (1953).

108. G. Manecke, *Z. Elektrochem.*, **58**, 363 (1954).

109. G. Manecke, *Z. Elektrochem.*, **58**, 369 (1954).

110. G. Manecke, *Z. Elektrochem.*, **67**, 613 (1955).

111. M. Ezrin and H. Cassidy, *Ann. N.Y. Acad. Soc.*, **57**, 79 (1953).

112. M. Ezrin, H. Cassidy, and I. Updegraff, *J. Am. Chem. Soc.*, **75**, 1610 (1953).

113. M. Ezrin and H. Cassidy, *J. Polym. Sci.*, **78**, 2525 (1956).

114. L. Luttinger and H. Cassidy, *J. Polym. Sci.*, **20**, 417 (1956).

115. R. Lundberg and H. Makowski, "A Comparison of Sulfonate and Carboxylate Ionomers," in *Ions in Polymers*, A. Eisenberg, Ed., (ACS Symposium Ser. No. 187), American Chemical Society, Washington, D.C., 1980.

116. H. Makowski and R. Lundberg, "Plasticization of Metal Sulfonate-Containing EDPM with Stearic Acid Derivatives," in *Ions in Polymers*, A. Eisenberg Ed., (ACS Symposium Sec. No. 187), American Chemical Society, Washington, D.C.,1980.

117. A. Eisenberg and H. Yeager, Eds., *Perfluorinated Ionomer Membranes*, ACS Symposium No. 180, American Chemical Society, Washington, D.C., 1982.
118. *Chemical Week*, 35 (November 17,1982).
119. C. Martin, T. Rhoades, and J. Ferguson, *Anal. Chem.*, **54**, 1639 (1982).
120. Surlyn Ionomers, E.I. DuPont de Nemours, Wilmington, DE.
121. I. Salyer et al., U.S. Patent 3,799,356 (1974).
122. H. Yasuda and A. Schindler, "Reverse Osmosis Properties of Ionic and Nonionic Polymer Membranes," in *Reverse Osmosis Membrane Research*, H. Lonsdale and H. Podall, Eds., Plenum, New York, 1972.
123. D. Dieterich, W. Keberle, and H. Witt, *Angew. Chem. Int. Ed.*, **9**, 40 (1970).
124. X. Marze, J. Quentin, and M. Ruad, *Ann. Mines*, 1 (May 1976).
125. ICI Americas Inc., Wilmington, DE.
126. A. LaConti,"Advances in Development of Sulfonated PPO and PPO Membrane Systems for Some Unique Reverse Osmosis Applications," *Ind. Eng. Chem.*, **44**, (1952).
127. C. Giori and V. Adamaites, *Polym. Prepr.*, **15**(1), 626 (1974).
128. G. Smets, J. Roovers, and W. Humbeek, *J. Appl. Polym. Sci.*, **5**, 149 (1961).
129. G. Mino, S. Kaizerman, and E. Rasmussen, *J. Polym. Sci.*, **38**, 393 (1959).
130. A. Myers et al., *J. Appl. Polym. Sci.*, **4**, 159 (1960).
131. N. Geacintov, personal communication.
132. W. Baldwin, D. Holcomb, and J. Johnson, *J. Polym. Sci.*, **A3**, 833 (1965).
133. W. Chen, R. Mesrobian, D. Ballantine, D. Metz, and A. Glines, *J. Polym. Sci.*, **23**, 903 (1957).
134. R. Kesting and V. Stannett, *Makromol. Chem.*, **65**, 247 (1963).
135. V. Stannett, personal communication.
136. R. Kesting, J. Newman, and K. Jackson, *Proc. Sixth Int. Symp. Fresh Water from the Sea*, **3**, 213 (1978).
137. H. Gregor, H. Jacobson, R. Shair, and D. Wetstone, *J. Phys. Chem.*, **6**, 141 (1957).
138. L. Kavanau, *Water and Solute Water Interactions*, Holden-Day, San Francisco, 1964.
139. F. Franks (Ed.), *Physico-Chemical Processes in Mixed Aqueous Solvents*, American Elsevier, New York, 1967.
140. H. Frank and W. Wen, *Disc. Faraday Soc.*, **24**, 133 (1957).
141. G. Nemethy and H. Scheraga, *J. Chem. Phys.*, **63**, 3382 (1964).
142. F. Franks and D. Ives, *Quart. Rev.*, **20**, 1 (1966).
143. H. Frank and M. Evans, *J. Chem. Phys.*, **13**, 507 (1945).
144. O. Samoilov, *Disc. Faraday Soc.*, **24**, 141 (1957).
145. O. Samoilov, "Structure of Aqueous Electrolyte Solutions and the Hydration of Ions," Trans. by D. Ives, Publication Consultants Bureau, New York 1965.
146. A. Vincent, M. Barsh, and R. Kesting, *J. Appl. Polym. Sci.*, **9**, 2363 (1965).
147. G. Miller, *J. Chem. Eng. Data*, **9**, 418 (1964).
148. F. Franks and D. Ives, *J. Chem. Soc.*, 741 (1960).
149. W. Gerrard and E. Macklen, *Chem. Rev.*, **59**, 1105 (1959).
150. H. Schwann, *Ann. N.Y. Acad. Sci.*, **125**(2), 344 (1965).
151. H. Berendsen and C. Migchelsen, *Ann. N.Y. Acad. Sci.*, **25**(2), 365 (1965).
152. G. Ling, *Ann. N.Y. Acad. Sci.*, **25**(2), 401 (1965).
153. J. A. Barrie, in *Diffusion in Polymers*, J. Crank and G. Park, Eds., Chap. 8, Academic, New York, 1968.
154. R. Kesting, U.S. Patent 4,220,477 (1980).

155. R. Kesting, U.S. Patent 4,280,970 (1981).

156. J. Wellons, J. Williams, and V. Stannett, *J. Polym. Sci.*, **A1**(5), 1341 (1967).

157. R. Petersen and L. Rozelle, *Ultrathin Membranes for Blood Oxygenators*, Report PBI-324 to NHLI (January 1974).

158. J. Caldwell and W. Jackson, Jr., U.S. Patent 3,514,422 (1970).

159. B. Breslau, A. Testa, B. Milnes, and G. Medjanis, "Advances in Hollow Fiber Ultrafiltration Technology," in *Ultrafiltration Membranes and Applications*, A. Cooper, Ed., Plenum, New York, 1980.

5 POLYMER SOLUTIONS

A polymer solution is a uniform (usually at the molecular level) dispersion of macromolecules in a solvent system which consists of one or more components, the strength of whose interactions with the polymer is greater than the strength of the competing interaction of the polymer molecules with one another. In Doolittle's[1] nomenclature this is expressed in terms of polymer–solvent (P–S) and polymer–polymer (P–P) interactions such that if P–S > P–P, then

$$P + S \rightarrow P—S \text{ (solution)}$$

The preparation of polymer solutions is one of the most widely used and versatile means of rendering synthetic polymers sufficiently tractible to permit their conversion from raw material pellet or powder forms into ultimate membrane configurations as flat film and fibers.

Macromolecules can be viewed as three-dimensional surfaces with unequally distributed clouds or smears of electrons. To the extent that electron density differs from point to point, there will be permanent dipoles or even ions present on these groups. It will be noted that this concept is very similar to the Lewis theory of acids and bases with the difference that for the present discussion, polarity is not restricted to electron pairs. Solvent molecules or ions can be viewed in a similar light and are capable of minimizing any imbalance in electron distribution on complementary polymer groups. On the molecular level the way in which this is accomplished is by diffusion of solvating species to solvatable sites. Once there the solvent species align themselves insofar as is sterically possible with respect to the polymer groups to minimize local differences in electron density. After a sufficient number of P–S bonds has been formed and P–P bonds broken, polymer mobility increases, first in chain segments in amorphous regions and then in complete chains. Diffusion of both polymer chains and solvent species continues and eventually results in a uniform polymer solution. It is important to realize that the natures of both polymer and solvent are modified in solution (P + S ≠ P—S) and that such modifications can and usually do affect steric relationships which exist in the re-emergent solid state once desolvation is complete. Polymer molecules in solution can be stretched and twisted, coiled to various extents, and even participate in the formation of protocrystallites depending upon the nature and extent of P–S interactions. Among the factors which influence polymer solubility and P–S interactions are chain flexibility, molecular weight, crystallinity (virtual cross-links), covalent or ionic cross-links, and the strength of interactions between polar groups. These properties also influence the *accessibility* of polymer sites or groups to solvent molecules or ions.

A description on the macroscopic level of dissolution is more complicated. During the first phase of polymer dissolution, the solvent molecules penetrate the polymer structure and form a swollen gel layer.[2] Proceeding from the solvent into the solid polymer, as many as four sublayers can be distinguished:

1. The hydrodynamic liquid layer which surrounds every solid encompassed by liquid.
2. A gel layer which contains swollen polymer material in a rubberlike state.
3. A solid swollen layer in which the polymer is in the glassy state.
4. A solid unswollen layer in which the fissures and voids in the polymer are filled with solvent molecules.

The thickness of this gel layer increases with temperature and polymer MW and decreases with the rate of stirring. The state of the polymer also has a strong bearing upon the type of dissolution. If an amorphous polymer is dissolved at a temperature higher than the melt temperature T_m, that is, above the temperature where the rubbery state exists, the surface layer consists only of the hydrodynamic liquid layer and dissolution is a simple mixing of two liquids. If the polymer is in the rubbery state, the first two sublayers will be present and the gel layer will thicken since the solvent molecules can penetrate the matrix faster than the polymer molecules can disentangle themselves and leave the gel layer. If, as is most common, the polymer is in the glassy state, then all four sublayers will be found. As a practical matter a polymer will dissolve most readily if a gel layer is not allowed to encapsulate a large mass of polymer, the interior of which has no contact with the solvent. To avoid the formation of slowly dissolving clumps, a small portion of the solvent should be added to the polymer initially and the mass mixed vigorously. This permits the individual polymer particles to remain in the glassy state without forming tacky gel particles while taking up the solvent. Once the maintenance of individual particles has been assured, the remainder of the solvent can be added and solution will occur at a rate proportional to the size of the fine particles rather than at the rate proportional to the size of large aggregates of particles.

In some cases, particularly at low temperatures, complete dissolution may occur without the formation of a gel layer. This may explain the acetone solubility of cellulose triacetate when dissolution is brought about at 190 K, whereas it is acetone insoluble at room temperature.[3] Stress cracking may be responsible. Crazing of the polymer matrix occurs, with fissures separating small blocks of the polymer which leave the surface rapidly. This is attributable to stored stress energy, frozen in the segment interval which is set in motion at the T_g and concentrated along the wider cracks and void systems.

With the exception of those solutions utilized in the preparation of ultrathin films, most polymer solutions utilized for casting membranes are concentrated. Generally speaking, the volume fraction Φ of the polymer in a solution at room temperature is less than 0.5; typically, it is 0.3–0.4 for solutions for dry spinning dense hollow fibers where high viscosities ($\sim 10^5$ cps) are required, ~ 0.2 for solutions for wet casting flat-sheet membranes ($\sim 10^4$ cps), and ~ 0.1 for solutions for dry casting flat-sheet membranes ($\sim 10^3$ cps). If Φ is greater than 0.5, or if the polymer is semicrystalline and devoid of strong polar interactions with the solvent, then the addition of heat to the system is required to effect solubility or at least sufficient mobility to ensure tractibility of the polymer mass. Such situations are encountered, for example, in gel and melt spinning of hollow fibers and in the extrusion of dense films from a gel or a melt. To obtain a concentrated polymer solution may require a single solvent or several solvents. In addition to solvent(s), a multicomponent mix or *solvent system* may contain other components, such as swelling agents and nonsolvents, as well as plasticizers and wetting agents. The various constituents of a solvent system may be schematically represented as a continuum of species possessing varying affinities for the polymer, that is, a polymer–solvent interaction spectrum (Fig. 5.1).

The implications of this spectrum are of importance to a qualitative understanding both of the preparation of the dense films from two-component polymer solutions and of phase-inversion membranes by the gradual desolvation of multicomponent polymer solutions. At one end of the spectrum are the solvents (dispersing agents), which can interact with the macromolecules, thereby influencing the extent of aggregation and the conformations of the individual macromolecules within the solution. At the opposite end are the nonsolvents (diluents), which are tolerated to

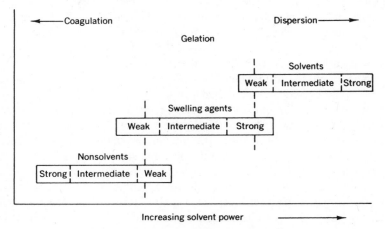

FIGURE 5.1. The polymer–solvent interaction spectrum.

a greater or lesser extent by both the polymer molecules and by the solvent itself. The stronger the nonsolvent, the less will be required to effect the precipitation of polymers from solution. Conversely, the stronger the solvent the more nonsolvent will be required to effect phase separation. In the middle of the spectrum are the swelling agents (weak precipitants, gelificants). Dense films imbibe swelling agents thereby increasing their volume; they do not, however, dissolve in them. In a multicomponent membrane casting solution it is useful to consider both swelling agents and nonsolvents as potential *pore formers*. Their presence in a polymer solution during the loss of a relatively more volatile true solvent effects a more or less gradual separation of the solution into two interdispersed liquid phases which subsequently become immobilized during the transition from sol to gel. The polymer–solvent interaction spectrum should be considered in a dynamic and qualitative sense rather than in a static and quantitative one. The position of any species on the spectrum will depend not only on the nature of the polymer but also upon the temperature, its own concentration, and the types and concentrations of the other species present. Substances which at low concentrations act as co-solvents may sometimes at higher concentrations act as nonsolvents. Examples are water and the lower alcohols in solutions of incompletely acylated cellulose esters. In such cases the hydroxyl-containing compounds may solvate the free hydroxyl groups of the cellulose ester. After the hydrogen-bonding requirements of the latter have been satisfied however, additional water or alcohol serves only to decrease the solvent power of the solvent system by dilution.

5.1 NATURE OF POLYMER–SOLVENT INTERACTIONS

The examples which follow exemplify specific group solvation which accompanies the formation of solutions of certain polymers. Cellulose acetate (CA) contains both acetyl (CH_3CO-) and hydroxyl ($-OH$) groups, which are dipoles with the following schematic electron distributions:

$$CH_3 \!-\! \overset{\overset{\displaystyle O^{\delta-}}{\|}}{\underset{\delta+}{C}} \!-\! \qquad \text{and} \qquad {}^{\delta+}H \!-\! O_{\delta-}$$

CA is soluble in aqueous solutions of certain lyotropic salts of the Hofmeister series such as magnesium perchlorate $Mg(ClO_4)_2$ because the following ion–dipole interactions take place[4]:

$$CH_3 \!-\! \overset{\overset{\displaystyle O^{\delta-}\cdots Mg^{+2}_{aq}}{\|}}{\underset{\delta+}{C}} \!-\! OCA \qquad\qquad \underset{\delta-}{H\!-\!O}\overset{\displaystyle Mg^{+2}_{aq}}{\vdots}\!-\!CA$$

In this case the small Mg^{+2}_{aq} cation because of its high charge density remains sufficiently electron deficient (electrophilic) that it seeks to associate with electron-rich (nucleophilic) oxygen atoms in both acetyl and hydroxyl groups. Spectropho-

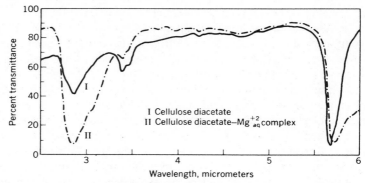

FIGURE 5.2. Infrared spectrum of dense cellulose acetate membrane in the presence and absence of aqueous $Mg(ClO_4)_2$ (from Kesting[4]).

tometric evidence for this effect may be seen in Figure 5.2. Both OH and CO stretching bands undergo the red shifts which are indicative of bond weakening owing to competition between electrophilic magnesium cations and the electropositive carbon and hydrogen atoms of the acetyl and hydroxyl groups for the electrons in the two dipoles. At the same time, the ClO_4^- counterion is of importance, albeit indirectly, because its large size and symmetrical shape are characteristic of a surface with a low and uniform charge density. Therefore, although ionic, it presents a more even and diffuse electron distribution than do the acetyl and hydroxyl dipoles and has little or no tendency to complex with the Mg_{aq}^{+2} ion. Were the salt $MgCl_2$ instead of $Mg(ClO_4)_2$, then complex ions of the type $MgCl_{aq}^+$ would form, thereby severely diminishing its electrophilic character relative to that of Mg_{aq}^{+2}. The crucial role of aqueous magnesium perchlorate in the casting solution developed by Loeb and Sourirajan[5] to produce their skinned hyperfiltration membrane was shown by Kesting[4] to be that of a swelling agent. Complexing of the hydrated magnesium ions with CA results in the inclusion of more water within the membrane gel than is the case in its absence. Zinc chloride, a traditional swelling agent for cellulose and cellulosic polymers, functions in a similar capacity.[6] Not only CA but also certain other polar polymers are known to be soluble in aqueous solutions of inorganic salts which are capable of forming complexes with polymer groups (Table 5.1). Although such salts are no longer universally employed in CA solutions in which organic swelling agents such as formamide[7] or nonsolvents such as the alcohols[8] have replaced them as pore formers in wet and dry phase-inversion processes, respectively, they are still of critical importance in the case of solutions of the more difficultly soluble aromatic polyamides of the type employed by Richter and Hoehn.[9] Fully aromatic polyamides and related polymers require not only the use of strong polar solvents (Table 5.2), but also inorganic salts such as LiCl.[10] In this case the Li^+ ion is believed to form a complex with the $-CONH-$ group

$$\overset{\delta^-}{O}\cdots Li^+$$
$$\underset{\delta^+}{\overset{\|}{C}}-\underset{H}{N}$$

TABLE 5.1 ION–POLYMER INTERACTIONS[a]

Polymer Type	Nucleophilic Group(s)	Model Compound(s)	Ion–Dipole Interaction Complex(es)	Spectral Changes in Solvent Saturated with $Mg(ClO_4)_2$
Cellulose acetate	$-OH$, $-OCOCH_3$	CH_3OH, CH_3COOCH_3	$CH_3-O\cdots M^+$ (H), $HC=O\cdots M^+$ with OC_2H_5	Red shift + broadening of OH bands; red shift of CO band from 5.85 to 5.92 μm
Cellulose	$-OH$	CH_3OH	$CH_3-O\cdots M^+$ (H)	Red shift + broadening of OH bands
Starch	$-OH$	CH_3OH	$CH_3-O\cdots M^+$ (H)	Red shift + broadening of OH bands
Polyacrylonitrile	$-C\equiv N$	CH_3CN	$CH_3-C\equiv N\cdots M^+$	Change in relative absorption of both halves of the primary $C\equiv N$ stretch doublet
Polypeptides	$-C(=O)-NH-$	$HC-N$ with O, CH_3, CH_3	$O\cdots M^+$, $H \rightarrow C-N(CH_3)_2$	Modification of first CH overtone

[a]From Kesting[4]; © 1965.

TABLE 5.2 PRINCIPAL SOLVENTS FOR POLYAMIDES AND RELATED POLYMERS[a]

Solvent	Structure
1. N,N-dimethyl acetamide (DMAC)[b]	$CH_3CON(CH_3)_2$
2. N-methyl pyrrolidone (NMP)[b]	$\overbrace{CH_2(CH_2)_2C}\overset{\overset{O}{\|\|}}{N}{-}CH_3$
3. N-methyl caprolactam (NMC)	$\overbrace{CH_2(CH_2)_3C}\overset{\overset{O}{\|\|}}{N}{-}CH_3$
4. Hexamethyl phosphoric triamide (HMPT)	$[(CH_3)_2N]_3PO$
5. Tetramethyl urea (TMU)	$(CH_3)_2NCON(CH_3)_2$
6. Dimethyl sulfoxide (DMSO)[b]	CH_3SOCH_3
7. N,N-dimethyl formamide (DMF)	$HCON(CH_3)_2$
8. Tetramethylene sulfone (TMS)	$\overbrace{CH_2(CH_2)_3SO_2}$
9. Formic acid (FA)[c]	$HCOOH$

[a]From Blais.[10]
[b]Most frequently used for barrier preparation.
[c]Restricted to aliphatic polyamides.

thereby disrupting the intermolecular hydrogen bonds which this group would otherwise form with similar groups on neighboring chains. Of course the amount of lyotropic salt which is necessary to produce the optimum void volume in the membrane as a whole and the requisite interchain displacement within the skin of both CA and aromatic polyamide membranes must be established empirically because 1:1 ratios between complexes of inorganic cations and polar polymer groups represent only values of potential maximum polymer solubility and bear no direct relationship to those levels at which optimum membrane performance characteristics are obtained.

The solubilities or enhanced solubilities of CA and the aromatic polyamides by virtue of ion–dipole interactions are certainly related to the solubility of CA in methylene chloride: methanol (3:1) described by Clement.[11] He postulated interactions between the Lewis-base carbonyl function of the acetate group and the Lewis-acid methylene group of methylene chloride

$$\overset{\delta^+}{C}{=}\overset{}{O}\underset{\delta^-}{\cdots}\overset{\delta^+}{CH_2}\begin{smallmatrix}{}^{Cl\delta^-}\\{}_{Cl\delta^-}\end{smallmatrix}$$

and hydrogen bonding between methanol and the free hydroxyl groups of CA

$$\overset{\delta^+}{H}\cdots\overset{\delta^-}{O}{-}CH_3$$
$$CA{-}\underset{\delta^-}{O}\cdots\overset{\delta^+}{H}$$

Although the solubility of nonpolar polymers such as the polyolefins is generally treated in black box or phenomenological terms, it can also be considered in terms of P–S interactions. Since in such cases the P–S interactions are admittedly less intense and less specific, input of thermal energy is required to effect solutions.

5.2 SOLVENT POWER

Several measures of the strength of P–S intractions in macromolecular solutions are available. Among these are solution viscosity, solution turbidity, nonsolvent (diluent) tolerance, various *cohesion parameters,* and the Lewis-acid–Lewis-base characteristics of polymer and solvent. The last two also provide the prospect of permitting a prediction of the solution properties of polymers in solvents from *a priori* knowledge of the properties of the components. Let us look at each of these in turn with the full realization that none of these scales, either singly or in combination with one another, will suffice to adequately quantify the behavior of polymer solutions in every situation.

Solution viscosity increases with increasing polymer molecular weight (MW) because as the latter increases so does chain length, and with it the hydrodynamic volume which this coiled chain occupies and hence the resistance of its solutions to shear and flow. Polymer chains in solution can take the form of rods, spheres, or coils, with the lattermost being the most prevalent for high-MW polymers. For a given chain length, the tightness of a coil and hence its size and viscosity can vary with the strength of competing P–P and P–S interactions. The viscosimetric behavior of polymer solutions can therefore be utilized to deduce solvent power and sometimes even the role of other components within the solvent system. It is necessary, however, to distinguish between the behavior of polymer coils in *dilute* solutions and their behavior in *concentrated* solutions.

The most widely utilized measure of the dilute solution viscosity of a polymer is its intrinsic viscosity $[\eta]$,

$$[\eta] = \lim_{c \to 0} \eta_{sp}/c$$

η_{sp} is defined as $(t - t_0)/t_0$, where t_0 and t are the times for the solvent and solution, respectively, to run through a capillary.[12] Because there is always competition between P–P and P–S interactions, it follows that in a good solvent, where P–S $\gg$ P–P, P–P will be minimal. In other words, a polymer chain will tend to avoid any intramolecular contact. In so doing it will tend to stretch out to the maximum extent possible. For this reason, if the value of $[\eta]$ for solvent A is greater than that for solvent B, A is a stronger solvent than b; likewise if a third component is added to a binary solution of polymer and solvent and $[\eta]$ increases, then the third component is acting as a co-solvent and the solvent system is a stronger solvent than the pure solvent alone. Conversely, if $[\eta]$ is found to decrease with the addition of a third component to a binary solution, then the third component is acting as a nonsolvent and the solvent system is a weaker solvent than the pure solvent by itself. Also [η] may first increase and then, with increasing additive concentration, decrease. This implies that the additive is acting in a concentration-dependent manner, first as a co-solvent and then as a nonsolvent.

The intrinsic viscosity obviously depends on the extensions of individual polymer chains. It is often related to the viscosity average molecular weight $\overline{M}_v$ by the Mark–Houwink equation[13]:

$$[\eta] = K\overline{M}_v^{\alpha}$$

The value of α depends on the strength of P–S interactions, ranging from 0.5 in poor (theta) solvents, through ~ 0.65 in fair solvents, to ~ 0.8 in good solvents. Values for $\overline{M}_v$ lie close to the weight-average molecular weight $\overline{M}_w$ because of the greater contribution of larger molecules to viscosity.

Although the literature contains many studies relating solvent power to intrinsic viscosities, such correlations appear less certain than those made on the basis of specific viscosities at higher polymer concentrations. Whereas in dilute solutions viscosity increases with the strength of the P–S interaction, with increasing polymer concentration a crossover point is reached where the reverse is true. Thus in *concentrated* solutions the viscosity is lower in good solvents than in poor solvents. This is rationalized by postulating tighter coiling in good solvents as concentration is increased, where, since P–S >> P–P, *intermolecular* contact between polymer chains is forbidden. The coils must as a result become smaller as they are crowded together with increasing concentration. The result is that each coil contributes progressively less to overall viscosity. In poor solvents, on the other hand, where *P*–S is slightly greater than P–P, intermolecular contacts are not so strongly forbidden and the increasing size of the polymer network owing to chain interpenetration results in increased viscosity.

Solution turbidity is a strong indication of solvent power. A nonturbid (clear) solution normally indicates that the dispersed particles are smaller than the wavelengths of visible light. It is, of course, possible (though improbable) that a close correspondence between the refractive indices of solvent and dispersed particles will yield the same result. Ordinarily, the less turbid the solution, the finer the particle dispersion and the stronger the P–S interaction. Solution turbidity can indicate the presence of impurities. Coarse particles (grainy matter) can be caused by a variety of factors and should always be removed by filtration. The required filter pore size will depend on the type of membrane end product required and must be determined empirically. The finer the pore size required in the membrane, the more critical the necessity for the removal of turbidity. Thus whereas slight turbidity may not interfere with the structural and functional characteristics of microfiltration and ultrafiltration membranes, it can disrupt the perfection required by gas separation and hyperfiltration membranes. In general the largest-pore-size filter which will remove the contaminant will have the longest life and therefore be most economical to use. Often, however, turbidity will be due to the presence of solute particles which are too fine to be removed by filtration. For example, cellulose acetates which are prepared from a mixture of cotton linters and wood pulp cellulose may yield turbid solutions. Such turbidity has been attributed to the presence of noncellulosic polysaccharides such as xylans and mannans which are found in cellulose derived from wood pulp.[14] For hyperfiltration and gas-separation membranes the utilization of cellulose acetate which is derived solely from cotton linters cellulose is recommended. The latter not only yield clearer solutions but also more permselective membranes. Still another origin of turbid solutions is incomplete solvation. The dissolution of CA in acetonitrile at room temperature yields a turbid solution which becomes clear upon heating to ~35°C. This solution retains its clarity upon cooling and remains clear indefinitely. Apparently, a potential-energy barrier must be overcome before the nitrile groups can form dipole–dipole complexes with the carbonyl and/or hydroxyl groups of CA. Possibly, steric hindrance inhibits complex formation at the lower temperature. Solution turbidity can increase with time thereby signaling an increase in P–P at the expense of the P–S interaction. This is especially common in the case of solutions of polar polymers which contain appreciable concentrations of nonsolvent. Such solutions not only become more turbid but, particularly if crystallization is possible, may eventually gel. The gels may in turn be reversible or irreversible on heating. Both turbidity and gelation can sometimes be avoided and useful solution life extended, by storage of the solution at a slightly elevated temperature (Table 5.3). Still another cause of turbidity is the presence of unsolvated ionic groups. The dispersion of ionomers in organic solvents often results in turbid and even grainy suspensions which can be converted to clear solutions by addition of water or other solvents of high dielectric constant such as methanol or formamide. Ionic groups usually require that the counterions be shielded from the charge of macroions before the latter can be sufficiently solvated to undergo dissolution at the molecular level.

In rare cases in which solute size is both uniform and in the same size range as the wavelengths of visible light, colored clear solutions are found. For example, when a slightly cross-linked, high-MW sample of polysulfone (PS) is dissolved in methylene chloride and trifluoroethanol (TFE) is added, a clear light-blue color reminiscent of glacial ice results. The appearance of this color in the solution signals the formation of an excellent, opaque white, skinless microfiltration membrane. If the PS MW is too low, then the blue color is not found in the solution and the membrane quality is poor.

TABLE 5.3 TIME DEPENDENCE OF CELLULOSE TRIACETATE SOLUTION VISCOSITY AS A FUNCTION OF TEMPERATURE[a]

Temperature (°C)	Vicosity (poises) at		
	0 h	72 h	144 h
17	726	5770	8800
30	236	—	253

[a]From Kobayashi et al.[15]

Although there are exceptions to the rule, the utilization of clear rather than turbid casting solutions is recommended. The former are usually more stable and provide greater leeway for the avoidance of imperfections such as streaks and drag-lines which begin at the leading edge of the casting hopper and are exacerbated by casting solutions which are too close to the point of incipient gelation.

Although it is theoretically advantageous to maximize the concentration of polymer in any solution which is utilized to cast membranes of films, processing considerations often dictate a compromise to concentrations at a somewhat lower level. One reason for this is that the processing of high-viscosity solutions is difficult. Another is the requirement for most membrane casting solutions that the solution possess excess solvation capacity so that pore-forming components can be added to the solution without bringing about precipitation or gelation. In a series of solutions with a given excess solvation capacity, nonsolvent tolerance can be used as a measure of solvent strength. The latter can be estimated by titration of the solution with nonsolvent until incompatibility (the appearance of turbidity or actual precipitation) results. The more nonsolvent required, the greater the solvent power. Nonsolvents appear to function by dilution of the solvent power of the solution thereby lowering the P–S interaction and permitting the occurrence of the P–P interaction. Determination of nonsolvent tolerance (dilution ratios) is not only simple, but also has the advantage of being directly related to membrane manufacture in that porosity (void volume) and/or (where a skin is present) skin thickness, with their profound effects on membrane permeability, are both proportional to the concentration of nonsolvent within phase-inversion casting solutions (Chapter 7).

In both solids and liquids strong attractive or cohesive forces exist between molecules. Hildebrand and Scott[16] developed an important relationship to express the magnitude of these cohesive forces and developed a scheme which would predict the miscibility of liquids in one another. Subsequently, by conversion of the forces in solids to equivalent liquid states, they were able to predict the solubility first of nonpolar amorphous polymers and subsequently of polar semicrystalline and even ion-containing types. The reader is hereby cautioned, however, that cohesion parameters have by no means been universally successful in this attempt. The concept becomes less applicable with increasing polymer polarity and loses credibility almost entirely when it is utilized to predict the compatibility of polymer blends.

In their original treatment the solvent cohesive energy density (CED) was taken as the heat of vaporization at constant volume, that is,

$$CED = (\Delta H - RT/V)$$

where ΔH = molar heat of vaporization, V is the molar volume of the liquid, R is the gas constant, and T is the temperature in °C (taken well below the boiling point of the liquid).

For the convenience of working with smaller numbers, CED is usually expressed in terms of the solubility parameter δ, where $\delta = (CED)^{1/2}$. Since CED is expressed in units of calories per cubic centimeter, δ has the units $cal^{1/2}/cm^{3/2}$. Recently, Barton[17] has suggested the adoption of SI units, joules$^{1/2}$ per centimeter$^{3/2}$, where $1\ cal^{1/2}/cm^{3/2} = 2.0455\ J^{1/2}/cm^{3/2}$.

If two substances have the same δ, enough energy is gained on mixing that solution can occur. As the difference between the δ values of the substance is increased, progressively more energy is required for mixing and less is gained by it. Eventually, the difference between the δ's becomes sufficiently large that mixing will not take place.

It soon became apparent that predictions of polymer solubility could be enhanced if the hydrogen-bonding capacities of polymer and solvent were taken into account. The first approach to accommodate this factor was to qualify δ with an estimate of hydrogen-bonding capacity in terms of weak, medium, and strong. Gordy and Stanford's[18] hydrogen-bonding index (HBI) was then employed. The HBI is defined as $\frac{1}{10}$ the shift in frequency in cm^{-1} of the 4-μm IR-band maximum when a given liquid is added to a solution of deuterated methanol in benzene. By plotting δ versus the HBI it was shown that all solvents for a particular polymer fall within a certain region. Crowley[19] further extended his concept to include the dipole moment μ as a measure of polarity. This combination of δ, HBI, and μ is the basis for the ANSI/ASTM D3132-72 standard test method for the solubility ranges of resins and polymers.[20]

Hansen and Skaarup[21] treated δ as a total parameter δ_T such that

$$\delta_T^2 = \delta_d^2 + \delta_p^2 + \delta_h^2$$

where δ_d, δ_p, and δ_h were partial parameters due to nonpolar-, polar-, and hydrogen-bonding parameters, respectively. They assumed the value of δ_d for a solvent to be equal to that of an isomorphous hydrocarbon. Dielectric constants, dipole moments, and refractive indices can be used to estimate δ_p, and δ_h can then be calculated. However, because there is no convenient means of plotting three variables in two dimensions, efforts have been made to reduce the three parameters to two. Bagley and Chen[22] combined δ_d and δ_p by defining $\delta_v = (\delta_d^2 + \delta_p^2)^{1/2}$ and plotted δ_v versus δ_h. Hansen himself found that in many instances δ_d could be neglected and two-dimensional coordinates based on δ_p and δ_h used to produce polymer solubility maps. A polymer is *likely* to be soluble in a solvent or mixture of solvents when its δ_p and δ_h values lie within the enclosed area on the diagram. For mixtures of solvents it is assumed that cohesive energies are additive when normalized for the volume fraction f_i of the constituent present. Thus, for a two-component system,

$$\delta_T = f_1 d_1 + f_2 d_2$$

Two-dimensional δ_p versus δ_h maps for CA and CTA (Fig. 5.3) were constructed by Klein et al.,[23] who utilized them to interpret the CA–acetone–formamide solutions of Manjikian et al.[7] They found that to produce useful skinned membranes for hyperfiltration desalination (1) the solution composition should be near the solubility boundary facing the nonsolvent gelation medium; (2) a volatile component should be such that its loss moves the solution composition out of the solubility envelope, (3) the solids content at the boundary should be high to increase the rate of transition from sol to gel, and (4) all components of the solution should be

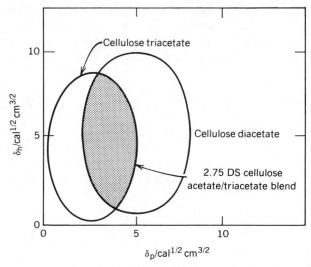

FIGURE 5.3. Hansen parameter δ_p versus δ_h solubility maps for cellulose triacetate, cellulose acetate, and (hatched area) a 2.75-DS CA/CTA blend (from Klein et al.[23] and Barton[17]).

miscible with the gelation medium. Klein et al. concluded, however, that solubility maps per se were incapable of establishing the requirements of a solution to form useful skinned membranes. Other external factors such as the kinetics involved in the exchange of solution components for the nonsolvent of the gelation medium were decisive in determining the thickness and structure of the skin in these membranes.

Hashimoto's[24] triangular fractional cohesion parameter solubility diagrams for CA and CTA are presented as examples of three-parameter maps in two dimension (Fig. 5.4). Fractional parameters are defined as

$$f_d = \frac{\delta_d}{\delta_d + \delta_p + \delta_h}; \quad f_p = \frac{\delta_p}{\delta_d + \delta_p + \delta_h}; \quad f_h = \frac{\hat{\delta}_h}{\delta_d + \delta_p + \delta_h}$$

Such fractional cohesion parameters have the advantage of spreading the points more uniformly over the triangular chart, but lack even the limited theoretical justification of Hansen's three-component parameters. Note moreover that the envelopes for CA and CTA almost overlap, thereby implying a closeness of solubility behavior which does not exist in reality. The two-parameter maps for CA and CTA are therefore more utilitarian in practice. Recognizing that few solubility parameter studies dealt with concentrated polymer solutions, Hoernschemeyer[35] plotted the Hildebrand parameter versus δ_h and obtained a map (Fig. 5.5) which he utilized to determine the influence of solvent type on the viscosity of concentrated (17.5 g/dL) CA solutions. He found that both for pure solvents and solvent mixtures the specific viscosity often attained a minimum value at a composition where the solubility parameter locus lay near the center of the solubility region of the polymer.

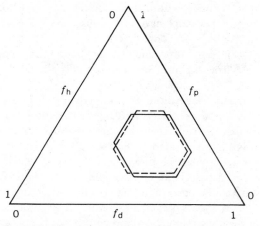

FIGURE 5.4. Triangular fractional cohesion parameter solubility diagram for cellulose acetate (solid line) and triacetate (broken line) (from Hashimoto[24] and Barton[17]).

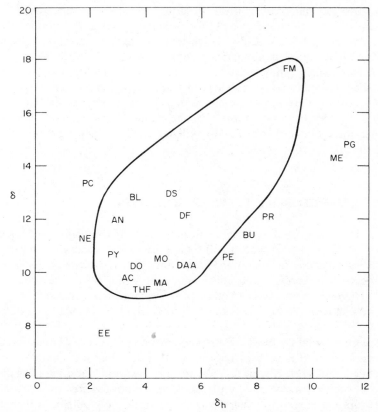

FIGURE 5.5. Solubility diagram for cellulose acetate (from Hoernschemeyer[35]; © 1974).

This map includes formamide within the solubility envelope for CA where it belongs, whereas several other maps have erroneously excluded it and other solvents. The truth is that solubility is to some extent subjective in that conditions such as concentration, agitation rate, time allowed for dissolution, temperature, and the plotting of different cohesion parameters can sometimes lead to conflicting conclusions.

The uneven electron surface representation advocated by the present author at the beginning of this chapter depicted P–S interactions in terms of interaction between electron-deficient and electron-rich groups on polymer and solvent molecules. This image is closely related to Fowke's[25] treatment of the P–P and P–S interaction in terms of Lewis-acid–Lewis-base theory in which an acid is an electron-pair acceptor and a base is an electron-pair donor. Most fluorinated or chlorinated molecules are acidic because of the electron-attracting tendency of the halogen atoms which leaves the carbon atoms to which they are attached in an electron-deficient state. For similar reasons, alcohol, carboxyl, phenol, and nitro groups are acidic as are the electrophilic carbon atoms of the carbonyl groups in ketones, esters, and carbonates and the carbon atoms of nitrile groups. Typical basic sites are the oxygens of carbonyl groups, ethers, and alcohols; the nitrogen atoms of amines, amides, and nitriles; divalent sulfur atoms; and π electrons. Polymers containing halogen or nitro groups, for example, poly(vinyl chloride) (PVC) or cellulose nitrate (CN), are acidic. Polyesters, polyamines, poly(vinyl pyrollidone) (PVP), and polymers with aromatic or olefin groups are primarily basic. There are also polymers which can be considered as amphoteric, namely poly(vinyl alcohol) (PVA), the polyamides (PAs), polyacrylic acid (PAA), and polyacrylonitrile (PAN). The solvent strength of acid–base interactions between polymer and solvent or two polymers in a blend can be estimated by the extent of spectral shifts in the IR[25] in a manner analogous to that originally utilized by Gordy and Stanford[18] for determining the HBI.

5.3 REPRESENTATIVE POLYMER SOLUTIONS

This section is designed to bridge the gap between the theory and practice of polymer solutions. The approach will be to present and discuss a number of practical casting solutions for important membrane-forming polymers. Individual polymer species will be discussed under major categories which include homopolymers, copolymers, and polymer blends. In the remainder of this chapter any number in brackets [ ] refers to the number of a casting solution in Tables 5.4, 5.6, 5.7, 5.11, and 5.12.

Cellulose is by far the most common polymer and thanks to its ubiquity in nature will always be so. It is perhaps only fitting therefore that membranes of cellulose and cellulose derivatives (Table 5.4) also occupy the leading position from the standpoints of both polymer volume utilization and membrane area production. Membranes from the parent polymer cellulose are principally employed in kidney dialysis and are prepared either by the extrusion of solutions of cellulose in cupra-

TABLE 5.4 CASTING SOLUTIONS OF CELLULOSIC HOMOPOLYMERS

Solution No.	Polymer(s)	Solvent(s)	Pore Former(s)	Manufacturing Process	Relative Void Volume	Ref.
1	Cellulose	Cuoxam	Glycerol	Wet	Dense	26
2	CN	Acetone	1-Butanol, isobutanol	Dry	Porous	27
3	CA	Acetone	$Mg(ClO_4)_2$(aq) $ZnCl_2$(aq)	Wet	Porous	5, 6
4	CA	Acetone	Formamide	Wet	Porous	7
5	CTA or CA	Sulfolane	PEG 400	Thermal/Wet	Dense	28
6	CTA	Acetone + dioxane	Maleic acid + methanol	Wet	Porous	29
7	CA	Acetone, dioxolane, methyl formate	Alcohol(s)	Dry	Porous	30
8	CA	Methyl formate, propylene oxide	SAIB	Dry	Dense	31
9	CA	Acetone, methanol	$CaCl_2$ + cyclohexanol	Wet	Porous	32
10	EC-PFB*	Cyclohexanone	—	Wet	Dense	33
11	EC-PFB*	Methylene chloride	Methanol	Dry	Porous	This volume

*Ethyl cellulose perfluorobutyrate

mmonium hydroxide (Cuoxam) [1] or by the posthydrolysis of CA or CTA membranes which are cast from solutions in organic solvents [5, 8]. Cuoxam solutions of cellulose are extruded both as hollow fibers and as flat sheets into aqueous salt or caustic solutions and then into aqueous acid and glycerol solutions. Glycerol serves both as a plasticizer and as a pore former. It acts in the latter role by preventing excessive densification and crystallization during drying. In spite of this such cellulose membranes are rather dense with void volumes ~20% in the dry state. The mechanical properties of cellulose membranes prepared form [1, 8] are almost identical and somewhat superior to those prepared from [5] (Table 5.5). The reason for this is primarily attributable to the nature of the pore former which is utilized in each case and only secondarily to cellulose origin, manufacturing process, or hydrolysis procedure. The basis for this assertion is that the spinning of solution [8] with PEG 400, a plasticizer, in place of sucrose acetate isobutyrate (SAIB), a nonplasticizing extender, yields CA membranes with physical properties closer to those of the membranes prepared from [5].

An extender such as SAIB provides potential porosity which is actualized on leaching during hydrolysis in methanolic NaOH without the structurally debilitating presence of a plasticizer. This effect is partially carried over to the hydrolyzed products, perhaps suggesting a degree of structural isomorphism between the

TABLE 5.5 PHYSICAL PROPERTIES OF CELLULOSIC HOLLOW FIBERS

Fiber Type	Swelling Modifier	Tensile Strength (psi) $\times 10^{-4}$	Tensile Modulus (psi) $\times 10^{-5}$
Cuprophan	—	3.04 ± 0.10	7.85 ± 1.88
CA (Run 93)	—	3.00 ± 0.29	6.53 ± 0.82
CA	16% Carbowax 400	1.49	3.58
CA	16% Carbowax 600	0.93	3.14
CA	16% Triacetin	2.25	5.78
CA	16% SAIB	2.69	6.26

prehydrolytic- and posthydrolytic structures of CA. Porosity in de-esterified CA membranes is also maintained by the inclusion of glycerol in hydrolysis and post hydrolysis steps.

Organsoluble cellulose derivatives have remained one of the favored classes of membrane polymer because of their great solubility in common solvents, their variety and broad applicability, and their low cost and universal availability. Cellulose nitrate (CN) was the first synthetic polymer and the first material utilized in the preparation of synthetic polymeric membranes. It is acidic and soluble in a large number of convenient organic solvents and is utilized in substantial quantities in the production of microfiltration membranes both by itself and in conjunction with other cellulosic polymers [62–64]. Although the membrane literature[34] indicates that methyl acetate is a practical solvent, the present author believes that the total unacceptability of technical-grade methyl acetate[35]—certainly today and probably in the past—suggests that it was never actually utilized in production. At any rate, the solvent properties of expensive reagent-grade methyl acetate are very similar to those of inexpensive technical-grade acetone, which is acceptable from both cost and purity standpoints. The pore former most widely utilized is 1-butanol which has the advantage of effectively removing during evaporation any water contained in the casting solution and limiting the quantity of water which can condense on the nascent membrane. Water is undesirable because it promotes the precipitation of low-MW polymer fragments known as *blushing*,[36] which manifests itself in the form of a fine powder on the membrane surface (Chapter 7, Section 7.8).

Unhydrolyzed cellulose acetate and cellulose triacetate membranes are of considerable importance as membrane materials for various pressure-driven separation processes as well as for electrophoresis and plasmaphoresis. A considerable variety exists between the various patented solutions and processes involved with the preparation of CA and CTA membranes. In this instance at least, these differences are related to the actual structural and functional requirements of the end products. The higher the void volume required in the final membrane, the greater the concentration of pore former required by the solution. Therefore, insofar as requisite pore former concentrations is concerned, solutions for dense films are less concentrated than those for skinned high-void-volume membranes which are less concentrated than those for skinless high-void membranes. Although in principle it should be possible to utilize the same basic solvent system with a given polymer such as CA and produce the entire range of membranes of low and high void volumes, in practice this is not always possible. Thus, for dense films, a two-component system—polymer and solvent without any pore former—for example, CA in acetone, suffices; for dialysis membranes low concentrations of plasticizers [5, 7, 8] are required, for skinned membranes (for gas separations, HF, UF) modest concentrations of swelling agent and/or nonsolvent pore formers are required [3, 4, 6, 7], and for skinless membranes for microfiltration, high concentrations of nonsolvents are required [7, 9]. In the case of solutions for hollow fibers for dialysis the requirement for low void volumes and the strength requirements of spinning dictate high polymer concentrations. This requirement can be met in two ways: the utilization of a thermal gel aided by a high boiling solvent, sulfolane, and a nonvolatile plasticizer, PEG 400 [5]; or the utilization of extremely strong and volatile solvents, propylene oxide or methyl formate, together with a neutral extender, sucrose acetate isobutyrate (SAIB) [8].The most critical solutions are those which are utilized in the spinning of skinless high-void-volume fibers for plasmaphoresis [7] (with methyl formate) and [9]. The strength of the solvent [7] permits the inclusion of high concentrations of both polymer and nonsolvent pore former. In [9], on the other hand, the solvent power of acetone is enhanced with methanol and $CaCl_2$ to retard gelation and a cyclic alcohol is utilized as a nonsolvent pore former. Acyclic diluents are generally more compatible than their linear counterparts of equivalent MW. In the former instance the extreme volatility of methyl formate (bp 30°C) allows it to flash off rapidly so that hollow-fiber strength develops rapidly and is sufficient to permit a drop height consistent with maintaining a high rate of spinning and takeup. The acetone-based solution, on the other hand, is not required to develop fiber strength solely through evaporative solvent loss because it is quenched in a nonsolvent gelation bath. One of these nonsolvent baths, namely methanol, is needed to extract the nonvolatile cyclohexanol nonsolvent.

A large number of cellulose esters and cellulose acetate mixed esters have also been successfully incorporated into solutions similar to those listed in Table 5.4. Indeed certain CA mixed esters—specially prepared from commercially available CA—and having a higher total DS than the CA starting material—are actually more soluble than the latter. Although not commercially available at present, the tremendous versatility of these CA mixed esters suggests that some may become so in the future. Among these are CA mixed esters bearing unsaturated groups which can be cross-linked in a postformation step.

The solubility of CA decreases as the acetyl content is increased from the usual acetone-soluble 2.5-DS material to the $\geqslant 2.75$-DS acetyl contents of CTA. Moore[14] held that CTA is more flexible in solution and exhibits behavior more typical of a random coil than does CA. In spite of this the latter is more soluble than the former. This is at least partially due to the lower crystallinity and hence greater accessibility of CA. However, it may also be related to the fact that CTA is primarily basic whereas CA is amphoteric and can, therefore, be solvated by both acidic and basic solvents. CTA is soluble in acetone if first cooled to 190 K and then warmed,[3] but is never cast from acetone solution in practice because the P–P interaction and its ready separation into crystalline gels is too pronounced in such solutions. Instead, a mixture of acetone and dioxane solvents, and maleic acid and methanol pore formers is utilized to prepare CTA membranes for hyperfiltration. An almost identical solution is more commonly encountered, for the same reasons, in the more widely utilized CA/CTA blend membrane (Table 5.12 [65]).

The choice of the pore formers is believed to be significant. CTA is basic and therefore probably solvated by maleic acid. Among the cellulosic polymers, the organosoluble organic cellulose esters have received far broader applications, over the entire range of separation processes—including gas separations, HF, UF, and MF—than have the inorganic ester CN and the cellulose ethers such as ethyl cellulose (EC). The reason for this may be in part related to the nonuniform substitution of the latter two which is in turn attributable to the heterogeneous reactions employed in their manufacture.[37] This lack of uniformity manifests itself in solution turbidity and suggests that these polymers should be restricted to those membranes where maximum order at the molecular level is not required. One such application is noncritical gas permeation of the type encountered in blood oxygenation. Of the various ECs and EC derivatives which were investigated, perfluorobutyrated (PFB)-EC proved to be most permeable and least thrombogenic.[33] Cyclohexanone [10] solutions of this polymer were utilized because they tend to spread out on water thereby forming thin, dense films which can be floated onto a polypropylene suport to produce composite membranes. Kesting [11] produced skinned membranes by dry casting high-MW grades (T-100 or T-200 EC starting materials) of this polymer from methylene chloride/methanol solutions. The solubility of EC and its derivatives in other alcohols dictates the choice of methanol as the nonsolvent pore former. The solubility characteristics of cellulose ethers are somewhat atypical and will be discussed further in the section on solutions for polymer blend membranes.

A class of membrane homopolymers which rivals the cellulosics in importance and surpasses them in variety is that of the polyamides (PAs) and related materials (Table 5.6). I strongly recommend the excellent review of this subject by Blais.[10] The most important solutions of aliphatic polyamides consists of nylon 6,6, nylon 610, nylon 11, or nylon 12 in 98% formic acid [12] and of blends of high-MW nylon 6,6 with the amorphous multipolymer binder nylon(6,6, 6,10, and 6) in 90% formic acid [13]. All of these solutions are utilized in the preparation of skinless microfiltration membranes of considerable and growing commercial importance. The anhydrous 98% acid [12] permits the dissolution of the polymer and even its storage at elevated temperatures for extended periods of time without degradation due to hydrolytic cleavage of the amide linkage. The lower viscosity at elevated temperatures is convenient for the preparation of letdown mixes in which nonsolvent water is added as a pore former. Solutions of nylon 6,6/nylon multipolymer in 90% formic acid, on the other hand [13], do not require the preparation of letdown mixes but must be monitored for the hydrolytic degradation which limits solution lifetime to about three weeks at room temperature. Solution [13] is the only case known to the author in which the complete evaporation of a two-component system results in membranes of sufficiently high volume as to be viable for commercial microfiltration. Ordinarily, the complete evaporation of a two-component solution, con-

TABLE 5.8   CHARACTERISTICS OF CA–PS GRAFT COPOLYMERS[a]

Sample	Combined Polystyrene (%)	[DS] of CA	$\overline{M}_v$ CA	$\overline{M}_v$ PS
P_3	45.0	1.84	54,000	119,000
P_4	44.1	2.25	111,000	119,000
P_5	28.0	1.84	54,000	32,000
P_6	18.2	2.25	111,000	32,000

[a]From Wellons et al.[62]; © 1967.

characteristics of these graft copolymers are listed in Table 5.8. Both short- and long-chain grafts of PS were made onto short and long CA chains. All four graft polymers and the four corresponding homopolymers were soluble in dimethylformamide (DMF). The tolerance of these graft copolymers and the two parent homopolymers for three nonsolvents is shown in Table 5.9. Grafting increased the tolerance of nonsolvents for the least soluble component. The polystyrene content of the graft had the strongest influence since the tolerance of the grafts to water and methanol was closer to that of polystyrene and their tolerance for toluene was much greater than that of CA for toluene.

The intrinsic viscosities of the grafts and homopolymers in mixed solvents and nonsolvents were utilized to study solution behavior. Two grafts with the same length of PS chains but with long- and short-chain CAs were examined in a number of solvent systems (Table 5.10). The addition of nonsolvent for one component resulted in the tighter coiling of this component thereby reducing the viscosity of the graft towards that of the other component. In the case of the acetone–toluene system, it is significant that although neither was a solvent for the graft copolymer, the mixtures were solvents across a wide range of compositions.

The compatibilization effect of graft copolymers for the blends of the homopolymers is shown in Figure 5.6. Whereas 75% of the uneven graft copolymers (P_3 or P_6) is needed to compatibilize equal parts of CA and PS, a much smaller per-

TABLE 5.9   NONSOLVENT TOLERANCE OF CA–PS GRAFTS AND CORRESPONDING HOMOPOLYMERS[a]

		Nonsolvent Volume (mL) to Precipitate 10 mL of 1% Solution in DMF								
		H_2O			MeOH			Toluene		
		CA	PS	Graft	CA	PS	Graft	CA	PS	Graft
Graft	P_3	6.36	0.28	0.30	29.5	3.6	4.0	21	∞	500
	P_4	2.85	0.28	0.30	45.1	3.6	3.6	37	∞	450
	P_5	6.36	0.40	0.50	29.5	5.1	7.4	21	∞	500
	P_6	2.85	0.40	0.55	45.1	5.1	7.4	37	∞	500

[a]From Wellons et al.[62]; © 1964.

TABLE 5.10 INTRINSIC VISCOSITIES OF CA–PS GRAFTS AND CORRESPONDING HOMOPOLYMERS IN MIXED SOLVENTS AND NONSOLVENTS[a]

Solvent 2	% Solvent 2	$[\eta]$ PS	$[\eta]$ CA	$[\eta]$ P_3 Graft
Graft P_3, solvent 1 = dimethylformamide				
Toluene	0	0.36	0.96	0.76
Toluene	20	0.37	0.95	0.78
Toluene	40	0.38	0.93	0.78
Toluene	60	0.39	0.90	0.78
Toluene	80	0.39	Insol.	0.78
Toluene	90	0.39	Insol.	0.52
Toluene	100	0.39	Insol.	0.48
Water	2	0.25	0.96	0.68
Water	20	Insol.	0.95	Insol.
Graft P_4, solvent 1 = acetone				
Toluene	0	Insol.	2.80	Insol.
Toluene	15	Insol.	2.76	1.66
Toluene	30	Insol.	2.44	1.53
Toluene	35	0.27	2.16	1.04
Toluene	45	0.32	Insol.	1.01
Toluene	60	0.37	Insol.	Insol.
Toluene	100	0.39	Insol.	Insol.

[a]From Wellons et al.[62]; © 1964.

centage of graft is required when the graft is even, that is, composed of long chains of CA and PS.

The synthesis of block and graft copolymers can be planned with the aid of cohesion parameters. Solvents can strongly influence structure, properties, rates of formation, and molecular weights of the copolymer. If a copolymer containing approximately equal amounts of both rubbery and glassy segments is in a solvent with a solubility parameter similar to that of the rubber, chains of the latter will be extended while the glassy chains are tightly coiled so the product will exhibit the properties of a filled rubber. Conversely, if in a solvent which interacts more strongly with the glassy segments, the resulting polymer will be a rubber-modified glass. Obviously, if one segment is present at a much higher concentration than the other, this fact will determine which will be the continuous phase.

Block and graft copolymers, polymer blends, and ionomers often show two separate peaks when the solubility or degree of swelling by solvents is plotted versus the Hildebrand parameter. Ionomers such as the sodium or zinc salts of ethylene–acrylic acid copolymers[63] are soluble in toluene:isobutanol, a solvent mix in which the isobutanol solvates the acrylic acid moieties and the toluene, the ethylene units. The sulfonated polysulfones [58], sulfonated poly(phenylene oxides) [57], and the trimethylammonium salt of CA bromoundecanoates (TMA CA 11-Bru) with greater than 0.2 DS of quaternary ammonium groups [66], all require solva-

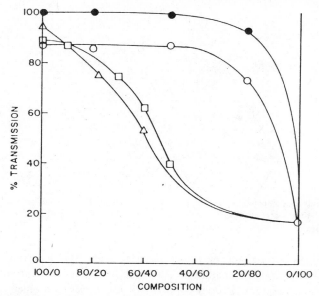

FIGURE 5.6. Light transmission versus composition of the graft copolymers and a 50:50 blend of CA and PS homopolymers: ($\triangle$) P_6 graft/homopolymer blend; ($\square$) P_3 graft/homopolymer blend; ($\circ$) P_3 + P_6 blend/homopolymer blend; and ($\bullet$) P_4 graft/homopolymer blend (from Wellons et al.[62]; © 1967).

tion of their macroions by media of high dielectric constant such as water, methanol, or formamide. Such solvents are needed to allow charge separation between the macroions and their associated counterions.

The solubility characteristics of weak polyelectrolytes such as poly(acrylic acid) (PAA) which can exist in ionized or unionized forms are interesting in that in the former condition they tend to be highly extended because of repulsion from like charges on neighboring chain segments and in the latter they assume the random coil conformations of neutral polymers.[12] Two oppositely charged polyelectrolytes can be blended in a single solution if water (or presumably other solvents of high dielectric constant) and inorganic salts are added to the organic solvent. In such cases the organic solvent solvates the polymer chain, the water permits separation of macroions from their counterions, and the inorganic salt serves to inhibit premature gelation by masking the oppositely charged macroions from one another[64] (Fig. 5.7).

Typical casting solutions of copolymers and ionomers are listed in Table 5.11. Ethylene–vinyl acetate (EVA) random copolymers are soluble in aqueous solutions of the lower alcohols [40] largely because of the basicity of VA and the acidity of the alcohols. Ethylene acrylic acid (EAA) random copolymers with a high ethylene content require a low-solubility-parameter hydrocarbon solvent [41]. The fact that toluene is somewhat basic probably helps with the solubilization of the acrylate ion and the high dielectric constant of isobutanol helps shield the acrylate ion from its Na^+ and Zn^{2+} counterions. Another factor is the equal volatilities of toluene and

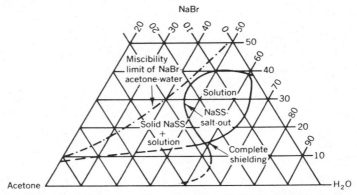

FIGURE 5.7. Solvent-composition phase diagram for "neutral" VBTA Cl–NaSS polysalt in the ternary system NaBr–acetone–water at 25°C (from Michaels[64]; reprinted with permission from *Industrial Engineering Chemistry,* © 1965 American Chemical Society).

isobutanol which permits the ratio to remain constant during drying and thereby avoids an unwanted phase separation. The solubilization of fully aromatic polyamides, even when they are random copolymers specifically designed to maximize accessibility and minimize crystallinity [42], requires both a strong solvent, DMAC, and a lyotropic salt, LiCl, to enhance solubility and avoid premature gelation. Random copolymers of AN with a small amount of methyl acrylate [43] are more soluble than the PAN homopolymers and can be cast from similar solvent systems. Their greater flexibility on the molecular level also results in less brittle membranes. Although random copolymers of aromatic polyamides are much more soluble than their fully aromatic counterparts, their conversion into high-void-volume membranes also requires both a strong solvent NMP and a lyotropic salt [44].

Block copolymers containing > 20% PEG 20M and the remainder BPA polycarbonate are soluble in dioxolane [45]. They can be wet cast into skinned membranes for dialysis if DMSO is added to their solutions as a pore former. The block copolyether–polycarbonate containing 5% of PEG 4000 was utilized for the preparation of MF membranes. Its superior solubility in MC compared to PC homopolymers is due to increased chain flexibility [46]. The utility of the acidic solvent methylene chloride and acidic pore former TFE or HFIP, respectively, suggests that PC is basic. Other polyesters are too. Additional block copolymers of this type are the basic copolyether–polyesters [47] which can be dry cast from the acidic solvent methylene chloride and the copolyether polyurethanes [48] which are amphoteric and can be dry cast from DMF.

Silicone–polycarbonate block copolymers are soluble in methylene chloride [49], an acid solvent which strongly solvates the basic (aromatic and carbonyl) groups of the PC blocks. If hexane is present, it preferentially solvates the siloxane blocks ($\delta = 7.5$). Therefore, films cast from mixed solvents are more rubbery than those cast from methylene chloride solutions.

Nylon 6,6 onto which PEI has been grafted is soluble in formic acid [50]. This is almost predictable behavior in that nylon 6,6 itself is soluble in formic acid and

PEI is basic. The fact that DMF is a solvent for CA which has been grafted by PEI or PS [51] is probably related to the fact that DMF is in the center of the solubility envelope for CA (Fig. 5.5) and hence one of its strongest solvents. DMF also has strong hydrogen-bonding and polarity components which should help with the PEI graft. The fact that PS is aromatic and therefore electron deficient may explain its solubility in the extremely nucleophilic DMF. There is a fairly substantial difference in δ's 9.5 and 12.5, respectively.

The solubility of the nylon random multipolymer 6/6, 6/10, to which vinyl acetate (VAc) has been grafted [52] is attributable to the structural irregularity of the ungrafted multipolymer (which is itself soluble in methanol) and to the basicity of PVAc which renders it soluble in the acidic methanol. This example was included to illustrate just how complex even neutral membrane polymers may become since in this case a random copolymer was further modified by grafting.

Solutions [53–60] illustrate the increasing importance of ionomers as membrane polymers and the unavoidable complexity of some of their solutions. Solutions [53,–56] illustrate the incorporation of charges or potentially charged monomers into random copolymers as a means of producing ionomers. In the case of [53, 56] this objective was accomplished by quaternization in a postpolymerization step. Solubility is maintained by utilizing only a small percentage of the ionic monomer and a strong solvent for the homopolymer which corresponds to the monomer which is present in the highest concentration. The solubility of the quaternized random copolymer of vinyl pyridine and butadiene [56] is attributable to polymer acidity and the basicity of THF. Since PPO and PS are both basic they are soluble in acidic solvents such as $CHCl_3$ and nitromethane. The sodium salt of the sulfonated derivative of PPO [57] is soluble in the same solvents as the parent polymer if the high dielectric constant cosolvent ethanol is added to solvate the sodium ion and shield it from the sulfonate counterion. Another approach, based on solubility parameters, has been employed by Kinzer et al.[76] who utilized two nonsolvents for the parent PS polymer, THF and formamide, to develop the proper solubilty parameter for the sulfonated derivative [58]. They have since developed solutions for sulfonated PES based on ethyl formate/formamide and MEK/formamide. The success of this approach is probably attributable to the ion-solvating capacity of the high dielectric constant formamide molecule. There are a number of other solvent systems for sulfonated polysulfone which has received and is continuing to receive a great deal of attention as an HF membrane polymer.

One of the most important new classes of membrane polymer is that of the *perfluorinated ionomers*. These are generally processed from the melt in their sulfonyl fluoride forms. However, for specialized applications the preparation of solutions is essential [59, 60]. The no-longer-available 970 eq wt material was soluble in ethanol [59] and could be dry cast to yield dense membranes. To obtain solutions of the 1100 and 1200 eq wt polymers it is necessary to dissolve them in an autoclave with superheated ethanol or isopropanol and water [60]. A dual solubility parameter has been found for insoluble Nafion membranes (Fig. 5.8).

Polymer blends are of increasing importance in both bulk-polymer and membrane applications. Unfortunately, no satisfactory list of criteria is as yet available

TABLE 5.11 CASTING SOLUTIONS OF COPOLYMERS AND IONOMERS

Solution No.	Polymer(s)	Polymer Class	Solvent(s)	Pore Former(s)	Manufacturing Process	Relative Void Volume	Ref.
40	EVA	Random co	Methanol + H_2O	—	Wet	Dense	65
41	EAA, Na^+ or Zn^{+2}	Random co	1-Propanol + H_2O Toluene + isobutanol	—	Dry	Dense	63
42	Ar–Ar PA from m-, o-, and p-phenylene diamines + iso- and terephthaloyl chlorides	Random co	DMAC	H_2O, LiCl	Gel/Wet	Porous	9
43	PAN–MA	Random co	75% HNO_3(−3°C)	Water	Wet	Porous	65
44	Poly (piperazine iso- and o-phthalamides)	Random co	NMP	LiCl	Wet	Porous	42
45	PEO-b, co-PC	Block co	Dioxolane	DMSO	Wet	Porous	66
46	PEO-b, co-PC	Block co	Methylene chloride	IPA + TFE or HFIP	Dry	Porous	60
47	PEO-b, co-PEster	Block co	Methylene chloride	—	Dry	Dense	67
48	PEO-b, co-PU	Block co	DMF	—	Dry	Dense	67
49	Silicone-co-PC	Block co	Methylene chloride or methylene chloride + hexane	—	Dry	Dense	68

No.	Material	Type	Solvent	Additive	Wet/Dry	Structure	Ref.
50	Nylon-6/6-g-PEI	Graft co	HCOOH	—	Wet/Dry	Dense	69
51	CA-g-PEI/CA-g-PS	Graft co	DMF	—	Dry	Dense	69, 62
52	Nylon (6/6, 6, 10)-g-VAc	Random co	Methanol	—	Dry	Dense	70
53	PA–MMA–K$^+$ vinyl oxybenzene sulfonate	Random co Ionomer	DMSO, DMF	—	Wet	Dense	71
54	PAN–QVP	Random co Ionomer	DMAC	PEG 400	Wet	Dense	72
55	PAN–Na$^+$ methallyl sulfonate	Random co Ionomer	DMF	—	Wet	Dense	71
56	PQVP-1,3-butadiene	Random co Ionomer	THF	—	Dry	Dense	73
57	NaSO$_3$–PPO	Ionomer	Chloroform + ethanol; nitromethane + ethanol	—	Dry	Dense	74
58	NaSO$_3$–PS	Ionomer	THF + formamide	—	Wet	Porous	75
59	Nafion eq wt 970	Ionomer	Ethanol	—	Dry	Dense	76
60	Nafion eq wt 1100, 1200	Ionomer	Superheated IPA + H$_2$O; superheated ethanol + H$_2$O	—	Dry	Dense	76

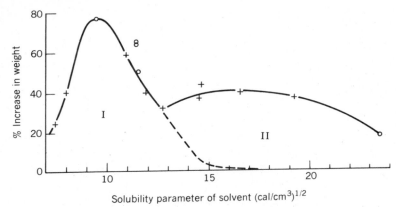

FIGURE 5.8. Solvent uptake of Nafion 120 versus solubility parameters of solvents (from Yeo[77]).

for the choice of suitable polymer pairs, although the acid–base concept which specifies interactions between acidic and basic copolymers appears to offer significant promise.[25] Occasionally, solubility parameter maps can be of some assistance in selecting polymers for blending. The δ_d and δ_p values of polystyrene and polyphenylene oxide which form a compatible (single T_g) blend lie very close to one another. Several factors influence the $\Delta\delta_{cr}$, the critical maximum difference between solubility parameters which permits a compatible blend. Among these are the molecular weight of both polymers (compatibility becomes less probable the higher the MW), the existence of specific interactions between groups on one polymer with those of another which favor compatibility, and steric properties such as tacticity which also influence the likelihood of specific interactions. Whereas true compatibility may be a requirement for membranes which are utilized in gas separations and hyperfiltration (the finest separations), it is less likely to be one for ultrafiltration and microfiltration membranes. Less compatible blends than those characterized by a single T_g are possible and yield coarser and more porous structures which are preferred in the separation of larger solutes or suspended particles. Cabasso et al.[78] have proven the efficiency in HF membranes of blends of CA and phosphorylated PPO in which specific group interactions are known. Kesting et al.[79] have formed DRY-RO membranes from blends of CA and/or CA mixed esters and isonomers such as the trimethylammonium salts of CA 11-bromoundecanoate and CA 6-bromocaproate [66]. Incidentally, the fact that blends such as these can yield highly permselective membranes is presumptive evidence that the skins of these membranes are in the glassy state since polymer blends are not generally believed to exist in crystalline forms.

Relevant to the preparation of solutions of polymer blends for preparing membranes are the following observations: (1) gross solution incompatibility such as the presence of coarse-grained turbidity and the separation into two liquid phases is never allowed; and (2) limited compatibility may occur over less than the full range of compositions.

Solutions [61–72] have received a considerable amount of attention in the preparation of blend membranes of two types: Membranes in which both polymers are to remain in the end-use application and those in which one polymer is present in an assisting role to form an interpenetrating polymer network and is removed by leaching prior to use (Table 5.12). Minor components with less than single T_g compatibility can serve a useful function as *gelation promoters*. The inclusion, for example, of the nylon 6,6; 6,10; 6 multipolymer in 90% formic acid solutions of nylon 6,6 [61] causes the solution to gel at an early stage in desolvation and hence yields skinless membranes of high porosity. In contrast, the complete evaporation of nylon 6,6 homopolymer solutions results in skinned membranes of low porosity which are characteristic of retarded gelation. This particular blend solution is the only case known to the author in which a solvent system which contains no nonsolvent pore former can be dried completely to yield an integral, skinless membrane of high porosity. Complete evaporation of such solutions typically yields dense membranes or skinned membranes of low porosity (Chapter 7). In this case intermolecular hydrogen bonding results in crystalline gels whose rigidity is sufficient to overcome the combined effects of plasticization and gravity which oridinarily lead to loss of porosity. Cellulose acetate functions similarly in CN/CA solutions [62, 63]. In the absence of CA, CN is almost too soluble to permit orderly inversion and gelation with the result that tears and pinholes are more common in CN homopolymer membranes than the blend membranes. Isotropic microfilters, that is, those with little homogeneity in depth, are produced by the slow evaporation of the relatively nonvolatile acetone–butanol solvent system [62]. On the other hand, highly anisotropic skinless CN/CA blend membranes, in which the pore size on the feed side of the membrane is greater than that on the product side, may be prepared from solutions in highly volatile solvents such as methyl formate [63].

The compatibility of acidic CN with basic EC has been documented. However, porous membranes are difficult to prepare from this blend because of the solubility of EC in the alcohols which are nonsolvent pore formers for the dry casting of porous MF membranes. Thus when solvent evaporation has proceeded to the point where the true solvent for CN is largely gone and hence CN forms micelles and then gels, the alcohols which are nonsolvents for CN continue to hold EC in solution. As the alcohols themselves depart they leave behind thin films of EC which form a film over pores in the membrane, rendering it useless. The solution to this dilemma was the substitution of another basic cellulose ether, cyanoethyl cellulose (CEC), presumably also compatible with CN-, but alcohol insoluble [64]. The result was that when CN formed micelles and gelled, so did CEC, and the resultant membrane consisted of porous open cells as required. Cellulose ethers confer strength and flexibility upon CN membranes. A commerically important solution is that utilized by Cannon to prepare CA/CTA blend membranes for HF [65]. In view of the difficulty of preparing these solutions and the limited choice of pore formers, this is probably not a truly compatible blend. It may be that maleic acid functions to solvate the basic CTA while methanol forms hydrogen bonds with the free hydroxyl groups of CA. Another useful cellulosic blend solution is that of the ionomeric CA mixed ester which contains the trimethylammonium salt of the 11-

TABLE 5.12 CASTING SOLUTIONS OF POLYMER BLENDS

Solution No.	Polymer(s)	Solvent(s)	Pore Former(s)	Manufacturing Process	Relative Void Volumes	Ref.
61	Nylon 6,6 (high MW) + nylon 6,6; 6,10; 6 multipolymer (low MW)	90% Formic	None	Dry	Porous	39
62	CN(RS) + CA	Acetone	1-Butanol, IBA	Dry	Porous	27
63	CN(RS) + CA	Methyl formate	IPA	Dry	Porous	80
64	CN(RS) + CEC	Acetone	1-Butanol, IBA	Dry	Porous	81
65	CA + CTA	Acetone + dioxane	Maleic acid + methanol	Dry	Porous	29
66	CA + TMA CA 11-Br undecanoate	Acetone + methanol; dioxolane + methanol	IBA	Dry	Porous	79
67	PBT + PVA	HFIP	PVA	Dry (PAPI)	Porous	57
68	PS + PVP	DMAC	PVP	Wet (PAPI)	Porous	45
69	Fully Ar PA of m-phenylene diamine + isophthaloyl chloride	DMAC	PVP	Wet (PAPI)	Porous	45
70	PVC + P (p-dimethyl-amino styrene)	Cyclohexanone	P (p-dimethylamino styrene)	Dry (PAPI)	Porous	58
71	PVC + PVME	Gel	PVME	Thermal (PAPI)	Porous	59
72	Isotactic PMMA + syndiotactic PMMA	DMSO + H$_2$O	—	Thermal	Porous	82

bromoundecanoyl group [66]. The mixed ester is blended with the parent CA starting material to improve reproducibility and to decrease cost. It is known that C_{10}–C_{12} chains provide maximum solubility but it is unknown whether the blend is truly compatible. The clarity of solutions (once a co-solvent such as methanol or water has been added to solvate the quaternary ammonium groups) and the glossiness of the skins of the resultant HF membranes do suggest a high degree of compatibility.

Solutions [67–71] are all used in PAPI processes. They exist as blends in the solution form and in the dense-film form before the assisting polymer has been leached. Basic PBT is soluble in the acidic solvent HFIP which tolerates the blending of basic PVA into the solution. Nevertheless, the blend film is not quite clear which suggests only limited compatibility. Polysulfone [68] and the fully aromatic polyamide based on m-phenylenediamine and isophthaloyl chloride [69] are soluble in DMAC. Each solution will tolerate the addition of PVP, provided the latter is anhydrous. There does not seem to be an a priori reason to expect more than physical compatibility—which I suspect is exactly what is required anyway when microporous membranes are the objective. The acidity of PVC and the basic characters of poly(p-dimethylaminostyrene) [70] and poly(vinyl methyl ether) [71], respectively, are consistent with a higher degree of compatibility. Cyclohexanone, a cyclic aromatic ketone, is basic both because it is a ketone and because it is aromatic which explains its capacity for dissolving PVC. The aromaticity of poly(p-dimethylaminostyrene) may be sufficient to explain its solubility in cyclohexanone in spite of the fact that both are basic. The final blend solution [72] is of interest because it involves the use of syndiotactic and isotactic PMMA. There is of course no certainty that stereoregular polymers will be soluble in solvents for their atactic counterparts.

The behavior of solutions of polymer blends is poorly understood at present. The development of a comprehensive theory which is capable of quantitative predictions of the varying degrees of compatibility of polymer blends in a spectrum of solvents remains a distant objective.

REFERENCES

1. A. Doolittle, *Ind. Eng. Chem.*, **36**, 239 (1944); **38**, 535 (1946); *J. Polym. Sci.*, **2**, 121 (1947).

2. K. Überreiter, "The Solution Process," in *Diffusion in Polymers*, J. Crank and G. Park, Eds., Academic, New York, London, 1968.

3. Cowie et al., *Makromol. Chem.*, **143**, 105 (1971).

4. R. Kesting, *J. Appl. Polym. Sci.*, **9**, 663 (1965).

5. S. Loeb and S. Sourirajan, U.S. Patent 3,133,132 (May 12, 1964).

6. R. Kesting, M. Barsh, and A. Vincent, *J. Appl. Polym. Sci.*, **9**, 1873 (1965).

7. S. Manjikian, S. Loeb, and J. McCutchan, U.S. Patent 3,344,214 (September 26, 1967).

8. R. Kesting, U.S. Patent 3,884,801 (May 20, 1975).

9. J. Richter and H. Hoehn, U.S. Patent 3,567,632 (1971).

10. P. Blaïs, "Polyamide Membranes," in *Reverse Osmosis and Synthetic Membranes*, S. Sourirajan, Ed., NRCC Publ. No. 15627, Ottawa, Canada, 1977.

11. P. Clement, *Ann. Chim. (Paris)*, **12**(2), 420 (1947).

12. F. Billmeyer, Jr., *Textbook of Polymer Science*, 2nd Ed., Wiley-Interscience, New York, 1971.

13. H. Mark, *Der feste Koerper*, Hirzel, Leipzig, 1938, p. 103; R. Houwink, *J. Prakt. Chem.*, **155**, 241 (1940).

14. W. Moore, "Concentrated Solutions," in *Cellulose and Cellulose Derivatives*, N. Bikales and L. Segal, Eds., Chap. XIV, Pt. IV, Wiley-Interscience, New York, 1971.

15. K. Kobayashi et al., French Patent 1,541,292 (October 4, 1968).

16. J. Hildebrand and R. Scott, *Solubility of Non-Electrolytes*, 3rd Ed., Reinhold, New York, 1950.

17. A. Barton, *CRC Handbook of Solubility Parameters and Other Cohesion Parameters*, CRC Press, Boca Raton, FL, 1983.

18. W. Gordy and S. Stanford, *J. Chem. Phys.*, **7**, 93 (1939); **8**, 170 (1940); **9**, 204 (1941).

19. J. Crowley et al., *J. Paint Technol.*, **38**, 269 (1966); **39**, 19 (1967).

20. ANSI/ASTM D3132-72, American Society for Testing Materials, Philadelphia, American National Standards Institute, New York, reapproved 1976.

21. C. Hansen and K. Skaarup, *J. Paint Technol.*, **39**, 511 (1967).

22. E. Bagley and S. Chen, *J. Paint Technol.*, **41**, 494 (1969).

23. E. Klein, J. Eichelberger, C. Eyer, and J. Smith, *J. Water Res.*, **9**, 807 (1975).

24. I. Hashimoto, Sen'i Gakkaishi, **34**, T469 (1978) (cited in Reference 17).

25. F. Fowkes, D. Tischler, J. Wolfe, L. Lannigan, and C. Ademu-John, *J. Polymer Sci. Polymer Chem. Ed.*, **22**, 547 (1984).

26. M. Isugu et al., U.S. Patent 3,888,771 (June 1975).

27. A. Goetz, U.S. Patent 2,926,104 (February 23, 1960).

28. B. Lipps, U.S. Patent 3,546,209 (December 8, 1970).

29. C. Mungle and R. Fox, U.S. Patent 4,026,978 (May 31, 1977).

30. R. Kesting, U.S. Patent 4,035,459 (July 12, 1977).

31. R. Kesting, U.S. Patent 4,219,517 (August 1980).

32. M. Mishiro et al., U.S. Patent 4,234,431 (November 18, 1980).

33. R. Peterson, U.S. Patent 4,210,529 (July 1, 1980).

34. K. Maier and E. Scheuermann, *Kolloid Z.*, **171**, 122 (1960).

35. D. Hoernschemeyer, *J. Appl. Polym. Sci.*, **18**, 61 (1974).

36. C. Smoulders, "Morphology of Skinned Membranes," in *Ultrafiltration Membranes and Applications*, A. Cooper, Ed., Plenum, New York, 1980.

37. E. Ott, H. Spurton, and M. Graffin, Eds., *Cellulose and Cellulose Derivatives V*, Pt. 2, Wiley-Interscience, New York, 1954.

38. D. Pall, U.S. Patent 4,340,479 (July 20, 1982).

39. R. Kesting, L. Cunningham, M. Morrison, and J. Ditter, *J. Parenteral Sci. Technol.*, **37**(3), 97 (1983); R. Kesting, U.S. Patent 4,450,126 (1984).

40. D. Pall and F. Model, U.S. Patent 4,340,480 (1982).

41. A. Castro, U.S. Patent 4,247,498 (January 27, 1981).

42. L. Credali and P. Parrini, U.S. Patents 3,743,597 (July 3, 1973), 3,696,031 (October 3, 1972).

43. F. Model, "PBI Membranes for Reverse Osmosis," in *Reverse Osmosis and Synthetic Membranes*, S. Sourirajan, Ed., NRCC Publ. No. 15627, Ottawa, Canada, 1977.

44. H. Hoehn, U.S. Patent 3,822,202 (July 2, 1974).

45. E. Klein et al., U.S. Patent 4,051,300 (September 27, 1977).

46. R. Mahoney et al., U.S. Patents 4,020,230 (April 26, 1977), 4,115,492 (September 19, 1978).

47. T. Nohmi, U.S. Patent 4,229,297 (October 21, 1980).

48. R. Hughes and E. Steigelmann, U.S. Patent 4,106,920 (August 18, 1978).

49. K. Yamamoto et al., U.S. Patent 4,071,454 (January 31, 1978).

50. R. Chapurlat et al., U.S. Patent 3,907,675 (September 23, 1975).

51. A. Michaels, U.S. Patent 3,615,024 (October 1971).

52. Y. Hashino et al., U.S. Patents 3,871,950 (March 18, 1975), 4,181,694 (January 1, 1980).

53. W. Benzinger, U.S. Patent 4,384,047 (May 17, 1983).

54. Y. Hashino et al., U.S. Patent 4,208,508 (June 17, 1980).

55. R. Cross, U.S. Patent 3,691,068 (September 12, 1972).

56. H. Hoehn and J. Richter, U.S. Patent Reissue, 30,351 (July 1980).

57. R. Kesting, U.S. Patent 3,957,651 (May 18, 1976).

58. G. Bourat and A. Fabre, U.S. Patent 3,751,536 (August 7, 1973).

59. B. Baum, R. White, and W. Holley, Jr., "Porous Tubelets for Desalination Barriers,' in *Reverse Osmosis Membrane Research*, L. Lonsdale and H. Podall, Eds., Plenum, New York, 1972.

60. R. Kesting, U.S. Patent 3,945,926.

61. R. Kesting and V. Stannett, *Makromol. Chem.*, **65**, 248 (1963).

62. J. Wellons, J. Williams, and V. Stannett, *J. Polym. Sci.*, *A-1*, **5**, 1341 (1967).

63. E. I. DuPont, *Surlyn Ionomer Resins*, Wilmington, DE.

64. A. Michaels, *Ind. Eng. Chem.*, **57**(10), 32 (1965).

65. S. Yamashita et al., U.S. Patent 4,234,837 (January 16, 1979).

66. W. Higley et al., U.S. Patents 4,075,108 (February 21, 1978) and 4,160,791 (July 10, 1979).

67. Y. Thakore, D. F. Shiek, and D. Lyman, "Chemical and Morphological Effects of Solute Diffusion through Block Copolymer Membranes," in *Ultrafiltration Membranes and Applications*, Plenum, New York, 1980.

68. R. Kambour, in *Block Polymers*, S. Aggarwal, Ed., p. 263, 1970; A. Noshay and J. McGrath, *Block Copolymers*, Academic, New York, 1977, p. 454.

69. R. Crowley, U.S. Patent 3,857,782 (December 31, 1974.

70. E. Steigelmann et al., U.S. Patent 4,047,908 (1977).

71. G. Christen et al., U.S. Patents 3,930,105 (December 30, 1975), 4,056,467 (November 1, 1977).

72. I. Salyer et al., U.S. Patent 3,799,356 (March 26, 1974).

73. H. Yasuda and A. Schindler, "Reverse Osmosis Properties of Ionic and Nonionic Polymer Membranes," in *Reverse Osmosis Membrane Research*, H. Lonsdale and H. Podall, Eds., Plenum, New York, 1972.

74. A. P. La Conti, "Advances in Development of Sulfonated PPO and Modified PPO Membrane Systems for Some Unique Reverse Osmosis Applications," in *Reverse Osmosis and Synthetic Membranes*, S. Sourirajan, Ed., Chap. 10, NRCC Publ. No. 15627, Ottawa, Canada, 1977.

75. K. Kinzer, D. Lloyd, J. Wightman, and J. McGrath, *Desalination*, **46**, 327 (1983).

76. C. Martin et al., *Anal. Chem.*, **54**, 1639 (1982).

77. R. Yeo, *Polymer*, **21**, 432 (1980).

78. I. Cabasso, J. Jagur-Grodzinski, and D. Vofsi, in *Polymer Alloys; Blends, Grafts, and Interpenetrating Polymer Networks*, D. Klempner and K. Frisch, Eds., Plenum, New York, 1977.

79. R. Kesting, J. Newman, K. Nam, and J. Ditter, *Desalination*, **46**, 343 (1983).

80. R. Kesting, U.S. Patent 4,338,972 (June 8, 1982).

81. R. Kesting, U.S. Patent 4,220,477 (September 2, 1980).

82. H. Tanzawa et al., U.S. Patent 3,896,061 (July 22, 1975).

6 DENSE MEMBRANES

Dense polymeric membranes, often referred to as polymeric films, constitute the high-polymer-density extreme of the spectrum of variously swollen structures which includes porous membranes and is bounded at the other end by liquid membranes. One consequence of the high polymer density which is characteristic of these types is the scarcity of voids which give rise to structure on the colloidal level.

On the microcrystalline level, the fringed-micellar model of microcrystalline structure has been tempered by the realization that crystalline polymers can sometimes crystallize from the melt and from solution in the form of thin lamellae, on the order of 100 Å thick, in which the molecules are folded back and forth on themselves.[1] This is a result of the fact that the transition from a randomly coiled configuration in the liquid state to an extended-chain configuration in the solid state is a kinetically hindered process.[2] An important consequence of this is that the mechanical and transport properties of the aggregated polymer state may be ascribed more to intramolecular and less to intermolecular cross-linking than was formerly considered to be the case. Thus contrary to classical theory, chain entanglements are not common in the aggregated polymer state (Fig. 6.1).

Dense films are subject to change with environmental conditions. Thus voids, although initially absent, may be induced by swelling and plasticization. An obvious corollary of this statement is that a given membrane may be either dense or porous depending on the nature of its environment. A cellulose film, for example, may be dense insofar as the permeability of water as a vapor is concerned but porous when utilized as a barrier between aqueous or waterlike solutions. Similarly, solutes may in a concentration-dependent manner be capable of evoking colloidal inhomogeneities in the membrane by virtue of strong interactions with the membrane and/or solvent.

In the following sections the natures of dense polymeric membranes prepared by solution, melt, and polymerization methods will be discussed. The special case of thin and ultrathin dense films will be treated in Chapter 7.

 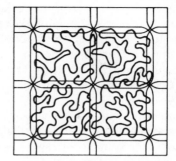

FIGURE 6.1. Classical and individual coil-structure modes of the aggregated polymer state (from Vollmert and Stutz[3]).

6.1 SOLVENT-CAST FILMS

Dense membranes from polymer solutions (cast films) are prepared by dissolution of a polymer substrate in a solvent medium, followed by the application of a liquid film onto a suitable substrate and complete evaporation of the solvent to form a dense film. The nature of both polymer and solvent are important in determining the morphologies of amorphous and semicrystalline films (Chapter 5).

The discussion which follows assumes the utilization of pure solvents and polymers which have been dissolved, filtered, reprecipitated, and dried prior to dissolution in the solution from which the membranes are to be cast. These assumptions are necessary because of the inordinately large effects which small amounts of solvent impurities[3] and/or insoluble gel clumps[4,5] may have upon the mechanical and transport properties of dense polymeric membranes.

In its most general sense a polymer solution may be defined as a dispersion of a polymer in a solvent system. The dispersion may be molecular (consisting of individual polymer molecules) or supermolecular (consisting of molecular aggregates). The nature of the dispersion can vary with polymer type and concentration, molecular weight, temperature, solvent system, and storage time. The solvent system may consist of a single solvent, a mixture of solvents, or a complex system incorporating various concentrations and combinations of solvents, swelling agents, and nonsolvents. Within any polymer solution there are a host of competing forces, referred to as polymer–solvent and polymer–polymer interactions, which tend on the one hand to increase dispersion and on the other to promote aggregation. In addition to changes in dispersion, configurational changes may also occur. Where strong polymer–solvent interactions are operative, molecular dispersion is favored over supermolecular aggregation; where strong polymer–polymer forces predominate, the reverse will be true. These facts are of importance because the disposition of polymer molecules, particularly at the point of incipient gelation, that is, just prior to the transition from sol to gel, is the single most important factor affecting their disposition within the primary gel.

It has long been established that the nature of the solvent mixture from which a dense polymeric membrane is formed has an important influence on the physical, mechanical, and permeability properties obtained.[6] Jones and Miles[7] found, for example, that the tensile strength and elongation of nitrocellulose films varies with the solvent from which they are prepared. Reasoning that the more amorphous the film, the greater the strength developed by stretching should be, they suggested that crystallinity increases in the solvent series methanol < ether–alcohol (2:1) < acetone. Cellulose triacetate can be crystallized into well-defined platelets only from nitromethane solution,[8] whereas polyacrylonitrile yields platelets from propylene carbonate solution and amorphous gels from the stronger solvents dimethylformamide and dimethylacetamide.[9] Solvent power for ethyl cellulose (as indicated by both osmotic and swelling data) increases in the series benzene < chlorobenzene < 2-nitropropane (Table 6.1). Film birefringency, density, and toughness, however, all decrease with increasing solvent power, indicating a corresponding decrease in structural regularity. Because of stronger polymer–polymer interaction, the formation of crystallites is more favored from poor solvents than from good ones.

Zubov et al.[11] utilized electron microscopy to follow the effects of polymer-solvent interactions upon the structure and properties of polystyrene films. Glass plates were coated with solutions of the atactic polymer in xylene, carbon tetrachloride, or a mixture of aromatic hydrocarbons. Both ordered structures and amorphous globules appeared, which varied with the solvent types. Film tensile strength and adhesion varied as a result. Katz and Munk[12] cast films of nitrocellulose, chlorinated rubber, and an alkyd from solvents of varying polarity onto several substrates. The polarity of the solvents exerted a definite influence on the permeability of the film. Kolonils[13] found that the casting of the cellulose acetate films from ethyl acetate resulted in a dense honeycomb, whereas chloroform produced a fine-grained structure. Thus for membranes which are cast from polymer solutions, solution properties are an important factor affecting membrane structure and function.

However, although the nature of the various polymer–solvent and polymer-polymer interactions exerts a great influence upon the structure of such dense membranes as are cast from polymer solutions, it is by no means the sole determining factor. The rate of desolvation and such additional environmental factors as solvent vapor pressure, relative humidity, and the size and concentration of atmospheric

TABLE 6.1   SOLVENT POWER OF SOLVENTS FOR ETHYL CELLULOSE[a]

Solvent	Polymer–Solvent Interaction Parameter μ	Volume Fraction of Polymer[b] after Swelling 48 h at 50°C and 48 h at 25°C
2-Nitropropane	−4.38	0.013
Chlorobenzene	+0.17	0.033
Benzene	+0.30	0.041

[a]From Haas et al.[10]
[b]Cross-linked with hexamethylene diisocyanate

particulate matter can also have pronounced effects upon the structural and transport properties of dense membranes.

The nature of the *desolvation process*[14] can have profound effects upon the permeability of the resultant membranes even insofar as a solution of a given polymer in a given solvent is concerned. In fact desolvation and polymer–solvent-interaction phenomena are closely akin to one another, in that they act to influence the nature of the bonding which occurs during the sol–gel transition.

Consider for a moment the implications of this transition. As solvent molecules depart from solvated polar groups, the latter become free to engage in intramolecular, and to a lesser extent, intermolecular, virtual cross-links. As these cross-links involve more and more sites, a point is reached at which aggregation predominates over disaggregation and a gel is formed. It is apparent that such factors as the configuration of the macromolecules in solution and the number of sites available for cross-linking at any given moment can influence the structure of the gel. In some cases desolvation can be varied to produce effects identical to those produced by variations in solvent systems. In others, however, particular polymer-solvent interactions may result in steric configurations which may not be duplicated by adjustment of desolvation conditions.

The relative humidity of the atmosphere above the desolvating polymer solutions may have a profound influence upon the properties of resultant membranes, particularly in the cases where water is a strong nonsolvent for the polymer. Porosity and permeability increase with increasing relative humidity. Both temperature and solvent volatility affect the rate of desolvation, which in turn influences the probability that polar groups will be in a position to cross-link with other groups on the same or adjacent molecules. Where the rate of desolvation is fast, a zipper effect may obtain, which will favor maximum density and crystallinity, although individual crystallite size may be small, owing to the simultaneous appearance of many nucleation sites and slow rates of growth onto nuclei. Backter and Nerurkar[15] investigated crystallization in films of polyvinyl alcohol prepared by evaporation of aqueous solutions. They found the rate of crystallization became rapid after a prolonged induction period and varied with the rate of evaporation, although ultimate crystallinity appeared to be practically independent of the evaporation rate. Slow desolvation can result in the formation of large and more perfect crystallites particularly when chain mobility is enhanced by the presence of plasticizers or solvent vapors.[16]

Substantial effects upon the permeability and permselectivity of the resultant dense membranes accompany differences in desolvation rate. The permselectivity of dense cellulosic membranes to NaCl was found, for example, to vary by more than an order of magnitude, increasing with the concentration of acetone vapor in the atmosphere above the desolvating solution[17]. This was true in spite of the fact that in both cases desolvation was allowed to proceed to completion. Caution must therefore be exercised in the assignment of diffusion coefficients to dense films. Such coefficients can be expected to vary greatly, depending upon the detailed nature of the fabrication process even for a given solvent. Where comparisons between dense membranes prepared from different polymers in different solvents are

TABLE 6.2 STRESS–STRAIN BEHAVIOR OF ETHYL CELLULOSE FILMS, 25°C, GLASS AND MERCURY CASTS[a]

Film From	Young's Modulus (psi) $\times 10^{-5}$	Ultimate Tensile Strength (psi)	Elongation (%)	Stress at Yield (psi)	Strain at Yield (psi)
80:20 Toluene–alcohol on glass	2.72	6.665	13.2	6.475	5.7
80:20 Toluene–alcohol on mercury	1.94	5.315	36.1	4.875	5.0
Benzene on glass	2.49	7.020	20.6	6.275	5.2
Benzene on mercury	1.87	5.540	50.5	4.510	4.3

[a]From Haas et al.[10]

made, evaluations of the relative permselectivities of the polymeric substrates may be particularly difficult.

Dense membranes can be subjected to certain postformation treatments which serve to modify their structural and performance characteristics. Thermal annealing, particularly at temperatures in excess of the glass transition temperature[18] can be utilized to increase both crystallite size and the extent of crystallinity. Annealing of amorphous films has the effect of diminishing the average interchain displacement. Increasing chain mobility by the inclusion of plasticizers or by subjecting the dense membrane to an atmosphere of solvent vapors can also promote crystallization even at room temperature. The crystallinity of polycarbonate films, for example, is increased by exposure to acetone vapor.[19]

The application of stress, particularly in the presence of plasticizers, has been utilized to increase crystallinity. Thus the crystallinity of poly(ethylene terephthalate), polycarbonate, and cellulose membranes was found to increase when subjected to stress under water amounting to 15% of ultimate tensile strength.[20]

As another example of the influence of stress may be cited the differences between the stress–strain behavior of ethyl cellulose films cast on glass and those cast on mercury (Table 6.2). Because the cast film adheres to the immovable substrate, its planar dimensions are determined at the time of casting. All subsequent shrinkage due to solvent loss is normal to the plane of the film and is similar to a force being applied normal to the sheet, thus forcing the chains into the plane of the film and accounting for the high birefringence of membranes cast on glass (Table 6.3).

TABLE 6.3 CROSS-SECTION BIREFRINGENCE OF ETHYL CELLULOSE FILMS GLASS CASTS[a]

Film From	$\Delta n \times 10^3$
2-Nitropropane	−5.75
Chlorobenzene	−6.65
Benzene	−7.40

[a]From Haas et al.[10]

The movable mercury surface, on the other hand, permits planar shrinkage and yields isotropic films which, although weaker, exhibit greater elongation at break.

The morphology of dense membranes in the glassy state is still poorly understood. For several cellulosic films the latest evidence from electron microscopy is consistent with a dense structure consisting of hemispherical subunits linked together in an irregular closely packed array.[21]

These nodular subunits were first observed by Schoon and Kretschmar,[21] who found them too small to accommodate an entire molecule and suggested that a single polymer chain forms a series of units comparable to a string of pearls. Yeh and Geil[18] observed similar structures, called *crystalline nodules* by Keith[22], in poly(ethylene terephthalate) which were about 75 Å in diameter and spaced with an average distance between centers of 125 Å. These nodules were credited with some paracrystalline order. When such membranes are annealed at temperatures close to the glass transition temperature (65°C), the nodules move relative to one another and aggregate into clusters about 5 to 10 nodules in diameter. At this point electron and X-ray diffraction indicate the presence of crystallinity. With continued heating the first recognizable indications of spherulitic growth appear. The nodules seem to align themselves in fibers in rows transverse to the fiber axis. Annealing at 154°C gives rise to spherulites composed of lamellae. Cold-drawing "amorphous" films (amorphous in the sense of being clear and not exhibiting crystallinity by X-ray diffraction) causes alignment of the nodules. Heating near the melting point causes the alignment to become better defined and the nodules to increase appreciably in size.

Sjöstrand[23] observed strikingly similar modes of random subunit packing in dense collodion membranes, and similar hemispheres have been observed in thin cellulose acetate dense films and in the skin layer of asymmetric membranes by Schultz and Asunmaa[24] (Fig. 6.2). The latter are now believed to represent densi-

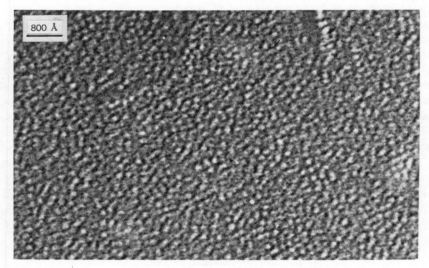

FIGURE 6.2. Electron photomicrograph of Pt–C preshadowed carbon replica of the surface of the active desalination layer of a Loeb–Sourirajan cellulose acetate membrane (from Schultz and Asunamaa[24]).

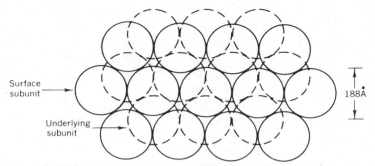

Surface subunit →

Underlying subunit →

188Å

FIGURE 6.3. Active desalination layer of cellulose acetate membrane idealized as an assembly of close-packed 188-Å diam spheres (from Schultz and Asunmaa[24]; © 1966).

fied micelles 180–190 Å in diameter (Fig. 6.3). Because of their small size and apparently amorphous nature they are not detected by X-ray diffraction.

6.2 MELT-EXTRUDED FILMS

Thermoplastic polymers can be melted and extruded through a die to produce molten structures which, after cooling and solidification, may be arbitrarily called either films or dense polymeric membranes, depending as much upon the intended application as upon any structural differences implied by the two terms. The structure of such membranes depends upon the nature of the polymer (intramolecular forces, chain rigidity, molecular weight, and branching) and, for a given polymer, upon the various kinetic factors involved in quenching and annealing.

Any discussion of dense membranes from polymer melts should begin with a description of the melts themselves. As a bulk polymer is heated, energy is added to the system which enables first groups and eventually both smaller and large chain segments to move. Since any liquid, including a polymer melt, can be viewed as a collection of holes moving about in matter, melting can be considered as the process of the occupation of holes which occurs when sufficient energy is present to overcome the attractive forces between the polymer molecules or portions thereof. As a hole is occupied, another is formed by default from the space which was vacated by the jump of the moving polymer chain into its new position. Because the amount of energy required to initiate flow levels off at molecular weights below that of the polymer itself, it is apparent that as the chain length increases, the molecules begin to move in segments rather than as a unit. For the case of polyethylene, for example, motion involves chain segments containing approximately 20–25 carbon atoms.[25] The factors controlling the melt viscosity, and hence the structure of dense polymeric membranes prepared from the melt, are[26]

$$\eta = \frac{f(M, \text{molecular architecture})}{j(V_f T)}$$

where η = melt viscosity,
     f = statistical factor,
     j = segmental jump frequency,
     M = molecular weight,
     V_f = free volume, and
     T = temperature.

The statistical factor f is related to the fact that, since the segments are joined by primary bonds, there must be a coordination of the movements of the individual segments before the molecule as a whole can progress under the action of a shearing force. The free volume V_f is the difference between the measured volume and that actually occupied by the polymer molecules. The temperature T controls the energy available to each segment and the number of holes in the melt.

As the incipient membrane emerges from the die, it solidifies in a manner depending upon the crystallization conditions (Fig. 6.4). As the melt becomes supercooled by 10°C or more, heterogeneous[27] nuclei initiate the growth of crystallites. Soon after growth has begun, the spherulites take on globular form and continue to expand in size until they impinge on the other spherulites, which then deform into polyhedra. The spherulites are polycrystalline and consist of fibrous crystals radiating outward, roughly from a common center. The fibers (the term denotes any crystal longer in one dimension than in either of the others) are ribbonlike lamellar crystals composed of molecules folded in much the same ways as in polymer single crystals grown from dilute solution. Crystallization takes place in two stages: the first results in the formation of spherulites composed of chain-folded lamellar crystals, and the second in a number of forms which grow in the interstices within the spherulites. Crystals of these various kinds are connected by tie molecules and by more substantial intercrystalline links, and between them there are disordered regions arising both from irregular folds at crystal surfaces and from polymer whose crystallization has been impeded by entanglements. On the molecular level, as chain mobility decreases, segments of polymer molecules begin to make intramolecular contact by chain folding. (Intermolecular contact with the formation of extended crystallites is kinetically hindered.) Where molecular regularity is of a high enough order, chain folding results in the formation of very small crystalline lamellae comparable to those encountered in the preparation of dense membranes from solution. If the molten polymer is quench crystallized, very small crystallites with extremely rough chain-folded surfaces are formed, together with row vacancies and some short chains in crystalites and numerous amorphous linking chains (Fig. 6.4a). If the primary (as quenched) membrane is annealed, crystallite size increases and interlamellar links become strained. In addition some low-molecular-weight chains may depart from the lamellar structure to yield an extended-chain phase (Fig. 6.4b). As the membrane is isothermally crystallized near its melting temperature, the lamellae tend to line up into spherulites; molecular fractionation to form an extended-chain phase may occur (Fig. 6.4c). Molecules longer than a certain length form folded-chain crystals, whereas those shorter than this critical length form extended-chain crystals. Anderson[28] showed that the mechanical properties of a given

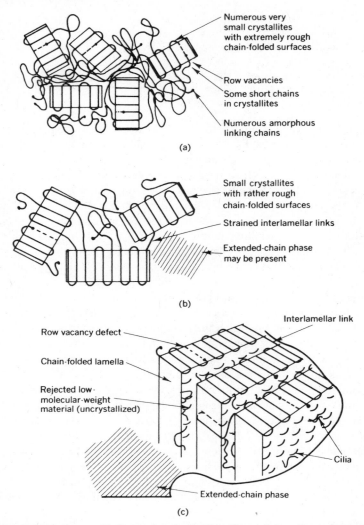

Numerous very
small crystallites
with extremely rough
chain-folded surfaces

Row vacancies

Some short chains
in crystallites

Numerous amorphous
linking chains

(a)

Small crystallites
with rather rough
chain-folded surfaces

Strained interlamellar links

Extended-chain phase
may be present

(b)

Interlamellar link

Row vacancy defect

Chain-folded lamella

Rejected low-
molecular-weight
material (uncrystallized)

Cilia

Extended-chain phase

(c)

FIGURE 6.4. Ultrastructure of bulk samples of polychlorotifluoroethylene linear chain polymer prepared under different crystallization conditions: (*a*) quench crystallized, (*b*) quench crystallized and annealed, and (*c*) isothermally crystallized near T_M (from Hoffman et al.[27]; © 1965).

dense membrane will be sensitive to the relative amounts of the various lamellar habits present. The presence of a large number of extended crystallites permits cleavage in a large number of directions without affecting molecular bonds, whereas samples containing the more common folded chains can undergo considerable deformation before cleavage.

Microbeam X-ray studies of partially spherulitic poly(ethylene terphthalate) have revealed that the spherulites are crystalline but the spaces between them are not.[29]

Several mechanisms have been suggested to account for the growth of spherulites. Schurr[30] favors an *auto-orientation*, in which a crystal nucleus formed initially tends to orient the surrounding amorphous chains perpendicularly, after which these chains crystallize and tend to orient their surrounding amorphous chains, and so on. Bryant et al.[31] state that fine crystalline fibrils grow out from a primary nucleus. Fine strands of crystalline order proceed outward along the fringes of the nucleus, thereby seeding further crystallizable domains and leading to a statistically random structure which is spherically symmetrical only in appearance. Keith and Padden[32] hold that fine crystalline fibrils grow outward into the melt as more or less discrete entities but in a densely packed array. The fact that the polymer within the interstices of the spherulites does not crystallize at the same rate as the spherulites themselves may indicate that fractionation is occurring; in other words, molecular species of varying order may be present.

The effect of molecular architecture upon crystallinity is exemplified by reference to polyethylene, a polymer whose degree of crystallinity varies with the number of branches it contains. High-pressure polyethylene (35–70% crystalline) contains on the average two branches per 100 carbon atoms, with extremes between 1 to 8. Low-pressure polyethylene (60–90% crystalline) can be completely linear but frequently contains 0.1–0.5 short branch per 100 carbon atoms. Crystallinity is determined by the number of branches on the chain (Fig. 6.5) because disruption of crystallinity occurs around branch sites. In addition to branching, the rate of quenching can significantly vary the degree of crystallinity, particularly for viscous high-molecular-weight samples.

Although the exact relationship between the various crystalline habits and

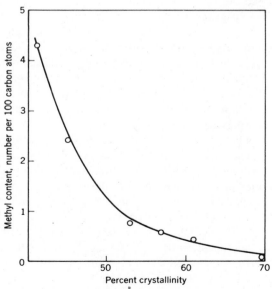

FIGURE 6.5. Crystallinity versus extent of branching in polyethylene (from Faucher and Reding[33]).

permeability is only beginning to be understood, it is generally true that crystallites restrict the motion of the chains in the amorphous regions (crystallite dislocations). This restriction will become greater as the volume fraction of the crystalline phase increases. In some cases orientation does not significantly change permeability until it induces crystallization.[34] The difference between the water-vapor permeability of oriented and unoriented poly(ethylene terephthalate) films of the same crystallinity is maximal at low crystallinity (10–15%) and becomes gradually less as the degree of crystallinity increases, until at 40–50% crystallinity no significant differences are discernible.

The ease and/or speed with which a dense membrane can be prepared from a polymer melt is inversely related to melt viscosity. Melt viscosity can be lowered by increasing the temperature, increasing the shear rate, decreasing the average molecular weight, and broadening the molecular-weight distribution. However, each of these methods for lowering the viscosity is in some way detrimental to the properties of the finished membrane. Increasing the temperature increases degradation. Decreasing the molecular weight and broadening its distribution can lead to the inclusion of extended-chain crystallites with their detrimental effects upon physical properties.

6.3 DENSE MEMBRANES FORMED DURING POLYMERIZATION

Whenever polymerization is attended by simultaneous cross-linking and consequent intractability of the resultant polymer, membranes of such polymers must necessarily be formed during polymerization. The most important class of membranes which fits into this category is the homogeneous ion-exchange type (Section 4.3). Intractability may be induced by means of cross-linking effected by various chain-transfer and coupling reactions. Although the majority of such types have not been generally considered in terms of any microcrystalline structure, it should be pointed out that they may differ from those formed from solutions or melts in that the former can exhibit all three of the limiting conformations of linear high polymers whereas the latter are usually excluded from participation in crystallites grown from extended polymer chains (Fig. 4.10).

The reason for this is that the randomly coiled molecular configuration (both in solution and in the melt) is for steric reasons kinetically hindered from uncoiling into the extended configuration prior to crystallization. In lieu of extending themselves, such molecules undergo chain folding into metastable crystals with small dimensions in the molecular-chain direction. Once crystallization has occurred, molecular extension and realignment are very difficult to effect. No such kinetic hindrance to forming extended-chain crystals exists for crystallization during polymerization unless long-chain portions are produced prior to the onset of crystallization. Well-crystallized extended-chain crystals exhibit densities close to those calculated from the unit cells derived from their X-ray diffractograms.

Crystallite formation can occur simultaneously with polymerization, or it can follow polymerization either before or after the polymerization is complete. The latter situation is comparable to ordinary crystallization from the melt except in those cases in which the macromolecules are formed in a special conformation and crystallization begins before randomization.

The preparation of amorphous structures is favored by rapid polymerization below the melting or dissolution temperature of the polymer, where chemical reaction rates are faster than crystallite nucleation rates. Interfacial condensations of the Schotten–Baumann type, in which a dibasic acid chloride reacts with a diamine or a diol, are typical of the reactions yielding amorphous structures. Such is the case for thin-film composite membranes of the FT-30 type formed from *m*-phenylenediamine and trimesoyl chloride.[35] Crystallization is also hindered in such cases because noncrystalline chain conformations are fixed by the formation of hydrogen bonds which constitute virtual cross-links. Crystallization does occur, however, when these bonds are ruptured by washing with hydrogen-bonding solvents such as water or alcohol.

Since less perfect crystals result when the polymerization proceeds faster than the crystallization, the most nearly perfect crystals are to be expected under crystallization conditions which are close to equilibrium. Lowering the polymerization temperature lowers crystallinity. Thus polyethylene with low crystallinity can be produced from diazomethane by polymerization and crystallization at $-50°C$, whereas at $20°C$ a highly crystalline product is formed.[36]

Crystallization during the polymerization of condensation (step-reaction) polymers begins when larger concentrations of oligomers nucleate the crystalline polymer phase. Once crystallization has begun, a change to chain-reaction polymerization occurs, so that the polymer chains continue to react with additional monomer at both ends. Vinyl monomers such as polytetrafluoroethylene are believed to form intramolecularly separate crystalline particles during the early stage of polymerization and crystallization, which during the later stages add to already existing "dead" crystalline nuclei.

REFERENCES

1. P. Geil, *Polymer Single Crystals*, Interscience, New York, 1963.
2. B. Wunderlich, *Adv. High Polym.Res.*, **5**(4), 568 (1967/1968).
3. B. Vollmert and H. Stutz, *Angew. Chem. Int. Ed. (England)*, **8**(5), 389 (1969).
4. H. Gregor and K. Sollner, *J. Phys. Chem.*, **50**, 53 (1946).
5. J. Neil, *J. Appl. Poly. Sci.*, **9**, 94 (1965).
6. A. Keller, *Phil. Mag.*, **2**, 1171 (1957).
7. G. Jones and J. Miles, *J. Soc. Chem. Ind.*, **52**, T251 (1933).
8. R. St. John Manley, *J. Polym. Sci.*, **47**, 149 (1960); *Nature*, **189**, 390 (1961).
9. J. Bischops, *J. Polym. Sci.*, **17**, 89 (1955).
10. H. Haas, L. Farney, and C. Valle Jr., *J. Colloid Sci.*, **7**, 584 (1952).
11. P. Zubov, V. Voronkov, and L. Sukhareva, *Vysokomolekul, Svedin*, **B10**(2), 92 (1968) from *Chem. Abstr.*, **68**, 96306g (1968).
12. R. Katz and B. Munk, *J. Oil Colour Chemists' Assoc.*, **52**(5), 418 (1969).

13.   V. Kolonils, *Kolloid Z. Z. Polym.*, **226**(1), 40 (1968).

14.   K. Maier and E. Scheuermann, *Kolloid Z.*, **171**, 122 (1960).

15.   A. Backter and M. Nerurkar, *Eur. Polym. J.*, **4**(6), 685 (1968).

16.   L. Boltenbruch, H. Schnell, and A. Prietschk, Belgian Patent 589,858 (1960).

17.   C. Saltonstall and W. King, personal communication.

18.   G. Yeh and P. Geil, *J. Macromol. Sci.*, **B1**, 235, 251 (1967).

19.   H. Schnell, *Chemistry and Physics of Polycarbonates*, Interscience, New York, 1964, p. 130.

20.   S. Kazakevich et al., *Fiz.-Khim. Mekh. Makm.*, (in Russian) **4**(5), (1968).

21.   T. Schoon and R. Kretschmar, *Kolloid Z. Z. Polym.*, **211**, 53 (1965).

22.   H.Keith, *Kolloid Z. Z. Polym.*, **231**, 430 (1969).

23.   F. Sjöstrand, cited in T. Kavenau, *Structure and Function in Biological Membranes*, p. 633, Holden-Day, San Francisco, 1965.

24.   R. Schultz and S. Asunmaa, *Recent Progr. Surface Sci.*, **3**, 291 (1970).

25.   W. Kauzmann and H. Eyring, *J. Am. Chem. Soc.*, **62**, 313 (1940).

26.   R. Raff and K. Doak, eds. *Crystalline Olefin Polymers*, Pt. 1, Interscience, New York, 1965.

27.   J. Hoffman, G. Williams, and E. Passaglia, *J. Polym. Sci.*, **C14**, 173 (1966).

28.   F. Anderson, *J. Polym. Sci.*, **C1**, 123 (1963); *J. Appl. Phys.*, **35**, 64 (1964).

29.   A. Keller, *Nature*, **169**, 913 (1952); *J. Polym. Sci.*, **17**, 291 (1955).

30.   G. Schurr, *J. Polym Sci.*, **11**, 385 (1953); **50**, 191 (1961).

31.   W. Bryant et al., *J. Polym. Sci.*, **16**, 131 (1955).

32.   H. Keith and F. Padden, *J. Polym. Sci.*, **39**, 123 (1959).

33.   J. Faucher and F. Reding, in Reference 26, p. 677.

34.   S. Lasosky and W. Cobbs, *J.Polym. Sci.*, **39**, 123 (1959).

35.   J. Cadotte, U.S. Patent 4,277,344 (1982).

36.   J. Nagill, S. Pollack, and D. Wymann, *J. Polym. Sci.*, **A3**, 3781 (1965).

7 PHASE-INVERSION MEMBRANES

Phase inversion refers to the process by which a polymer solution (in which the solvent system is the continuous phase) *inverts* into a swollen three-dimensional macromolecular network or gel (where the polymer is the continuous phase). In thin-film form designed for use as a barrier layer, such a gel constitutes a *phase-inversion membrane*.

7.1 MECHANISM OF PHASE INVERSION

Phase inversion either begins with a molecularly homogeneous single-phase solution (Sol 1) which at some point prior to gelation undergoes a transition into a heterogeneous solution of molecular aggregates consisting of two interdispersed liquid phase (Sol 2), or it begins directly with Sol 2.

In other words, there are two possible reaction sequences for phase inversion:

(1) Sol 1 $\rightharpoonup$ Sol 2 $\rightharpoonup$ Gel or
(2)   Sol 2 $\rightharpoonup$ Gel

The essence of phase inversion is the appearance in a polymer solution of two interdispersed liquid phases followed by gelation (Fig. 7.1).

The micellar structure which exists in the primary gel, that is, the gel which exists immediately following the Sol 2 $\rightharpoonup$ gel transition, differs only infinitesimally from that of Sol 2 just prior to gelation. Therefore, since the structures of Sol 2 and of the primary gel are virtually identical, Sol 2 is conceded structural as well as temporal primacy over the gel.[2] In other words, the structure and function of the final phase-inversion membrane is primarily controlled by adjustments to the Sol 2 structure and only secondarily by modification of the primary gel once the latter has formed.

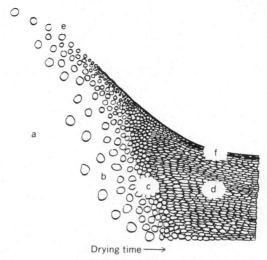

Drying time ⟶

FIGURE 7.1. Mechanism of formation of phase-inversion membranes: (*a*) Sol 1; (*b*) Sol 2; (*c*) primary gel; (*d*) secondary gel; (*e*) air–solution interface; (*f*) skin (adapted from Kesting[1]).

The dispersed phase of Sol 2 consists of spherical droplets or micelles which are coated with polymer molecules. The composition in the interior of the micelles and in the continuous phase will differ from case to case and depends upon the nature of whatever variation of the phase-inversion process is being employed. The reader may find it helpful at this junction to consider the phenomenological model originally developed by Cahn[3] to describe the two-phase structures of metal alloys and more recently also used in conjunction with polymer blends. This model explains the appearance of isotropic interdispersed domains in terms of the decomposition of the *spinodal*, that is, the metastable region of the polymer volume fraction versus temperature curve and yields some insight into the reasons why *uphill* diffusion, that is, diffusion against the concentration gradient, occurs in phase-inversion solutions. Sol 2 is present when some factor either promotes separation into two phases from one phase and/or prevents two phases from recombining into a single phase. It is expedient to entitle this factor *incompatibility*, and to discuss the subdivisions of the phase-inversion process in terms of the various reasons for incompatibility. In the sections to follow, four phase-inversion processes are discussed: the *dry* process, the *wet* process, the *thermal* process, and the *polymer-assisted* process.

7.2 THE DRY PROCESS

The *dry* or *complete evaporation* process is the oldest and easiest to interpret of the phase-inversion processes. It can be followed by reference to a typical cellulose nitrate (CN) casting solution (Fig. 7.2*a*). Final membrane thickness is only a frac-

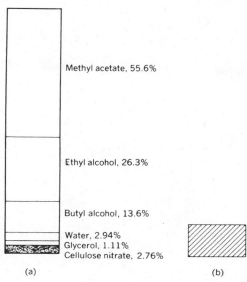

Methyl acetate, 55.6%

Ethyl alcohol, 26.3%

Butyl alcohol, 13.6%

Water, 2.94%
Glycerol, 1.11%
Cellulose nitrate, 2.76%

(a)   (b)

FIGURE 7.2. (*a*) Solution components and (*b*) resultant membrane with 85% void volume depicted as layers whose thickness corresponds to their concentrations (from Maier and Scheuermann[4]).

tion of the as-cast thickness owing to solvent loss and the resultant increase in the concentration of polymer per unit volume. However, because of the inclusion of voids, it is substantially greater than the thickness of a dense membrane containing an equivalent amount of polymer (Fig. 7.2*b*). The weight and loss of depth as a function of evaporation time of another solution are shown in Table 7.1 and Figure 7.3. This system can be used to exemplify the various macroscopically observable stages involved in the formation of membranes by the dry process.

1. Loss of volatile solvents and the inversion of a clear one-phase solution into a turbid two-phase (Sol 2) solution. Alternatively, the solution may be a turbid Sol 2 type to begin with. Ease of processing and reproducibility are enhanced if the solution begins with a Sol 1 or at least a Sol 2 which is somewhat removed from the point of incipient gelation. Thus in most cases it is desirable to formulate a Sol 1 which does not invert into a Sol 2 until some time after it has been cast.

2. Gelation. This is accompanied by a diminution in the reflectivity of the cast solution.

3. Contraction of the gel with or without syneresis. In the case of skinless membranes, syneresis causes expressed liquid to appear at the air–solution interface which can be at both surfaces if the membrane is cast on a porous support. In the case of membranes which are skinned at the air/solution interface, syneresis occurs downward into the porous support or, where no such support exists, it does not occur at all. In such a case drying can be a slow process since it will require the diffusion of vapor rather than liquid through what may be a relatively impervious skin layer.

TABLE 7.1 DECREASE IN CASTING–SOLUTION WEIGHT AND THICKNESS WITH DRYING TIME[a,b]

Evaporation Time (min)	Solution Weight (g)	Thickness of Nascent Membrane (μm)
0	10.5	650
0.40	9.5	—
0.83	9.0	500
1.58	8.5	—
2.08	8.0	450
2.8	7.5	350[c]
4.0	7.0	300
5.16	6.5	280
6.67	6.0	250
8.25	5.5	220
10.50	5.0	200
13.16	4.0	170
20.33	3.5	155
24.16	3.0	150
31.0	2.5	135
35.5	2.0	—
43.0	1.75	125
47	1.50	—
54	1.25	115
74	1.25	115
130	0.99	—
900	0.82	100

[a]From Maier and Scheuermann.[4]
[b]Evaporation at 21 ± 1°C in a 62 ± 2% relative humidity environment. Original casting solution: 5% cellulose nitrate, 54.2% methyl acetate, 23.7% ethyl alcohol, 12.3% butyl alcohol, 3.3% water, and 1.5% glycerol.
[c]From this point on values refer to the thickness of the gel exclusive of the layer of expelled liquid.

4. Capillary depletion. Here the largely nonsolvent liquid encompassed by the gel departs leaving behind empty capillaries. As this happens the membrane opacifies, usually with the formation of beautiful snowflake patterns that gradually fill in until the entire membrane becomes opaque. The reason for opacification is light scattering by the micrometer-sized empty voids. On the other hand, those membranes which contain voids that are less than 0.5 μm in diameter can be opalescent or clear. Subtle differences in void size can sometimes also be discerned from the turbidities of the dry gels once they have been wet by water.

5. Loss of residual nonsolvent (final drying). Depending on such factors as the volatility and concentration of the residual liquids left in the membrane after takeup, the amount of membrane on the roll, and storage temperature, final drying can vary between 2 weeks and 6 months. It is also possible to take up the membrane in an essentially dry condition by passing it over heated rollers. In either case, no

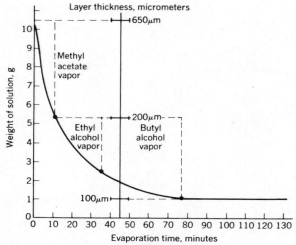

FIGURE 7.3. Decrease in casting solution weight and thickness with evaporation time (from Maier and Scheuermann[4]).

membrane should be handled unless it is completely dry because it is subject to shrinkage and warpage while in a plasticized condition.

The sequence of events on the colloidal level which corresponds to the five macroscopically observable stages has been deduced both from the nature of the gel network in the finished membrane[1-5] and from the ghosts of the nascent membrane, that is, the frozen and lyophilized nonvolatile remnants of the membrane in its various formative phases[6] (Fig. 7.4). The polyhedral cell structure of the final membrane gel is considered to be an immobilized and flattened version of the sol precursors which exist in the solution immediately prior to the sol → gel transition. As the loss of volatile solvent progresses, the solvent power of the solution, that is, its capacity for retaining the polymer in a Sol 1 decreases. If only polymer and solvent are present then at least three situations are possible.

1. Separation into two liquid phases may not occur prior to gelation. This would be the case if polymer and solvent are infinitely miscible. Even after gelation the solvent will continue to act as a plasticizer, which when combined with the effect of gravity can lead to a collapse and densification of the gel, resulting ultimately in a dense film.

2. Phase separation may occur prior to gelation if there is only limited solubility of the polymer in the solvent. However, even in this case residual solvent can act as a plasticizer and dense or nearly dense (low porosity) films may result.

3. In those cases in which P–P interaction is unusually strong, such as, for example, in the evaporation of solutions of nylon 6,6 in 90% formic acid,[7] gelation will occur with the formation of strong virtual (perhaps crystalline) cross-links. Such a gel can overcome the combined effects of plasticization and gravity so that

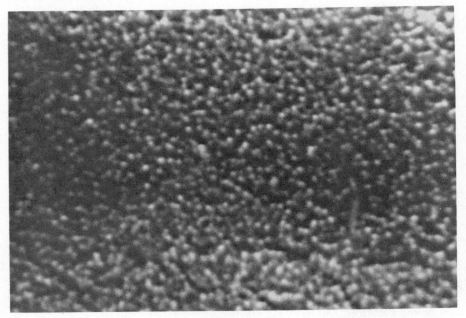

FIGURE 7.4. SEM photomicrograph of lyophilized micelles in the nascent skin of a dry cellulose acetate hyperfiltration membrane (from Kesting[6]); © 1973).

its porosity is maintained throughout complete evaporation. In the lattermost case, after phase inversion has occurred and prior to gelation, the sol structures exhibit long-range order. Virtually any disruption of this order or *nucleation* in the sol by, for example, rapid agitation or even fine filtration will result in a final membrane with larger pores sizes than would have otherwise resulted from gelation of the undisturbed sol.

Both the interior of the micelle and the continuous phase of a two-component system consist of polymer-poor regions whereas the micellar wall consists of polymer-rich regions (Fig. 7.5). In the latter, P–P interaction predominates over P–S interactions. Most dry-process casting solutions, however, consist of three or more components: polymer, volatile solvent, and pore former(s) from the nonsolvent side of the polymer–solvent interaction spectrum (Chapter 5). The nonsolvent should be substantially less volatile than the solvent—a practical rule of thumb is a 30–40°C minimum difference in boiling points between the two. Although Sol 1 is homogeneous on the colloidal level (Fig. 7.6*a*), compatibility decreases as evaporation proceeds. Eventually, the solvent power of the remaining solvent system is insufficient to maintain Sol 1, and inversion into Sol 2 occurs (Fig. 7.6*b*). Most of the polymer molecules distribute themselves about the micelles which have been formed, so that relatively few (perhaps 0.5%) are left dispersed in the liquid matrix containing the micelles. The interior of the micelle in this case consists of a liquid with a high concentration of the nonsolvent components of the casting solution. In the typical dry process the fundamental reason for incompatibility which leads to

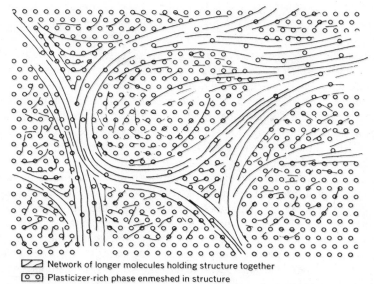

☐ Network of longer molecules holding structure together
☐ Plasticizer-rich phase enmeshed in structure

FIGURE 7.5. Theoretical structure of a plastic containing a nonsolvent plasticizer (from Spurlin, et al.[8]); © 1946).

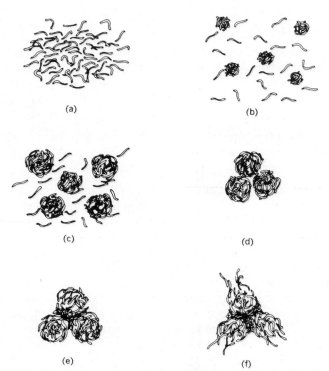

(a)

(b)

(c)

(d)

(e)

(f)

FIGURE 7.6. Membrane formation process (adapted from Maier and Scheuermann[4]).

phase inversion, gelation, and the maintenance of gel porosity in spite of forces which act to collapse the gel, is the presence of nonsolvent in the casting solution and/or strong P–P interaction forces. In other words, in the dry process incompatibility is an *internal* characteristic of the system. Inasmuch as solvent loss continues after phase inversion, the spherical micelles approach one another (Fig. 7.6c), eventually making contact in the initial phase of gelation (Fig. 7.6d). As the gel network contracts, the micelles deform into polyhedra and the polymer molecules diffuse into the walls of neighboring micelles causing an intermingling of polymer molecules at the interface (Fig. 7.6e). Finally, if the walls are sufficiently thin, for example, when a high initial concentration of components other than polymer and solvent causes the formation of numerous micelles with a large total surface area, contraction causes a tearing of the walls which then retract and form the hoselike skeleton which constitutes the gel network (Fig. 7.6f). A similar phenomenon occurs during the bursting of soap bubbles[9] and the formation of open-celled polyurethane foams. It can happen, however, that micelles are covered with such a thick coating of polymer that rupturing of cell walls is somewhat hindered or entirely inhibited. In such a case either mixed open-cell and closed-cell or closed-cell structures result. The principal factors which determine the porosity and pore-size characteristics of dry-process membranes are:

1. Polymer volume concentration in Sol 2 which is inversely proportional to gel porosity.
2. Ratio of nonsolvent volume/polymer volume in Sol 2 which is directly proportional to gel porosity.
3. Difference in boiling points between solvent(s) and nonsolvent(s) which is proportional to porosity and pore size.
4. Relative humidity which is proportional to porosity and pore size.
5. The presence of more than one polymer with less than perfect compatibility increases porosity.
6. The presence of a high-MW polymer tends to increase porosity because high-MW polymers tend to be less compatible and thus gel at an earlier stage.

Because dry-process solutions employ nonsolvent pore formers, the capacity of the solvent system to tolerate high concentrations of polymer is severely limited. In spite of this the casting solution must be sufficiently viscous to permit its handling during flat-sheet and tubular casting or hollow-fiber spinning operations. This dilemma is resolved by utilizing high-MW polymers which although slightly less soluble than their low-MW counterparts do contribute significantly more to solution viscosity. However, most engineering plastics are available only in the low- to intermediate-MW ranges because they are designed for melt processing applications such as injection molding. This can and often does limit the application of the dry process. Methods for circumventing this obstacle include the preparation of special grades of high-MW polymers,[10] the utilization of viscosity enhancers

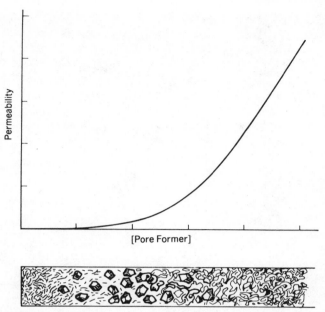

FIGURE 7.7. Relationship between cell type, skin thickness, concentration of pore former, and permeability in phase-inversion membranes.

such as a second polymer[11, 12] or finely divided colloidal silica, and casting at low temperatures.

A summary of the effects upon cell structure and porosity, and hence on the permeability of dry-process membranes which can result from variations in polymer and/or nonsolvent concentrations in the casting solution, is schematically depicted in Figure 7.7. In the absence of any nonsolvent pore former or strong P–P interaction, phase inversion does not take place (Sol 1 → gel) so that a dense high-resistance membrane or film is formed. Such a structure as a first approximation consists of a single dense skin layer. With low concentrations of nonsolvent, membranes possessing closed cells, low porosity, and substantial resistance to material transport are encountered (Fig. 7.8). However, the thickness of the dense skin layer is substantially diminished. At intermediate concentrations of nonsolvent a mixture of open and closed cells is formed (Fig. 7.9). The dense skin layer has thinned considerably and a thin transition layer is discernible which consists of closed cells which are intermediate in polymer density between those of the dense skin layer and those of the porous open-called substructure which is found in the bulk of the membrane. Permeability is low but measurable at this point. At high concentrations of nonsolvent a bilayered structure is found which is comprised of a thin skin and a porous substructure that consists entirely of open cells (Fig. 7.10). There is a break in the curve of permeability versus porosity at that concentration at which mixed open and closed cells give way to open cells. As the concentration of non-

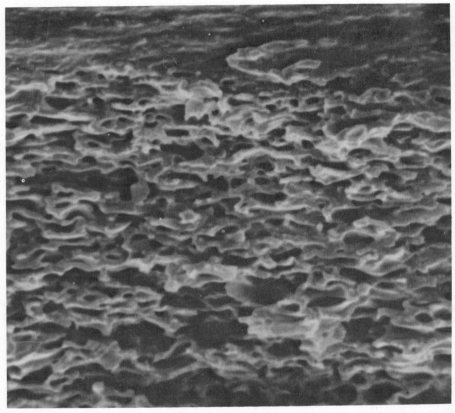

FIGURE 7.8. SEM photomicrograph of a cross section of a dry cellulose acetate membrane with closed cells.

solvent is increased beyond this point, skin thickness decreases and permeability increases. Eventually, however, the skin becomes sufficiently thin that its integrity is breached in places and the porous substructure becomes visible (Fig. 7.11). Ultimately, at very high concentrations of nonsolvent, the dense skin layer is absent altogether and both surface and interior regions consist only of open cells with tears in the walls, that is, the porous surfaces of microfiltration membranes are encountered. As nonsolvent concentration is increased even beyond this point, cell and pore size increase further as does permeability. Eventually, even the integrity of the porous substructure cannot be maintained (Fig. 7.12). If structures with pore sizes greater than about 5 μm are desired, processes other than phase inversion are employed.

The interrelationship between Δ bp (bp nonsolvent − bp solvent) of the two solvents acetone (bp 56°C) and dioxolane (bp 75°C) and the levels of a single nonsolvent isobutanol (bp 110°C) required to produce equivalent skinned mem-

FIGURE 7.9. SEM photomicrograph of a cross section of a dry cellulose acetate membrane with a transition layer.   intermediate

branes (as deduced from their equivalency in permeability and permselectivity) is shown in Table 7.2

Both solvent and nonsolvent evaporate simultaneously. Therefore, if a critical solvent–nonsolvent ratio must be reached before gelation occurs, a less volatile solvent requires a higher initial concentration of a given nonsolvent to reach this ratio at a given porosity than will a more volatile solvent. In like manner, the concentration of nonsolvent in the casting solution required to achieve a given porosity is inversely related to its volatility.

Bjerrum and Manegold[13] were among the first to observe the influence of the composition of the atmosphere above the desolvating solution upon membrane structure and function. The presence of a high concentration of solvent vapor retards gelation, whereas high temperatures and air velocities will hasten it. *Skinning* is enhanced by high air-flow rates and high polymer concentrations. The effect of atmospheric moisture is to hasten gelation which in turn acts to increase average pore size and permeability (Table 7.3). The inclusion of water in the casting solution has a very pronounced effect in those cases in which water plays a nonsolvent role

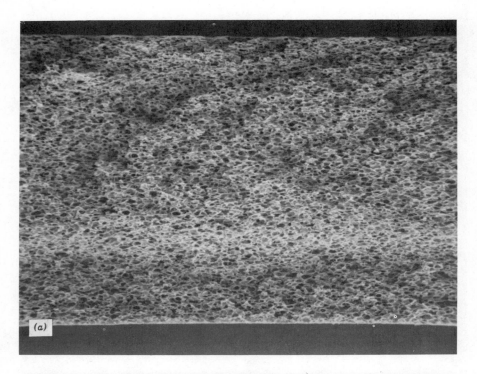

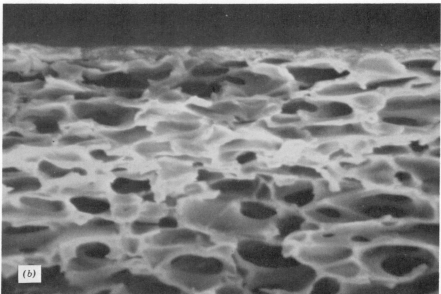

FIGURE 7.10. SEM photomicrograph of a cross section of a dry cellulose acetate membrane with open cells: (*a*) entire cross section; (*b*) partial cross section at skin surface.

high

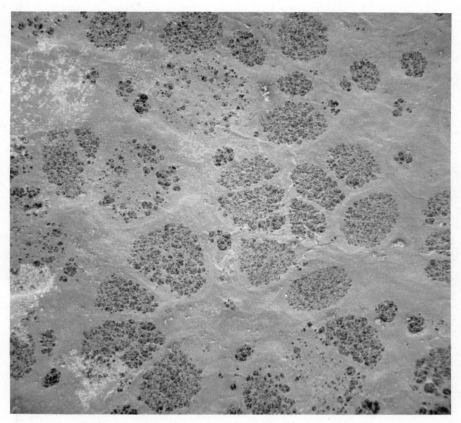

FIGURE 7.11. SEM photomicrograph of the top surface of a dry cellulose acetate membrane with a nonintegral skin. *too much*

(Table 7.4). In hydrophobic sols water acts both to hasten gelation and to increase the size of voids in the gel structure. This is attributable to two factors: (1) a high degree of incompatibility with the solvated polymer component of the casting solution; and (2) high surface tension. Both factors act to cause water to separate from the remainder of the solution and nucleate comparatively large micelles which then result in coarse microgels. The presence of a microgel structure in membranes from polymers such as the cellulosics and the polyamides which possess some affinity for water confers the important property of wet–dry reversibility on these membranes.[6] This is so because the magnitude of the capillary forces which come into play upon drying depends on the internal surface area of the membrane which in turn depends on the cell size. Microgel membranes possess large (1–10 μm diameter) cells which means that such membranes have a relatively small internal surface area and as a result will not lose porosity during drying. Ultragel membranes, on the other hand, have small (0.5–0.5 μm diameter) cells and consequently possess a larger internal surface area. Ultragels are therefore more likely to lose

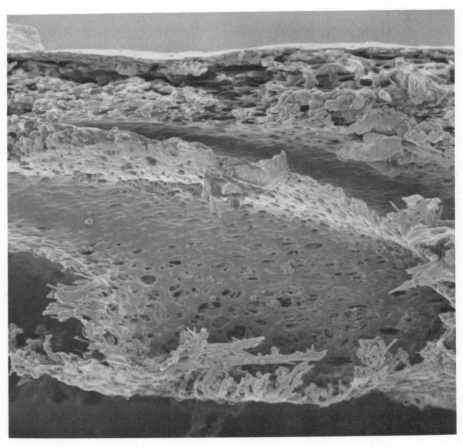

FIGURE 7.12. SEM photomicrograph of a cross section of a dry cellulose acetate membrane with a nonintegral substructure (from Kesting et al.[14]; © 1965).

TABLE 7.2 EQUIVALENT NONSOLVENT CONCENTRATIONS IN ACETONE AND DIOXOLANE SOLUTIONS[a] FOR DRY-RO BLEND MEMBRANES OF CA AND THE TMA SALT OF CA 11-BROMOUNDECANOATE

Solvent	Boiling Point (°C)	Δbp (°C)	Nonsolvent Concentration (g IBA/formulation)	Permeability[a] (gal/ft^2 day)	Salt[b] Rejection (%)
Acetone	56	54	38	5.6	97.9
Dioxolane	75	35	54	5.5	97.8

[a]Total polymer concentration, 10% wt/vol; polymer ratio, 6/1 JLF-68CA/TMA salt of CA 11-bromoundecanoate (made from E-383-40 CA with 0.3-DS (quaternary ammonium groups); methanol, 10 g/formulation.
[b]0.5% NaCl feed at 400 psi and 25 ± 1°C.

TABLE 7.3 EFFECT OF RELATIVE HUMIDITY UPON PERMEABILITY AND PORE SIZE[a]

Relative Humidity at 20°C (%)	Filtration Time[b] (sec)	Average Pore Diameter (nm)
80	25–40	~600
60	40–60	~500
40	60–80	~400

[a]From Maier and Scheuermann[4]
[b]For 500 mL H_2O/12.5 cm^2 at 70 cm Hg

TABLE 7.4 INFLUENCE OF CASTING SOLUTION WATER CONCENTRATIONS UPON PORE SIZE AND PERMEABILITY OF COLLODION MEMBRANES[a]

H_2O Concentration in Casting Solution (%)	Filtration Time[b] (S)	Average Pore Diameter (nm)	Casting-Solution Viscosity at 20°C (cps)
3.3	40	600	2011
0.4	800	30	1813
0.0 (trace)	4000	15	1600

[a]From Maier and Scheuermann[4]
[b]For 500 mL H_2O/12.5 cm^2 at 70 cm Hg.

porosity during drying and less likely to be wet → dry reversible. Since the dry process tends to employ more dilute solutions and less compatible pore formers (both of which characteristics promote the formation of microgels) than does the wet process, the former is more likely to produce microgels than the latter. However, there are many exceptions to this rule and it is possible both to produce microgels by a wet process and ultragels by a dry process.

7.3 THE WET PROCESS

The wet or combined evaporation–diffusion technique is that variation of the phase-inversion process in which a *viscous* polymer solution is either (1) allowed to partially evaporate after which it is immersed into a nonsolvent gelation bath where whatever is left of the solvent–pore-former system is exchanged for the nonsolvent or (2) is immersed directly into the nonsolvent gelation bath for the exchange of the solvent system for nonsolvent. The end products of the wet process are water-swollen membranes; moreover, the water content of membranes—the equivalent of porosity in the dry process—is a prime determinant of its functional performance characteristics. *It is therefore fundamental to consider the effects of such variables as casting-solution composition and environmental parameters in terms of their effects upon membrane-water contents.*[14, 15]

A wet-process solution must be relatively viscous ($\geqslant 10^4$ cps) at the moment of immersion in the nonsolvent so that it will retain its integrity throughout gelation. When it is too fluid, the primary gel will be subject to disruption by the weight of the nonsolvent and the uneven forces brought about by the various currents which come into play during immersion. The requirement for high viscosity and hence high polymer concentration is in most cases inconsistent with the attainment of high porosity via the inclusion of nonsolvent pore formers. Therefore, when they are required, pore formers which are utilized in wet-process casting solutions are frequently chosen from the swelling agent—weak solvent side of the polymer–solvent interaction spectrum (Chapter 5). Moreover, the presence of pore formers within the casting solution prior to its immersion into a nonsolvent gelation bath is not a requirement of every wet-process solution. In many instances, particularly when nonvolatile solvents with a strong affinity for the nonsolvent in the gelation bath are utilized, the phase-inversion sequence Sol 1 → Sol 2 → gel is evoked by the simple act of immersion into nonsolvent. In such a case the nonsolvent bath represents an *external* source of incompatibility and a two-component solution (polymer + solvent) becomes in effect a three-component solution (polymer + solvent + nonsolvent pore former) as a result of the diffusion of the nonsolvent into, and the solvent out of, the nascent membrane gel.

The effect of the strong nonsolvent, water, may be influenced by other components of the casting solution. The presence of lyotropic salt swelling agents from the Hofmeister series causes the aggregation of water molecules about the electrophilic cations, thereby considerably modifying the properties of the water so affected.[14] The result of this interaction is to change the role of water from that of a nonsolvent to that of a swelling agent (Table 7.5). Other polar nonsolvents such as the aliphatic alcohols function in much the same manner as water, except that their nonsolvent tendencies are less pronounced. The role of water in the atmosphere and in the solution to effect gross structural irregularities will be discussed later in this chapter.

The effects of increasing the concentration of the weak nonsolvent pore former, ethanol, in a casting solution containing CA and acetone is to increase the porosity of the resultant membranes (Table 7.6, Fig. 7.13). (Because of their excellent solubility, certain cellulosic polymers can be so formulated that their solutions represent exceptions to the rule that wet-process solutions require highly compatible pore formers). As the concentration of ethanol is increased, the values of δ_p decrease slightly and the values of δ_h increase appreciably, which has the effect of bringing the solution closer to the point of incipient gelation, that is, to the perimeter of the solubility envelope. Since a solution which contains a high concentration of nonsolvent can be presumed to be of the Sol 2 type close to gelation, its immersion into a nonsolvent bath and subsequent gelation will be accompanied by less gel contraction than would occur if the solution were further removed from the perimeter of the solubility map. The result is that porosity and permeability increase as the concentration of pore former increases. Because the pore former is

TABLE 7.5 CASTING SOLUTION WATER-CONCENTRATION EFFECTS[a]

Water Concentration (g/formulation)[b]	Swelling-Agent Concentration (g $ZnCl_2$/formulation)[b]	Description of Membrane	Wet Thickness of Unheated Membrane (mm) $\times 10^2$	Gravimetric Swelling Ratio of Unheated Membrane (wet wt/dry wt)	Rate of Water Transport (mL/cm^2 day)[c]		Salt Retention (%)
					Deionized-Water Feed	0.6 M NaCl Feed	
0	0	Brittle, opaque (microgel)	5.8	1.47	<1	—	—
5	0	Brittle, opaque (microgel)	6.4	1.77	<1	—	—
10	0	Brittle, opaque (microgel)	7.1	1.99	<1	—	—
15	0	Brittle, opaque (microgel)	8.0	2.35	<1	—	—
0	5	Clear (ultragel)	8.7	2.53	24	16	90.3
5	5	Opalescent (ultragel)	9.0	2.79	34	22.8	97.2
10	5	Opalescent (ultragel)	9.2	2.85	72	48	98.5
15	5	Opalescent, opaque (ultra gel–microgel)	9.6	2.92	136	82	96.2

[a]From Kesting et al.[15]; © 1965.
[b]Formulation: cellulose acetate, 22.2 g; acetone, 66.7 g (doctor–blade gap, 0.25 mm).
[c]Rate of water transport and salt retention at 102-atm pressure for heated membranes (86°C for 5 min).

TABLE 7.6 EFFECT OF SWELLING AGENT (ETHANOL) ON THE
MEMBRANE-WATER CONTENT[a]

| Membrane Code No. | Mixed Solvent | | Calculated δ Values of Mixed Solvents | | Membrane-Water Content (wt %) |
	Ethanol (mol %)	Acetone (mol %)	δ_p	δ_h	
CA-24	20	80	4.99	4.23	50.7
CA-23	30	70	4.93	4.69	50.3
CA-22	40	60	4.86	5.20	53.4
CA-25	46.6	53.4	4.83	5.47	61.2
CA-21	50	50	4.79	5.75	65.8

[a]From Chawla and Chang[16]; © 1975.

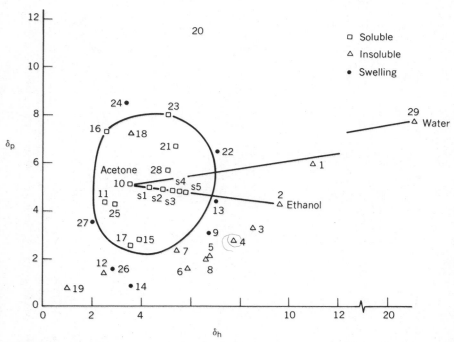

FIGURE 7.13. Solubility diagram for cellulose acetate. Solvents used: (1) methanol, (2) ethanol, (3) propanol, (4) butanol, (5) 1-pentanol, (6) 1-octanol, (7) 2-octanol, (8) cyclohexanol, (9) benzyl alcohol, (10) acetone, (11) methyl ethyl ketone, (12) diethyl ether, (13) ethylene glycol monoethyl ether, (14) dioxane, (15) tetrahydrofuran, (16) furfural, (17) ethyl acetate, (18) ethyl sulfate, (19) toluene, (20) formamide, (21) N,N-dimethyl formamide, (22) diethylene triamine, (23) dimethyl sulfoxide, (24) acrylonitrile, (25) pyridine, (26) chloroform, (27) 1,2-dichloroethane, (28) acetic anhydride, and (29) water (from Chawla and Chang[16]).

254

of a nonsolvent type, solution compatibility decreases with increasing ethanol concentration. This leads ultimately to increased diameters in the micelles of Sol 2
and consequently to greater opacity in the final membrane. It is worthy of note that
it is only the insufficient Δbp of 23°C between acetone and ethanol which prevents
this solution from being a candidate for the dry process. If methyl formate (bp
30°C) or propylene oxide (bp 35°C) had been employed as solvents in conjunction
with ethanol as the pore former, this solution could have served in either a wet- or
a dry-process mode. If acetone had been utilized as a solvent in conjunction with
propanol (bp 97°C) or isobutanol (bp 108°C) as pore formers, the same would of
course apply.   *water ?*

  The effects of increasing the concentration of the solvent-type pore former,
formamide, upon the porosity, optical, and permeability properties of CA membranes are found in Table 7.7. In the first place, the utilization of this solution in
the dry-process mode leads to the formation of a dense film. Since both acetone
and formamide are solvents, the loss of the more volatile acetone leaves behind a
high-boiling solvent, formamide, which plasticizes the CA gel as it evaporates. The
fact that [η] increases and both solution and membrane turbidity decrease with
increasing concentration of formamide in the acetone–formamide solvent system
suggests that solvent power increases as well. Concurrent increases in thickness,
porosity, and permeability are attributable to the strong hydrogen-bonding capacity
of formamide coupled with its strong affinity for solvating CA. After immersion,
desolvation of such solutions is slow rather than rapid because water can associate
with formamide by hydrogen bonding, thereby lessening water's role as a strong
nonsolvent. The net result appears to be that the Sol 2 → gel transition occurs at
a reduced rate during which the aggregating mass is more amenable to the infusion
of a higher concentration of nonsolvent than would otherwise be possible in the
case of a more abrupt Sol 2 → gel transition.

  The gelation bath temperature also exerts an important influence upon the struc-

TABLE 7.7 PROPERTIES OF SOLS AND GELS FROM ACETONE–FORMAMIDE
SOLUTIONS OF CA[a]

| Formamide Concentration (mol %) | Sol[b] Properties | | Gel Properties | | | |
	[η]25°C	Turbidity at 546 μm (x 10^2)	Turbidity at 546 μm (x 10^{-1})	Thickness (μm)	Wet wt / Dry Wt	Water Flux[c] (cm/day)
0	0.895	1.6	38.7	36	1.71	No flow
10	0.942	0.9	33.2	43	2.10	No flow
20	0.948	0.6	—	46	3.01	26
30	0.963	0.5	20.7	74	3.44	128
40	—	0.45	7.8	86	3.90	384
50	—	0.45	1.4	94	4.40	1320

[a]Adapted from Kesting and Menefee.[17]
[b]15 g E-398-10 CA/100 mL solution.
[c]Distilled-water feed at 40.8 atm and 25°C.

ture and function of HF membranes (Table 7.8). Increasing temperature hastens the onset of gelation which in turn results in increased void size, degree of swelling, and permeability, and decreased permselectivity.

Increasing the evaporation (drying) time prior to immersion in the nonsolvent medium causes a decrease in cell size and porosity and hence a decrease in permeability (Table 7.9). Permselectivity first increases and then decreases owing to stress imposed on the skin layer and possibly also to some swelling and rehardening of the skin as the solvent concentration in the nonsolvent bath increases.

The higher the affinity of the gelation medium for the components of the casting solution, the more gradual will be the Sol 2 → gel transition and the greater will be porosity in the final membrane. Thus the gelation of a CA solution in methanol will lead to a membrane of higher porosity than the gelation of the same solution in water. Methanol has greater affinity for CA than does water. Conversely, water is a stronger nonsolvent for CA than is methanol. Because the immersion of a casting solution in a strong nonsolvent such as water often leads to a skinned membrane it may be expedient, when a skinless membrane is desired, to immerse the casting solution into a nonsolvent solution which contains some solvent. Likewise, when a skinned membrane is available by any process, the skin may often be removed by immersing it into a nonsolvent/solvent solution. A closely related phenomenon known as *clearing* is utilized to collapse an opaque microporous electrophoresis membrane into a clear dense film so that the electrophoretogram can be read on an optical densitometer without changing the spacial relationships between the various protein fractions. Here the reverse of the dry casting process is employed. Instead of utilizing a volatile solvent and a nonvolatile nonsolvent to gradually decrease compatibility, a volatile nonsolvent and a nonvolatile solvent are employed to gradually increase the affinity of the clearing solution for the membrane substance as drying progresses. Gravity does the rest as the softened but intact gel slowly collapses.

The structure which is at hand immediately following the Sol 2 → gel transition in the dry process is known as a primary gel. It is seldom isolated as such because with continued evaporation (± syneresis) and drying the completely consolidated membrane, known as the secondary gel, is ordinarily the only product which is encountered or of interest. This is not usually the case for the wet process, however. Here, after the viscous solution has been gelled by immersion and the solvent system has been removed from the gel, a primary gel membrane which is stable as such for an indefinite period is the result. Such a membrane is easily distinguished from the secondary gels which result after the primary structures have been subjected to various postformation treatments.

As was the case for the dry process, the control of primary gel structure by environmental and especially casting-solution variables permits far greater latitude in the regulation of ultimate structural and performance characteristics of wet phase-inversion membranes than does the modification of primary into secondary gels. Because the properties of the primary gel determine to a large extent those of its secondary counterpart, the former should be considered as the more fundamentally characteristic and important structure in any consideration of the effects of varia-

TABLE 7.8 GELATION–BATH TEMPERATURE EFFECTS[a,b]

Gelation–Bath Temperature (°C)	Membrane Appearance	Intrinsic Viscosity [η] of Cellulose Acetate in Acetone–Water (66.7:100)[c]	Wet Thickness of Unheated Membrane [(mm) × 10²]	Gravimetric Swelling Ratio of Unheated Membrane (wet wt/dry wt)	Rate of Water Transport mL/cm² day[d]		Salt Retention (%)
					Deionized-Water Feed	0.6 M NaCl Feed	
0	Opalescent	0.985	9.2	2.85	84	50	98.6
10	Opaque	0.940	14.0	3.80	83	50	97.1
25	Opaque	0.05	22.8	5.80	90	58	90.1
40	Opaque	0.745	31.0	6.98	118	74	81.1

[a]From Kesting et al.[15]; © 1965.
[b]Casting-solution composition: cellulose acetate, 22.2 g; acetone, 66.7 g; water 10.0 g; ZnCl$_2$, 5.0 g (doctor–blade gap, 0.25 mm).
[c]Measured at the corresponding gelation–bath temperature.
[d]Rate of water transport and salt retention at 102 atm pressure for heated membranes (86%C for 5 min).

TABLE 7.9 DRYING TIME EFFECTS[a,b]

Drying Time (min)[c]	Description of Membrane	Wet Thickness of Unheated Membrane (mm) × 10²	Gravimetric Swelling Ratio of Unheated Membrane (wet dry/dry wt)	Rate of Water Transport (mL/cm² day)[d]		Salt Retention (%)
				Deionized-Water Feed	0.6 M NaCl Feed	
1	Opaque–opalescent (microgel blending into ultragel)	13.9	2.88	116	72	98.0
3	Opalescent (ultragel)	12.2	2.98	84	50	98.6
5	Opalescent (ultragel)	10.2	2.65	86	54	98.8
10	Opalescent–clear (ultragel)	8.5	2.41	80	50	96.3
20	Clear (ultragel)	5.8	1.75	72	50	75.1
30	Clear (ultragel)	5.3	1.60	50	36	71.5

[a]Kesting et al.[15]; © 1965.
[b]Casting-solution composition: cellulose acetate, 22.2 g; acetone, 66.7 g; water, 10.0 g; ZnCl₂, 5.0 g (doctor-blade gap, 0.25 mm).
[c]Drying time—interval between casting at −11°C and immersion into gelation bath (0°C).
[d]Rate of water transport and salt retention at 102 atm pressure for heated membranes (86°C for 5 min).

tions in fabrication parameters, for example, casting-solution composition, upon performance characteristics. Once a primary gel has been formed, it may be utilized as such (particularly for low-pressure applications), or it may be subjected to various physical and/or chemical treatments for conversion into a secondary gel which may be more suitable for a given end use.

Physical modifications of primary gel structures can be effected either to increase or to decrease the porosity (degree of swelling, void volume, water content, etc.), pore size, permeability, and permselectivity. The technique utilized to produce porous membranes from dense films can be used to effect an increase in porosity. In this variation of Brown's[18] technique (Chapter 8), an already porous primary gel is immersed in a swelling medium. To set the secondary gel in its more expanded condition the swelling medium is removed, either by exchange with a nonsolvent (nonsolvent–swelling-agent miscibility is essential) or by simple evaporation.

Since this technique adds another step to the fabrication process and is complicated by the leaching of low-molecular-weight polymer from the primary gel by the swelling medium, it is usually circumvented by the reformulation of the casting solution to produce a primary gel with an initially higher void volume. It is frequently encountered, however, as an undesirable factor in the permeation of certain organic solutes which interact with and swell the membrane, thereby altering initial pore characteristics and permeability. Of much greater practical importance are physical alterations of the primary gel structure to effect decreases in porosity. The most important means to this end are thermal annealing, pressurization, and solvent shrinking.

Annealing a porous membrane (particularly one which contains a nonsolvent capable of functioning to some extent as a plasticizer) results in a diminution of void volume and permeability and, because pore size is generally decreased as well, an increase in permselectivity. The reason for this can be seen on the molecular level where the introduction of thermal energy causes translational motion of the macromolecules, with the result that polar groups on the same and/or on neighboring molecules will approach one another closely enough to form virtual cross-links by dipole–dipole interactions. These cross-links tend to decrease chain mobility and, in a nonsolvent medium, are irreversible because of the inability of the nonsolvent to solvate and therefore intervene between the polar groups so enjoined.

Annealing has some effects which are continuous and some which are discontinuous. A continuous effect is the loss in water content and void volume with increasing temperature (Fig. 7.14). Water is lost from the primary gel during annealing, both because of the formation of virtual cross-links and because of the decrease in hydrogen bonding and cluster size in the water itself. An example of a discontinuous effect is the dramatic increase in permselectivity (salt retention) which is observed when cellulose acetate membranes are heated above 68.6°C, the glass transition temperature (Fig. 7.15). In fact, not one but two discontinuities are found on the permselectivity versus annealing temperature curve for cellulose acetate desalination membranes. The first signals an increase, and the second a decrease, in permselectivity. The increase, on the basis of the previously cited structural

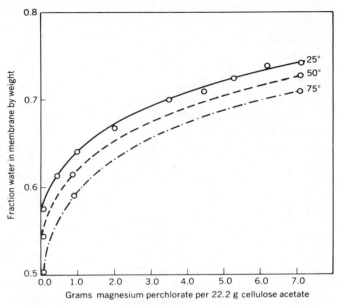

FIGURE 7.14. Membrane-water content as a function of temperature (from Kesting et al.[15]; © 1965).

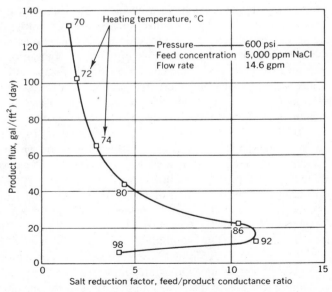

FIGURE 7.15. Permeability versus permselectivity for Loeb–Sourirajan membranes annealed at various temperatures.

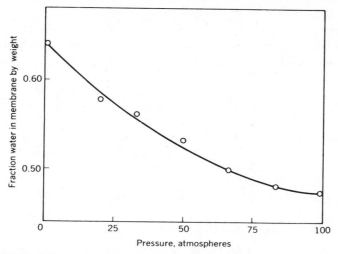

FIGURE 7.16. Membrane-water content as a function of pressure (from Kesting et al.[15]).

interpretation, may be attributed to the attainment of the critical interchain spacing or pore diameter, whereas the decrease may be related to disruption in the uniformity of these spacings owing to closer alignment of polymer chains in the glassy state in some regions at the expense of strain-induced removal of polymer chains from one another in others.

Whereas heating causes shrinkage in three dimensions, the application of pressure causes shrinkage primarily in one dimension, namely in the plane perpendicular to the surface. Two stages may be distinguished in the shrinkage of porous membranes under pressure (Fig. 7.16): (1) The rapid loss of void volume by the porous substructure which occurs at comparatively low pressures; and (2) the slower, more gradual loss of void volume by the comparatively dense skin layer. Inasmuch as the skin layer more closely approaches the structures of the bulk polymer, it is to be expected that significant compaction of this layer will require pressures in excess of the compressive yield point.

7.4 THE THERMAL PROCESS

A significant recent development in the technology of phase-inversion membranes is the invention of the thermal process by Castro.[19] The thermal process is applicable to a wide range of polymers, which because of their poor solubility, are otherwise inaccessible to the phase-inversion approach. In essence, the thermal process utilizes a latent solvent, that is, a substance which is a solvent at elevated ($\sim 220°C$) temperatures and a nonsolvent at lower temperatures, and thermal energy to produce a Sol 1 which on cooling inverts into a Sol 2, and on further cooling, gels. The reason for the incompatibility which evokes Sol 2 is loss of solvent power by the removal of heat. Both liquids and solids can serve as latent solvents. If, however, a solid is employed, it must be a liquid at the temperature at which Sol 2 appears. Because the latent solvents are nonvolatile substances, they are removed from the final gel by extraction with a liquid which is a solvent for the latent solvent and a nonsolvent for the polymer.

Although the thermal process is the most universally applicable of all the phase-inversion processes in that it can be utilized for the widest range of both polar and nonpolar polymers, its use for membrane applications will probably concentrate most heavily on the polyolefins, particularly on the polypropylenes. A large number of substances can function as latent solvents (Table 7.10). They usually consist of one or two hydrocarbon chains terminated by a polar hydrophilic end group. They, therefore, exhibit surface activity which may explain their ability to form the emul- sionlike Sol 2 micelles at elevated temperatures. One latent solvent which is worthy of special mention because of its broad applicability is *N*-tallowdiethanolamine (TDEA). The thermal process has a number of unique features:

1. Spherical shape of the cells in the final gel matrix. Although all phase-in-version membranes possess spherical micelles in their nascent Sol 2 condition, only thermal process Sol 2 solutions retain a speherical micellar shape in the final open-cell gel structures (Fig. 7.17). The diameters of the cells lie between 1 and 10 μm and the apertures or pores between them have diameters of between 0.1 and 1 μm with a narrow pore-size distribution. The Sol 2 micelles of dry- and wet-process

TABLE 7.10 THERMAL PROCESS POLYMERS AND LATENT SOLVENTS[a]

Polymer[b]	Latent Solvent(s)	Extrusion Temperature (°C)
LDPE	Saturated long-chain alcohols	—
HDPE	TDEA	250
PP	TDEA	210
PS	TDEA, dicholoro benzene	200
PVC	*trans*-Stilbene	190
SBB	TDEA	195
EAA	TDEA	190
Noryl (PPO/PS)	TDEA	250
ABS	1-Dodecanol	200
PMMA	1,4-Butanediol, lauric acid	210
Nylon 11	Sulfolane	198
PC	Menthol	—

[a]From Castio.[19]
[b]Commonly employed abbreviations.

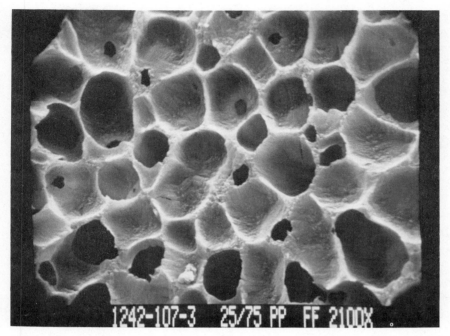

FIGURE 7.17. SEM photomicrograph of cross section of thermal process membrane with speherical cells (from Castro[19]).

membranes deform into polyhedra and flatten out by the time they have been fully formed.

2. Only the thermal process is capable of yielding isotropic microporous structures in thick sections. Wet and dry processes tend to become increasingly anisotropic as the thickness of the membrane gel is increased. This unique property renders thermal process gels suitable for use in controlled-release applications in which the gels can be cooled, ground up, extracted, and filled with, for example, volatile insect repellents.

If the solution is cooled slowly (8–1350°C/min), Sol 2 micelles appear. If, however, cooling is too rapid ($\geq$2,000°C/min), a continuous lacelike noncellular polymer network is apparent in the SEM photomicrograph (Fig. 7.18). This lacelike network may represent the frozen Sol 1 structure which for kinetic reasons is unable to assume the Sol 2 configuration before it becomes immobilized.

If the solution is cooled by casting on a metal belt, the bottom surface, that is, the side adjacent to the belt, will be skinned, whereas the top surface, the side adjacent to the air–solution interface, will be skinless. This is just the opposite of what usually occurs in wet and dry processes. The skin thickness is approximately equal to the thickness of a single-cell wall and can either be integral (totally nonporous) or nonintegral (some porosity) depending on the particular conditions. The

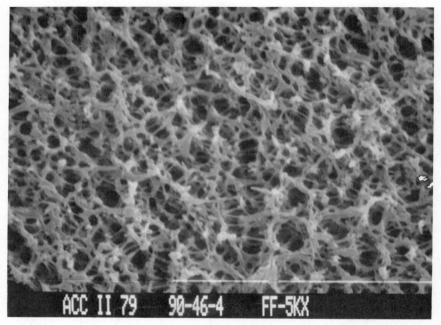

FIGURE 7.18. SEM photomicrograph of rapidly (2000°C/min) cooled 75% porosity polypropylene illustrating lacelike (possibly Sol 1) structure (from Castro[19]).

presence of an integral skin is desirable if the membrane is to be utilized for gas separations, HF, UF, and so on. If, on the other hand, no skin is desired it may be removed by allowing the membrane to be exposed briefly to a suitable solvent such as hexane in the case of polypropylene.

The latent solvent is removed from the membrane gel after the latter has attained sufficient strength to allow further processing. Typical leaching agents include volatile liquids such as isopropanol, methyl ethyl ketone, tetrahydrofuran, ethanol, and heptane.

7.5 THE POLYMER-ASSISTED PHASE-INVERSION (PAPI) PROCESS

The *polymer-assisted* variation of the phase inversion process (PAPI) utilizes a solution consisting of a solvent and two physically compatible polymers to cast a dense film with a morphology known as an interpenetrating polymer network (IPN). After complete (dry PAPI) or partial (wet PAPI) solvent evaporation, the IPN film is immersed in a liquid, usually water, which is a solvent for one of the polymers and a nonsolvent for the other. The insoluble network which remains after leaching is a skinless microporous PAPI process membrane. As originally conceived, the polymer which was to be leached was viewed as filling the role of a nonsolvent pore former, albeit one of higher than usual MW. As such it *assisted* the membrane polymer to assume the Sol 2 micellar structure prior to gelation. Physical, *but not single* T_g, compatibility is required for practical PAPI blends. Too compatible a blend presumably leads to Sol 1 structures which gel as such without assuming the necessary Sol 2 configuration. PAPI membranes are usually skinless, isotropic with a narrow pore-size distribution, of intermediate porosity ($\sim 50\%$), and characterized by good to excellent mechanical properties. The choice of both membrane and leaching polymers for the PAPI process is governed by the rules which are applicable to polymer blends which, unfortunately, given our present level of understanding, means that it is made on a largely empirical basis.

A potential application for PAPI process membranes is to serve as microporous supports for thin-film composites. This is particularly attractive in the *reverse-sequence* (RS) method for forming defect-free thin-film composites (see Section 7.7), where the thin film can be deposited prior to leaching the assisting polymer, thereby providing a dense impermeable surface which is ideal for coating. After a thin film of preformed-, or *in situ-*, formed polymers from fluid solutions has been deposited and/or cured, the support layer of the composite membrane can be made porous by leaching.

7.6 INHOMOGENEITY IN DEPTH

Prior to 1960, only isotropic or slightly anisotropic phase-inversion membranes were known. Today there are two types of inhomogeneity in depth which are of importance: *skinning* and *anisotropy*.

Skinning or asymmetry refers to a structure in which a thin ($0.1-0.25$ μm in depth) dense skin layer is integrally bonded in series with a thick (~ 100 μm) porous substructure. The skin layer determines both the permeability and permselectivity of the bilayer, whereas the porous substructure functions primarily as a physical support for the skin. The first skinned membrane, the wet-process cellulose acetate type developed by Loeb and Sourirajan[20] in 1960 for desalination by hyperfiltration, is universally acknowledged as the instrument which heralded the advent of the golden age of membranology. In the Loeb and Sourirajan, or *integrally-skinned*, membrane, skin and substructure are composed of the same material. Differences in density between the two layers are the result of interfacial forces and the fact that solvent loss occurs more rapidly from the air–solution and solution–nonsolvent bath interfaces than from the solution interior. Certain aspects of the skin layer remain a hotly debated issue. The earliest EM studies failed to discover any specific details of the microstructure, from which it was deduced that the skin was in an amorphous glassy state, similar to that of a solvent-cast bulk film. As we shall see, this conclusion was only partly true. Schultz and Asunmaa[21] discovered the presence of (~ 200 Å in diameter) spherical micelles in the skins of cellulose acetate membranes which had been etched with argon ions. Similar micelles were subsequently found by Kesting[6] in the skins of dry-process mem-

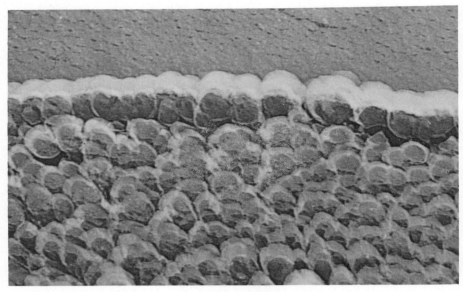

FIGURE 7.19. Top edge of cross section of polyamide–hydrazide skinned gel membrane (from Panar et al.[22]; reprinted with permission from *Macromolecules*, © 1973 American Chemical Society).

branes of cellulose acetate. Panar et al.[22] then discovered that the micellar morphology was quite general to integrally-skinned membranes not only of cellulose acetate but of polyamide–hydrazides and polyamides as well.

In the paragraphs that follow, the findings of the classical study of Panar et al. are cited at length in their own precise and descriptive terms.

The surface structure as shown in Figure 7.19 for a polyamide–hydrazide membrane is formed from a closely packed monolayer of micelles of about 400- to-800-Å diameter. The substrate (i.e., the structure immediately below the skin or uppermost layer of micelles...) is composed of similar spherical units randomly oriented with 75- to-100-Å voids between spheres. In the surface layer these structural units are compressed and distorted so that few voids appear. The skin is thus a denser form of the same "micellar" structure which forms the bulk of the membrane.

The surface must, however, be considered as a distinct mechanical entity. During the fracture procedure, the skin is frequently separated from the bulk. In Figure 7.19 it is evident that there are more pronounced cracks between surface and the bulk than between micelles of the surface. In Figure 7.20 we see a section of surface layer free of its substrate. The grainy surface in this micrograph is the fracture surface of the water surrounding the gel membrane. The skin has been broken off the substrate during fracturing and has adhered to the water "pot." The fact that the surface layer can be separated demonstrates the relatively poor fusion of the micelles directly below the surface. The substrate is itself poorly fused and relatively spherical.

The picture of the partial fusion of the structural units in a "skinned" gel membrane requires that if the entire structure were exposed to strong surface tension forces, as

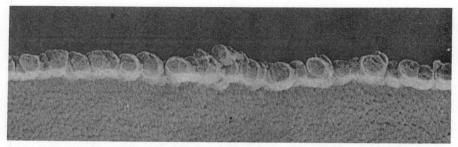

FIGURE 7.20. Surface skin of polyamide–hydrazide membrane (from Panar et al.[22]; reprinted with permission from *Macromolecules*, © 1973 American Chemical Society).

in the drying of the membrane, the micelles would be expected to fuse, just as a layer of freshly applied latex paint fuses to a solid. Consistently, Figure 7.21 shows the fracture surface of dried polymide–hydrazide gel membrane of the type discussed above. The fusion of the micelles to form a typically homogeneous, bulk phase is clearly evident.

The micellar structure of "skinned" membranes having a surface monolayer as the functional portion appears to be general. Polyamides (Fig. 7.22) and cellulose acetate (Fig. 7.23) gel membranes exhibit the same structure, as do freeze-dried polyamide–hydrazides (Fig. 7.24). The freeze-drying procedure, by obviating surface tension forces, apparently permits the micellar structure to be retained in the dry state.

The closely packed micellar morphology of the surface layer supports mechanistic hypotheses which assume permeation through free volume (dynamic pores) rather than through static pores. It seems reasonable that most permeation takes place through the anomalously high free volume in the zones between micelles when the deformed spheres are imperfectly fused.

Note, however, that these intermicellar low-density regions can themselves be considered as the static pores in Sourirajan's[23] model.

FIGURE 7.21. Fracture surface of air-dried polyamide–hydrazide membrane (from Panar et al.[22]; reprinted with permission from *Macromolecules*, © 1973 American Chemical Society).

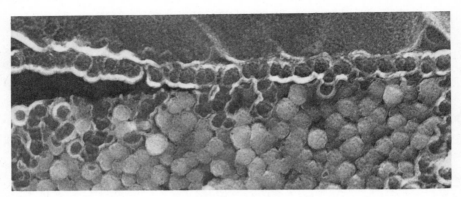

FIGURE 7.22. Skin structure of polyamide skinned membrane (from Panar et al.[22]; reprinted with permission from *Macromolecules*, © 1973 American Chemical Society).

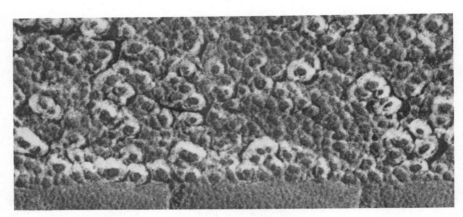

FIGURE 7.23. Skin structure of cellulose acetate membrane (from Panar et al.[22]; reprinted with permission from *Macromolecules*, © 1973 American Chemical Society).

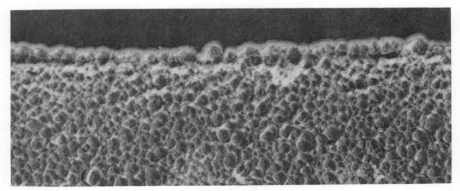

FIGURE 7.24. Skin structure of freeze-dried polyamide–hydrazide membrane (from Panar et al.[22]; reprinted with permission from *Macromolecules*, © 1973 American Chemical Society).

Areas in the surface which are imperfectly ordered, such as the gap visible in the right side of Figure 7.20 and the central region in Figure 7.23 represent defects because they occur too infrequently to be the significant morphology. These static pores would permit salt passage. The substrate presents comparatively little impedance to flow because water can move through the 100-Å gaps between the spheres.

The micellar structures of which the membranes are found appear to derive, at least for polyamide–hydrazides, from the structure of the solution from which the membrane is cast.

Note, these are the Sol 2 micelles in the present author's terminology. They are present in every phase-inversion membrane, even in those which are skinless.

The casting solution not only exhibits the same micellar structures as the bulk (Fig. 7.25) but the micelles form a surface monolayer under the influence of surface tension. Figure 7.26 is the fracture cross section of an air–liquid interface of the dimethylacetamide solution of the polyamide–hydrazides used to cast the membranes discussed above. The angle of view represented in Figure 7.26 is at 45° to the fracture cross section so that one sees the fracture surface at one side of the photograph and a section of the air interface at the other. The air interface appears structureless probably because of the effect of surface tension on the solvent. The micelles are only visible where a fracture has emphasized the mechanical differences between micelles and solvent. The micellar structure is not consistently visible on the surface of a dry *skinned* membrane, e.g. a freeze-dried gel membrane of a polyamide–hydrazide. When not visible the micellar nature can be made clear by a brief oxidative etching of the surface.

The apparent disappearance of the micellar structure was also noted by Schultz and Asunmaa[21] and Kesting[6] who employed etching with argon ions to bring the micellar structure to the fore in the skins of wet, and dry, CA membranes, respectively.

FIGURE 7.25. Fracture surface of casting solution of polyamide–hydrazide membrane (from Panar et al.[22]; reprinted with permission from *Macromolecules*, © 1973 American Chemical Society).

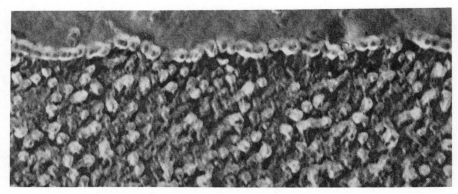

FIGURE 7.26. Air–solution interface of polyamide–hydrazide casting solution (from Panar et al.[22]; reprinted with permission from *Macromolecules*, © 1973 American Chemical Society).

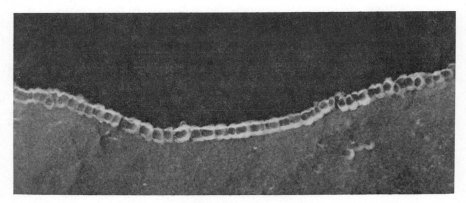

FIGURE 7.27. Air–solution "skin" of polyamide–hydrazide casting solution (from Panar et al.[22]; reprinted with permission from *Macromolecules*, © 1973 American Chemical Society).

The brief period between forming the air interface and freezing the solution to a rigid glass (about 60 s at 25° in this case) is apparently sufficient to result in a surface layer of some mechanical integrity as can be seen in Figure 7.27 in which a section of surface layer has, during fracturing, broken free.

The rapid appearance of the surface monolayer after casting is consistent with the demonstrated importance of air exposure prior to "gelation" in determining membrane properties.

The "skinned" membrane is thus seen to be related to the structure of the freshly cast polymer solution. The various procedures devised to prepare a high flux membrane appear to have been optimized to retain the solution structure in the solid phase. One may consider this trapped solution morphology as a functional definition of the "skinned" membrane of the type first described by Loeb and Sourirajan. This viewpoint clearly differentiates such membranes which have yielded the highest reverse osmosis fluxes from those fabricated with a thin dense layer of normal solid morphology.

The question of whether the micelles in the casting solution exist at room temperature, or form during the rapid freezing to a glass, requires comment. In the latter case, the freezing process must be considered to produce changes analogous to those occurring during "gelation," and the discussion above would refer to an undescribed solution orientation which leads to the micellar structures on freezing. The prior existence of the micelles is, however, suggested by the constancy of their sizes when a solution is frozen either very slowly or extremely rapidly in the form of a capillary film.

Immediately below the skin layer of integrally-skinned membranes is found a substrate[22] or transition[24] layer with a density intermediate between that of the skin and that of the porous substructure. This consists of less closely packed micelles than those in the skin layer and is composed of closed cells, and mixed open and closed cells. The depth and structure of the transition layer in wet-process membranes are functions of the various fabrication parameters rather than immutably fixed quantities. Although they are commonly found in wet-process membranes which employ concentrated polymer casting solutions, they are encountered in dry-process membranes, which employ more dilute casting solutions, only under specialized conditions. Trudelle and Nicolas[25] utilized light reflection, differential refractometry, and densitometry and found that the skin has a water content of 38% by weight, whereas the substructure contains 61.8%. They found that the water content increases steadily from the surface inward, quite quickly in the surface region and more slowly, but not negligibly, in the deeper regions (Fig. 7.28). The transition layer was estimated to be 19 μm (out of a total membrane thickness of 140 μm). Significantly, it was found that annealing the membrane resulted in increased asymmetry. The water content was decreased to a greater extent in the skin than in the substructure layer (Table 7.11).

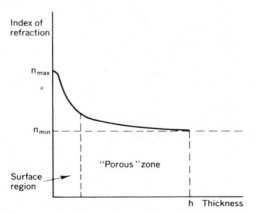

FIGURE 7.28. Variation of the index of refraction with penetration into the top surface of a Loeb–Sourirajan type of cellulose acetate membrane (from Trudelle and Nicolas[25]).

TABLE 7.11 EFFECT OF THERMAL ANNEALING UPON THE INDICES OF
REFRACTION AND WATER CONTENTS OF SKIN AND SUBSTRUCTURE
LAYERS OF LOEB–SOURIRAJAN-TYPE MEMBRANES[a]

	Skin Layer		Substructure Layer	
Membrane	Index of Refraction	%H$_2$O	Index of Refraction	%H$_2$O
N.T.[b]	1.409	46	1.3799	63
T.[c]	1.420	38.6	1.3823	61.8

[a]From Trudelle and Nicolas.[25]
[b]Unheated membrane.
[c]Membrane heated at 80°C for 5 min.

The conclusion which can be drawn from the above studies is that the skin layer in integrally-skinned hyperfiltration membranes consists of a single layer of consolidated Sol 2-type micelles and that this structure differs in kind from that of a solvent-cast film which is the result of the direct conversion of Sol 1, a molecular dispersion of macromolecules, to the gel without the intermediate step of Sol 2 micelle formation. Thus supermolecular structure of a very specific nature is present in the skin of integrally-skinned hyperfiltration membranes which is not present either in specially prepared thin films of the type utilized by Krishnamurthy and McIntyre[26] to mimic the skin layer or in the thick dense films utilized by Lonsdale[27] to determine the "intrinsic" salt rejections of various polymer HF membrane candidates. The ultrathin film X-ray studies of Krishnamurthy and McIntyre and the thick solvent-cast film DSC (differential scanning calorimetry) studies of Morrow and Sauer[28] were previously cited[29] as evidence for the existence of crystallinity in the skin layer of integrally-skinned membranes. However, from the vantage point of our present state of knowledge of the skin structure, it now appears that these two studies were largely irrelevant since they dealt with structures which were unrelated to those of the skins in integrally-skinned membranes. Furthermore, several additional facts argue against the necessary existence of crystalline order in the skins of HF membranes:

1. Interfacial thin-film polycondensates whose mode of formation makes crystallinity highly improbable are known to function as excellent HF membranes.

2. Richter and Hoehn[30] deliberately designed their aromatic polyamides to be amorphous and yet they are highly permselective.

3. The trimethyl ammonium salts of CA 11-bromoundecanoate both by themselves and as blends with CA[31] result in integrally-skinned membranes with the highest permselectivity in a series of quaternized omega haloesters of various chain lengths. Since the inclusion of long aliphatic chains into cellulose (Table 7.12) is known to result in lower T_m values and hence in lower crystallinity, it follows that crystallinity in the skins of integrally-skinned membranes is not only unnecessary but may actually be detrimental.

TABLE 7.12 APPARENT MELTING TEMPERATURES OF VARIOUS HIGHER ALIPHATIC ACID TESTERS OF WOOD PREPARED BY THE TFAA OR THE CHLORIDE METHOD[a]

Sample (acyl)	N (carbon atoms)	Melting Temperature (°C)	
		TFAA	Chloride
Butyroyl	4	300	310
Valeroyl	5	235	305
Caproyl	6	250	260
Capryloyl	8	210	245
Capryl	10	205	290
Lauroyl	12	195	240
Myristoyl	14	200	—
Palmitoyl	18	195	195
Stearoyl	20	—	220

[a]From Shiraishi.[32] Reprinted with permission from *CHEMTECH*, © 1983 American Chemical Society.
[b]Measured under a pressure of 3 kg/cm^2.

To accomodate both micellar skins and interfacial thin films into a single structural model consistent with the functional behavior of skins of HF membranes is quite obviously not an easy task. Since the skin is not crystalline, it is amorphous. It cannot be a rubber so it must be a glass. Since, however, the glassy state is a nonequilibrium condition, a good deal of uncertainty regarding the nature of this state and how it relates to HF membrane performance remains.

In view of the above, the "solution–diffusion" theory of demineralization by HF membranes should be critically reexamined. Consider the following:

1. The structures of the skins of integrally-skinned and thin-film composite membranes are known to be different in kind from those of the thick dense films of the type which are utilized to determine *intrinsic* salt rejections.

2. Chan et al.[33] have calculated that HF membranes have pores with two different pore-size distributions.

3. Katoh and Suzuki[34] have developed a new EM technique which has enabled them to establish the presence of pores within integrally-skinned CA HF membranes.

4. Panar et al.[22] have obtained EM evidence (Figs. 7.20 and 7.23) that imperfections in the form of imperfectly fused micelles exist in the skins of integrally-skinned membranes. It seems probable that these gaps represent the larger of the two pore-size distributions calculated in (2) and seen in (3) above.

5. Mathematical treatments[35] of various models suggest that pore defects combined with predictions from solution–diffusion theory yield the best agreement between theory and experiment. With these thoughts in mind it is time to restrict predictions from the hyperfiltration variation of the solution–diffusion model to some specially prepared defect-free dense films which are unrelated to any functional membrane,[36] whether of the integrally-skinned, or of the thin-film composite, types.

Perhaps because more attention has centered on hyperfiltration membranes, the fine pores present in their skins were observed prior to their discovery in the functionally larger-pored skins of ultrafiltration (UF) membranes. Recently, however, pores of ~ 30 Å have been observed by Zeman[37] in the skins of a UF membrane with a MW cutoff of 10^5 (Fig. 7.29). Their density, uniformity, and diameters leave no doubt that these are actually the pores which are functional during UF. Furthermore, since our ability to actually "see" the intermicellar defect pores (the population of larger-size pores) in the skins of HF membranes extends to the 10-Å range, it is not unreasonable to expect that at some point we shall be able to extend this ability to the population of smaller-sized pores, whose existence is predicted by Sourirajan's pore theory.[23]

The list of polymer membrane materials is virtually endless insofar as possible chemical varieties are concerned. However, the number of fundamental physical structures into which they may be formed is much more limited. For present purposes, a distinction is made between skinned membranes and skinless ones. However, in view of the substantial and growing evidence cited above for the existence of pores in HF and UF membranes, even this is done with trepidation. Further subdivision results in three types of skinned membrane: (1) integrally-skinned ultragels; (2) integrally-skinned microgels; and (3) nonintegrally-skinned microgels, the thin-film composite membranes. Such skinned membranes are utilized in gas

FIGURE 7.29. SEM micrograph ($\times 10^5$) of polysulfone UF membrane with MW cutoff 100,000 (from Zeman[37]).

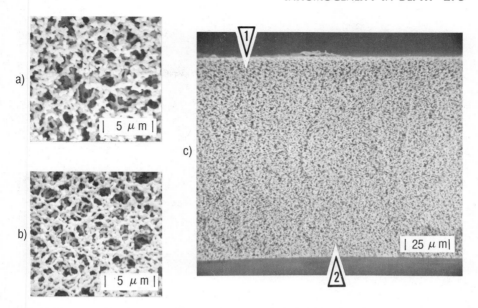

FIGURE 7.30. SEM micrograph of an isotropic 0.45-μm membrane: (*a*) surface at 1; (*b*) surface at 2; (*c*) cross section (from Kesting et al.[7]).

separations, hyperfiltration, and ultrafiltration. On the other hand, two types of skinless membranes are discernible and both are utilized in microfiltration and related applications: (1) isotropic (actually, slightly anisotropic) microgels and (2) highly anisotropic microgels.[38] The former are the conventional microfiltration membranes of commerce. Both surfaces of the conventional membrane are quite similar in appearance (Fig. 7.30*a,b*) and the cross section (Fig. 7.30*c*) reveals only a slight degree of anisotropy with little difference in pore and cell size from one surface to the other. In contrast, a considerable difference is apparent between the pore sizes at opposite surfaces of the highly anisotropic microgel (Fig. 7.31*a,b*). SEM photomicrographs reveal a degree of anisotropy (DA), that is, a difference in pore size from one surface to the other, of approximately 5. The cross section indicates the presence of two integral layers, the thicker of which contains the coarser cells (Fig. 7.31*c*).

A simple but graphic illustration of the morphology of anisotropy is also afforded by a wick test. In this procedure, a hollow-point pen is brought into contact with either surface of (in this instance) a 0.45-μm microfiltration membrane. Circular ink stains of unequal diameter then appear on opposite sides of the filter. Photomicrographs of the cross sections of the stained membranes indicate wicking patterns attributable to differences in capillarity. A low degree of anisotropy leads to minimal wicking (Fig. 7.32*a*); an intermediate degree of anisotropy to modest wicking (Fig. 7.32*b*); and a high degree of anisotropy to maximum wicking (Fig. 7.32*c*). In the last of these an integral bilayer is indicated, one part consisting of

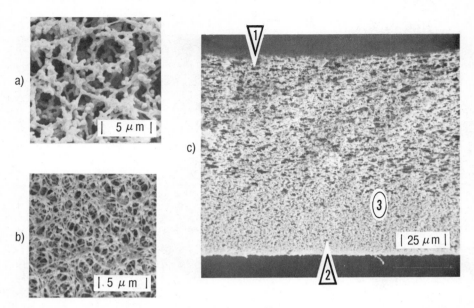

FIGURE 7.31. SEM photomicrographs of a 0.45-μm highly anisotropic membrane: (*a*) surface at 1; (*b*) surface at 2; and (*c*) cross section (from Kesting et al.[7]).

FIGURE 7.32. Photomicrographs of ink-stained cross sections of 0.45-μm membranes with various degrees of anisotropy: (*a*) low DA; (*b*) intermediate DA; (*c*) high DA; and (*d*) artist's conception of (*c*) (from Kesting et al.[7]).

fine cells (in which capillarity is quite pronounced) and the other of much coarser cells (Fig. 7.32d).

This gradation of pore sizes from one surface to the other confers the filtration capacity of a prefilter–filter combination upon these integral bilayers and accounts for their significantly higher dirt-holding capacity. Because the fine-pored layer of highly anisotropic membranes is only a fraction of the overall membrane thickness, its cells and pores are deliberately made smaller than those of a similarly rated isotropic membrane filter. For example, a 0.2-μm-rated highly anisotropic membrane may have pores in its fine layer of ~ 0.15 μm or even 0.1 μm and pores in its coarse layer as large as 10 μm. In the latter case the DA equals 100.

Highly anisotropic membranes are produced by manipulation of casting-solution parameters such as solvent volatility and environmental factors such as temperature and relative humidity which influence the kinetics of phase inversion, gelation, syneresis, and capillary depletion. The surface of the membrane which constitutes the air–solution interface during its nascent phases becomes the fine-pored side of the finished membrane. During filtration, however, it is the coarse-pored side of the membrane which is positioned to face the feed solution. When this is done, the throughput of the highly anisotropic membrane is much greater than that of the conventional isotropic type (Fig. 2.21). The throughput is greatly diminished, however, when the fine-pored surface faces the feed, although it is still roughly equivalent to that for the coarser-pored surface of the conventional membrane.

7.7 COMPOSITE MEMBRANES

The combination of two or more membranes in series results in a composite membrane. The permeability contant P for a composite membrane may be expressed[39]

$$\frac{1}{P} = \sum_{1}^{n} \frac{x_i}{l} \frac{1}{P_i}$$

where P = overall permeability constant,
   n = number of layers of membrane, each of thickness x, and
   l = total thickness of the composite structure.

The permeability constant of the composite membrane is therefore represented by the harmonic average of the permeability constant of the individual layers, the respective weights being x_i/l, the ratio of layer thickness to the total. Although composite membranes include layers of dense films or even liquid layers in series with films, in this case the term is being limited to those series, at least one of whose members is a phase-inversion membrane which itself can be either of the integrally-skinned or of the skinless variety.

In the case of composite membranes consisting of skinless porous substrates and dense films, permeability and permselectivity may be determined solely by the resistance of the dense films. Different membrane polymers may therefore be employed for the thin barrier layer and the thick support structures which thereby permits the attainment of a combination of properties which are not available in a single material. Such membranes were initially developed for desalination by hyperfiltration where they are known as thin-, or ultrathin-, film composites or nonintegrally skinned membranes. The other type of composite membrane is utilized for gas separations. It is a composite consisting of an integrally-skinned or asymmetric membrane, the defects of whose skin layer are coated by the addition of a second, albeit more permeable, skin. The inventors of the latter have entitled their device "resistance-model" membranes.

In the standard sequence for the formation of thin-film composites, a preexisting microporous membrane is utilized as a support onto which a thin barrier-layer film is deposited. It is hoped that the nature and preparation of the microporous support layer is sufficiently familiar to the reader by this time that little more need be said other than that it is a skinless microgel with approximately 0.1–μm pores and today in most cases consists of polysulfone.

The microporous support layer can be combined with the thin film to form the thin-film composite membrane in a variety of ways:

1. A dilute solution of a preformed polymer can be separately cast, preferably from a surface-active "spreading" solvent such as cyclohexanone, onto water, thereby yielding a thin film which is then laminated to the porous support layer. This approach was originally developed by Carnell and Cassidy[41] and was the first one utilized by Francis and Cadotte[41] in 1964 to coat microporous CA membranes.[1] Petersen[42] also utilized this approach to cast thin films of ethyl cellulose perfluorobutyrate which were then laminated to Celgard® and Tyvek® supports for use as blood oxygenation membranes.

2. Application of either a preformed polymer or a prepolymer solution by dip coating, wicking, or some other transfer procedure directly onto the porous support where it is subsequently dried or cured. This procedure has the advantage that the difficult handling of thin films is avoided. It has the disadvantage that only solutions which do not interact with the porous support may be utilized in the application of the barrier layer. This is, of course, not a disadvantage if, as is frequently the case, the coating can be applied from an aqueous solution. This approach was first utilized by Cadotte in the application of an aqueous solution of polyethyleneimine (PEI) and the subsequent reaction with tolylene diisocyanate (TDI) in hexane to yield the cross-linked polyurea which constituted the NS-100 membrane. When PEI or an ethoxylated derivative of PEI were coated and reacted with a hexane solution of isophthaloyl chloride, the resultant thin film was a polyamide (NS-101 or PA-300). The reaction of furfuryl alcohol with sulfuric acid resulted in a sulfonated polyfuran known as NS-200. Recently, interest in coating and subsequent cross-linking of a water-soluble sulfonated polysulfone has been revived.

3. Plasma polymerization involves the buildup of a dense layer from the deposition of monomers produced in an RF plasma.[43] Such monomers are unrelated to those encountered in free radical polymerization. Hydrophilic coatings can sometimes be produced from materials which, prior to their conversion to a plasma, were hydrophobic and vice versa. No commercial use is being made of this approach at present.

4. Interfacial polycondensation of reactive monomers on the surface of the porous support. The basic interfacial polycondensation technique is known[44] to have a number of unusual but desirable features among which the lack of strict requirements for monomer purity and reagent stoichiometry are perhaps the most outstanding. Cadotte[45] has applied this concept to the preparation of the commerically successful FT-30 desalination membrane. A polysulfone support is coated with an aqueous solution containing at least 0.01% of *m*-phenylene diamine. The coated membrane is then brought into contact with a hexane solution containing trimesoylchloride. Reaction is swift and terminates when the completed interfacial thin film inhibits further reaction. Only slightly more than two of every three carboxylic acid chloride groups condense so that the lightly cross-linked final membrane polymer closely approximates the formula

$$\left[HN - \langle\!\!\!\!\bigcirc\!\!\!\!\rangle - NHOC - \langle\!\!\!\!\bigcirc\!\!\!\!\rangle - CO \atop COOH \right]_n$$

The high permeability of the FT-30 membrane is attributable to the presence of the hydrophilic carboxyl groups. An SEM photomicrograph of the cross section shows that the barrier layer is itself composed of several layers of varying density to a total depth of approximately 0.25 μm (Fig. 7.33). This thickness adds physical strength in the form of flexibility and abrasion resistance which earlier and thinner thin-film barrier layers did not possess. The fact that none of the thin-film composites, including FT-30, exhibits any appreciable resistance to oxidative degradation by chlorine is believed to be coincidental with the chemical characteristics of the particular thin films which have thus far been investigated rather than with the nature of thin-film composites per se. The *cross-linked polyether*[46] membrane, for example, is so subject to degradation even by dissolved oxygen, that the feed must be continuously dosed with sodium bisulfite. On the other hand, thin-film composites consisting of cross-linked thin films of sulfonated polysulfone are expected to be chlorine resistant.

The finer the particles in a mixture the more difficult it is to obtain an appreciable difference in the rates at which they permeate a membrane. Thus separation of bacteria from a suspension is routinely achieved with log reduction values

$$\frac{(\log_{10} \text{ colony-forming units (CFUs) in feed})}{\text{(CFUs in product)}}$$

of between 7 and 11, whereas in HF salt reduction factors

$$SRF = \frac{\text{salt concentration in feed}}{\text{salt concentration in product}}$$

12014 HVO 0.5 μ

FIGURE 7.33. SEM micrograph of cross section of FT-30 membrane (from Cadotte[45]).

of 100 or less, that is, LRVs of ~ 2, are the norm. In the case of many gas separations, even lower separation factors are observed. Furthermore, because the fluidity of gases is so high, the rate of gas permeation of unseparated gas mixtures through porous imperfections in the skin layer of an integrally-skinned membrane can be so great as to overwhelm by dilution of separated gas with unseparated gas mixture, the effects of whatever separation was achieved. Thus even CA/CTA blend membranes which are capable of SRFs of 100 in a desalination application, frequently encounter difficulty when utilized, after suitable drying, as gas-separation membranes. Henis and Tripodi[47] noticed this effect while investigating integrally-skinned hollow-fiber membranes of polysulfone and solved the problem by applying a coat of silicone and/or other elastomers to effectively seal any imperfections in the skinned polysulfone membrane against a high flow rate of unseparated feed gas mixture. This approach should be applicable to virtually any skinned membrane. The sealing polymer coat need not itself be capable of separating gaseous mixtures so long as it is sufficiently permeable as not to significantly reduce the permeability of gases through the separating membrane, while at the same time sufficiently impermeable as to constitute a barrier to the bulk flow of unseparated gases.

5. The *reverse-sequence* (RS) method of forming thin-film composites is an approach which, in its more common manifestation, involves the preparation of defect-free thin films in series with thick dense (but potentially porous) films. The conversion of what amounts to a thin dense film: thick dense-film bilayer to a thin film: microporous membrane composite is effected by the immersion of the bilayer into a bath where certain constituents are leached from the thick dense-film portion of the bilayer, while leaving the thin dense film intact. The result is a thin-film composite in which the preparation of the thin film preceded the preparation of, or at least the development of porosity in, the microporous layer. In the standard sequence, the microporous membrane is prepared first and then subsequently coated with a thin dense film. The latter procedure leads to defects in the thin film because of the presence of occasional large pores in the surface of the microporous substrate. In contrast the casting substrate utilized in the RS method is a smooth nonporous surface which is amenable to the preparation of defect-free thin films. In another variation of the RS method, a thin dense film is cast on a smooth nonporous surface such as glass or polished stainless steel and a microporous membrane or potentially microporous thick film is then cast onto the thin film. The composite is then released from the casting substrate by immersion in water.

7.8 STRUCTURAL IRREGULARITIES

A by no means exhaustive list of commonly encountered structural irregularities in phase-inversion membranes includes: irregular gelation, wavemarks, macrovoids, and blushing.

Irregular gelation can take many forms. Streaks or draglines appear at the top side of the membrane, that is, at that surface which constitutes the air–solution interface during the early nascent phases of membrane formation. If a solution is close to the point of incipient gelation when it passes under the casting knife as it is first exposed to air, virtually any imperfection, such as a scratch on the blade surface, can serve to nucleate a gel particle. This particle can adhere to the blade and cause the formation of streaks when additional gel particles are nucleated in the solution as it passes under the original site. Such premature gelation is prevented by elimination of nucleation sites by, for example, polishing the casting blade, and by postponing gelation until the solution is a safe distance downstream from the casting blade. The latter is usually accomplished by the installation of a *quiet zone* immediately downstream from the hopper. An impermeable plate, a few centimeters in length, is positioned close to the surface of the casting solution to maintain the concentration of solvent vapor at a sufficiently high enough level as to prevent gelation until the solution has passed underneath the plate. Another type of irregular gelation is visible as subsurface imperfections which are most easily seen by light-box inspection. Although such subsurface gel particles are often only a cosmetic defect, they can at times be more serious and do represent a potential locus of failure as, for example, when roll stock is convoluted for inclusion in pleated cartridges. Such subsurface gels are usually attributable to inadequate filtration of the casting solution. Good manufacturing practice requires that a casting solution be constantly recirculated through a filter after it has been mixed. It then receives a final filtration just prior to its being fed into the casting hopper. Any particle, however minute, represents a potential nucleation site for random gelation. This can sometimes be seen as in the form of irregular snowflake patterns during capillary depletion. Ordinarily, snowflake patterns appear in an even line in the transverse direction across the entire width of the membrane. If they appear on one side earlier than on the other, the membrane is thinner in that area. If snowflakes suddenly appear randomly where no others are found, this may indicate a thin spot or even a pinhole. Both defects are apparent on light-box inspection.

As the name implies, wavemarks are encountered only in wet-process membranes. The casting solution enters the nonsolvent bath at which point a skin is formed as the solution is immersed. Interfacial tension causes the water to adhere to the leading edge of the membrane until continued motion of the nascent membrane into the nonsolvent medium causes the water to break away and establish another leading edge. Wavemarks represent a thickening of the membrane at the crest of the wave and, depending on various factors such as casting solution fluidity, can achieve varying amplitudes at the crests. The problem is overcome by adjusting the angle of entry of the casting solution into the nonsolvent bath.

Large (10–100 μm in the longest dimension) subsurface voids, tear dropped, spherical, ellipsoidal, or fingerlike in shape, are known as *macrovoids*. They represent, at the very least, weak spots within the gel matrix, and at worst, that is, when located near the high-pressure surface of the membrane in pressure-driven applications, areas of potential rupture. It was once thought that fingerlike cavities represented volume elements of low resistance which contributed to overall permeability. Such cavities are never advantageous and should always be avoided whenever possible. There are two basic reasons for the appearance of macrovoids and an examination of SEM photomicrographs of membrane cross section usually permits the responsible cause to be established. When the walls of the cavity are made up of open cells which are identical to the structure of the undisturbed gel matrix, then the cavities are the effect of trapped pockets of solvent vapor which has built up in subsurface domains faster than it can diffuse out. If the surrounding matrix gels before the solvent vapor can depart, then a skinless cavity wall remains. The presence of a high concentration of solvent prevents the skinning of the cavity walls. Such voids can be eliminated by minimizing the buildup of solvent vapor by lowering solution viscosity and/or by lowering the environmental temperatures. Another cause of similar voids is the physical entrapment of air bubbles owing to leaks into the hopper area or to too rapid influx of casting solution into the hopper.

When the cavity walls are skinned, this is a clear indication that nonsolvent intrusion from the gelation bath has taken place. Such intrusions can occur irregularly when certain voids close to the surface rupture for any reason prior to the gelation of the matrix surrounding the void. Thus a given membrane can contain voids with skinned—as well as voids with skinless—interior surfaces. The fingerlike cavities, on the other hand, appear on a regular basis with predominantly

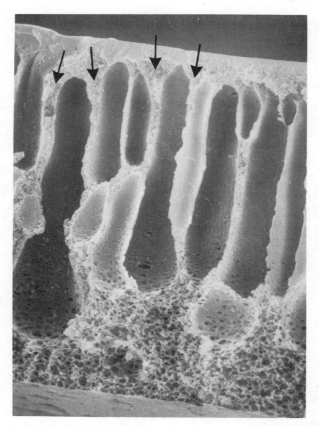

FIGURE 7.34. SEM photomicrograph of cross section of membrane with skinned fingerlike cavities. Arrows denote actual permeation pathways (from Cabasso[11]).

skinned interior surfaces. Cabasso has shown that no permeation of product takes place through these cavities which therefore represent useless dead space (Fig.7.34). They appear to be due to skinning of somewhat too-fluid solutions followed by "osmotic shock" to the formed skin, which subsequently ruptures, permitting the rapid intrusion of water into the still-fluid substructure where gelation eventually takes place in the matrix surrounding the fingerlike intrusions of water (Fig.7.35). The surface pores which result from the rupture of the skin remain, and are the reason why such UF membranes are unable to quantitatively remove bacteria whereas nominally coarser MF membranes can. They can be avoided, or at least minimized, by the addition of solvent to the gelation medium to lessen the osmotic shock and by utilizing more viscous casting solutions.

The presence of a fine white powder on one or both surfaces of a polymer film or membrane following wet or dry casting is known as *blushing*. Microscopical examination shows that the powder consists of spherical particles characteristic of polymer lattices. Obviously, the presence of these particles is, or can be, deleteri-

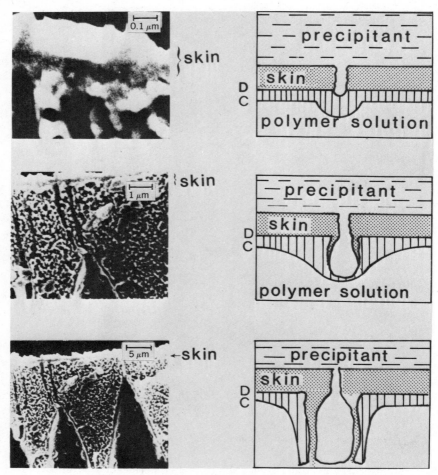

FIGURE 7.35. Formation of fingerlike cavities (from Strathmann[48]).

ous because the potential exists for their inadvertent sloughing off and addition to the product stream, the purification of which was the principal reason for utilizing a supposedly integral membrane in the first place. These particles consist of the most soluble polymer components within the casting solution which is utilized in the preparation of the membrane or film. They are the constituents which remained in solution the longest which in turn means that they are soluble in the solvent system even after it has become depleted of most of the more volatile true-solvent constituents. Furthermore, it is likely that they consist of lower-MW fragments since the latter are more soluble than are those of higher MW.

Water is, in most instances, a strong nonsolvent because of which it is a causative factor in the occurrence of blushing. After phase inversion and gelation and prior to capillary depletion, the membrane gel is still filled primarily with the less volatile nonsolvent components of the casting solution. Only residual amounts of true solvent remain. Therefore, any polymer remaining in solution is in a metastable state. The presence of water either in the solution or in the atmosphere acts to precipitate the remaining polymer. Blushing can appear at either or both surfaces. When it appears on the surface which is exposed to air during drying, it is the result of pore former (nonsolvent) liquid which has been expressed through this surface in the gel contraction known as syneresis. The expressed liquid contains the last traces of polymer in solution. Their precipitation as latex particles at this point depends on relative humidity and the composition of the syneresed liquid. The presence of blushing on the bottom surface, on the other hand, depends on the composition of the liquid which is still within the gel near the bottom surface.

There are several ways to prevent or minimize blushing:

1. Addition of a high-boiling (retarder) solvent to the mix. This is the approach taken for polymer films. It must be utilized judiciously for membranes because it can lead to loss of porosity. The presence of a solvent as the last liquid component to leave the scene prevents precipitation.

2. Elimination of water from the casting solution and/or reduction of relative humidity.

3. Employment of pore formers such as 1-butanol, which carry the water with them as they evaporate.

REFERENCES

1. R. Kesting, *Pure Appl. Chem.*, **50**, 633 (1978).
2. R. Kesting, in *Cellulose and Cellulose Derivatives, Part V*, N. Bikales and L. Segal, Eds., Chap. XIX, Wiley-Interscience, New York, 1971.
3. J. Cahn, *J. Chem. Phys.*, **42**, 93 (1963).
4. K. Maier and E. Scheuermann, *Kolloid Z.*, **171**, 122 (1960).
5. J.-G. Helmke, *Kolloid Z.*, **135**, 29 (1954).
6. R. Kesting, *J. Appl. Polym. Sci.*, **17**, 1771 (1973); U.S. Patent 3,884,801 (1975).
7. R. Kesting, L. Cunningham, M. Morrison, and J. Ditter, *J. Parenteral Drug Assoc.*, 98 (May/June 1983).
8. H. Spurlin, A. Martin, and H. Tennent, *J. Polym. Sci.*, **1**, 63 (1946).
9. A. Scheludko, *Kolloid Z.*, **155**, 39 (1957).
10. R. Kesting, U.S. Patent 3,945,926 (1976).
11. I. Cabasso, in *Ultrafiltration Membranes and Applications*, A. Cooper, Ed., Plenum, New York, 1980.
12. R. Kesting, U.S. Patent 3,957,651 (1976).
13. N. Bjerrum and E. Manegold, *Kolloid Z.*, **42**, 97 (1927).
14. R. Kesting, *J. Appl. Polym. Sci.*, **9**, 663 (1965).
15. R. Kesting, M. Barsh, and A. Vincent, *J. Appl. Polym. Sci.*, **9**, 1873 (1965).
16. A. Chawla and T. Chang, *J. Appl. Polym. Sci.*, **19**, 1723 (1975).
17. R. Kesting and A. Menefee, *Kolloid Z. Z. Polym.*, **230**(2), 341 (1969).

18. W. Brown, *Biochem. J.,* **9**, 591 (1915).

19. A. Castro, U.S. Patent 4,247,498 (1981).

20. S. Loeb and S. Sourirajan, U.S. Patent 3,133,132 (1964).

21. R. Schultz and S. Asunmaa, *Recent Progr. Surface Sci.*, **3**, 291 (1970).

22. M. Panar, H. Hoehn, and R. Herbert, *Macromolecules*, **6**, 777 (1973).

23. S. Sourirajan, *I & EC Fund.*, **2**, 51 (1963).

24. G. Gittens, P. Hitchcock, and G. Wakely, *Desalination*, **2**, 315 (1973).

25. Y. Trudelle and L. Nicolas, Paper presented at Paris Membrane Symposium, (December 1966).

26. S. Krishnamurthy and D. McIntyre, in *Reverse Osmosis Membrane Research*, H. Lonsdale and H. Podall, Eds., Plenum, New York, 1972.

27. H. Lonsdale, in *Industrial Processing with Membranes*, R. Lacey and S. Loeb, Eds., Wiley-Interscience, New York, 1972.

28. L. Morrow and J. Sauer, Report to OSW on Grant 14-01-0001-2130 (1969).

29. R. Kesting, *Synthetic Polymeric Membranes*, McGraw-Hill, New York, 1971.

30. J. Richter and H. Hoehn, U.S. Patent 3,567,732 (1971).

31. R. Kesting, J. Newman, and J. Ditter, *Desalination*, **46**, 343 (1983).

32. N. Shiraishi, *CHEM TECH*, 366 (June 1983).

33. K. Chan, L. Tinghui, and S. Sourirajan, *I & EC Prod. Res. &Dev.*, **23**, 116–125, 124–133 (1984).

34. M. Katoh and S. Suzuki, in *Synthetic Membranes*, A. Turbak, Ed., (ACS Symposium Ser. No. 153), American Chemical Society, Washington, D.C., 1981.

35. L. Applegate and C. Antonson, in *Reverse Osmosis Membrane Research*, H. Lonsdale and H. Podall, Eds., Plenum, New York, 1972.

36. W. Pusch, personal communication.

37. L. Zeman, in *Material Science of Synthetic Membranes*, D. Lloyd, Ed., American Chemical Society (Symposia Series Vol. 269), Washington, D.C., 1985.

38. R. Kesting, S. Murray, J. Newman, and K. Jackson, *Pharm. Tech.*, **5**(5), 52 (1982).

39. R. Bhargawa, C. Rogers, V. Stannett, and M. Szwarc, *TAPPI*, **40**, 564 (1957).

40. P. Carnell and H. Cassidy, *J. Polym. Sci.*, **55**, 233 (1961).

41. P. Francis and J. Cadotte, OSW R&D Report 177 (1964).

42. R. Petersen, U.S. Patent 4,210,529 (1980).

43. H. Yasuda, in *Reverse Osmosis and Synthetic Membranes*, S. Sourirajan, Ed., NRCC Publ. No. 15627, Ottawa, Canada, 1977.

44. P. Morgan, *Condensation Polymers*, Interscience, New York, 1965.

45. J. Cadotte, U.S. Patent 4,277,344 (1981).

46. K. Kuritara, U.S.Patent 4,366,062 (1982).

47. J. Henis and M. Tripoldi, U.S.Patent 4,230,463; *J. Membrane Sci.*, **8**, 233, (1981); *Science*, **220**, 11 (1983).

48. H. Strathmann, *Trennung von Molekularen Mischungen mit Hilfe Synthetischer Membranen*, Steinkopf Verlag, Darmstadt, 1979.

8 OTHER POROUS MEMBRANES

Virtually all polymeric membranes, even the relatively dense types, contain some porosity or void volume. However, it is usually not until the porosity becomes comparable to the volume occupied by the polymer, so that microvoids interconnect, that the resultant membranes are considered porous. The prototype of porous membranes is, of course, the phase-inversion membrane which was encountered in Chapter 7. However, where the high porosity which is presently attainable only by phase-inversion techniques is not required, a number of viable alternatives exist.

8.1 SWOLLEN DENSE FILMS

The total immersion of dense membranes into a swelling system followed by exchange of this system for a nonsolvent medium is one method for the production of porous membranes. Brown[1] originated this swelling technique to prepare reproducible collodion membranes for biological separations. The method consists of the preparation of dense films by the air drying of nitrocellulose solutions to completion. The dense films are then immersed in solutions of ethanol and water. After subsequent washing with water, the resultant membranes exhibit permeabilities which are directly proportional to the concentration of alcohol, that is, to the solvent strength of the swelling medium. The swelling parameters representative of collodion membranes prepared by this method are found in Table 8.1.

Even a cursory glance at Table 8.1 suffices to indicate that this method is not without its limitations. The swelling ratio, which is relatively insensitive to alcohol concentration up to 90% suddenly becomes extremely sensitive at values above that. The loss of polymer substrate owing to the solvent action of the swelling medium also becomes appreciable at higher concentrations. Complications may

TABLE 8.1 SWELLING PARAMETERS FOR COLLODION MEMBRANES PREPARED BY THE METHOD OF BROWN[a]

Alcohol Concentration in Swelling Medium (% of volume)	Wet Weight[b] $\times 10^2$	Dry Weight[b] $\times 10^2$	Swelling Ratio (wet wt/dry wt)	Polymer Loss during Swelling(%)
0	106.0	100.0	1.06	0.0
50	111.0	99.3	1.18	0.7
70	115.6	98.6	1.17	1.4
80	122.4	98.2	1.25	1.8
90	141.9	97.0	1.46	3.0
92	160.5	93.8	1.71	6.2
94	208.5	89.1	2.34	10.9
96	387.8	74.8	5.18	25.2

[a]From Brown.[1] Reprinted by permission from *Biochemical Journal*, © 1915 the Biochemical Society, London.
[b]Normalized with respect to the air dry weight of the original membrane.

also arise with respect to the preparation of the initial dense membrane. Because of the structural variations which are possible between bulk structures of a given polymer, considerable care must be exercised in the preparation of the intermediate (primary) films to ensure constancy in the final (secondary) membranes. Gregor and Sollner[2] found it necessary to utilize high-purity solvents, distilled water, a room whose temperature was constant at 20 ± 0.5°C, and a controlled-humidity atmosphere in order to obtain dense films of oxidized nitrocellulose which could then be reswollen to yield reproducible porous ion-exchange membranes.

The immersion of cellophane into alcohol–water solutions results in membranes whose degree of swelling increases with increasing water content. As a result of solvent exchanging 95% alcohol for alcohol–water, the membranes become permeable (proportional to the degree of swelling in alcohol–water) with alcohol-miscible liquids such as benzene (Table 8.2). The dimensions of cellulose filters vary with the hydrogen-bonding capacity of the swelling medium (Table 8.3). Changes on the macrolevel are, of course, indicative of changes on the colloidal level (Table 8.4). The void volume and average pore area increases with the increasing hydrogen-bonding capacity of the swelling medium. Pore density, on the other hand, decreases with increasing swelling because of the macroextension of the membrane. The effective pore area is not significantly affected by swelling.

Closely related to Brown's swelling technique is the phenomenon of conditioning polyethylene membranes by immersing them in the isomeric xylenes before using them in permeability experiments.[5] Not only is permeability increased, thanks to the swelling action of the xylenes, but the permselectivities increase as well. This may be indicative of a template effect, in which microvoids are formed whose size and shape correspond to those of molecules or molecular aggregates of which the swelling medium is composed.

TABLE 8.2 PERMEABILITY OF BENZENE THROUGH CELLOPHANE "600" SWOLLEN TO VARIOUS EXTENTS IN ALCOHOL–WATER AND SOLVENT EXCHANGED WITH ALCOHOL[a]

Swelling Solution (% alcohol in water)	Thickness (mm) Initial	In 95% Alcohol	Pressure (kg/cm^2)	Benzene Permeability (mL effluent/pressure hr cm^2)
Dry	0.0475	0.0475		
75	0.060	0.058	36	1.8
50	0.075	0.070	31	3.4
25	0.082	0.075	35	3.9
0 (distilled water)	0.095	0.082	38	13.0

[a]From McBain and Kistler.[3] Reproduced from *The Journal of General Physiology*, by copyright permission of The Rockefeller University Press.

TABLE 8.3 DIMENSIONS OF A CELLULOSE FILTER IN VARIOUS SWELLING MEDIA[a]

	Length (μm)	Width (μm)
Water	1216	573
Ethanol 70%	1171	562
96%	1159	547
100%	1148	537
Ethanol–benzene (1:1)	1144	537
Benzene	929	428
Benzene–methyl methacrylate (1:1)	929	428
Methyl methacrylate	919	428

[a]From Spandau and Kurz[4]; © 1957.

TABLE 8.4 PORE STATISTICS OF CELLULOSE MEMBRANES IN VARIOUS SWELLING MEDIA[a]

Swelling Medium	Average Pore Area $(\mu m^2) \times 10^2$	Pore Density (pores/cm^2) $\times 10^{-8}$	Effective Pore Area (%)
Water	4.85	2.98	16.9
Ethanol	3.76	3.15	17.4
Benzene	2.73	3.37	16.5
Methyl methacrylate	2.85	3.39	17.1

[a]From Spandau and Kurz[4]; © 1957.

8.2 STRETCHED SEMICRYSTALLINE FILMS

In many respects the Celgard® process, in which semicrystalline films or fibers are extruded from the melt and porosity induced by simply stretching the finished articles in the solid state, represents the ideal insofar as the manufacturing of microporous membranes is concerned. No solvents are required. Polypropylene (PP), the polymer chosen for extensive commercialization, is among the lowest-cost membrane substances and is available in a large number of specialty grades. Furthermore, production rates are believed to be high. Although the process is limited to certain slitlike pore sizes—generally ~ 0.2 μm in length and 0.02 μm in width—and porosities $\sim 40\%$ and thus lacks the extremely broad range of pore sizes and void volumes encompassed by phase-inversion membranes, nevertheless, for many applications such structures are extremely useful.

The Celgard® process is comprised of a number of interrelated steps:

1. Extrusion of film or fiber under conditions of relatively low melt temperature and high melt stress. In other words, the takeup speed is considerably greater than the extrusion rate. Under these conditions the PP molecules align themselves in the machine direction in the form of microfibrils which are believed to nucleate the formation of folded-chain row lamellar microcrystallites perpendicular to the machine direction.[6, 7]

2. The row lamellae are consolidated by annealing at a temperature just below the T_m. Segmental motion is permitted which results in crystallite growth and densification as well as folding of the polymer chains at crystallite surfaces but prohibits melting which would tend to relax the lamellae and allow them to assume the more random spherulite formation which obtains under unstresssed conditions. The lamellae are separated from one another by amorphous regions composed of atactic blocks or otherwise noncrystalline material in the 50% crystalline polymer. At this juncture the precursor films or fibers remain dense but exhibit different stress–strain properties (Fig. 8.1) and greater elasticity (Fig. 8.2) than comparable objects prepared from unstressed and unannealed PP. The morphology of the row lamellar precursor films or fibers are shown schematically in Figure 8.3.[8-10] In what amounts to controlled crazing, the dense precursor objects are stretched (50–300%) at a temperature above the initial annealing temperature but below the T_m. This deforms the amorphous regions between the lamellae into fibrils and results in a porous interconnecting network of slitlike voids in the machine direction (Fig. 8.3b). The dimension of the pores are defined by the drawn fibrils. They are 0.4 μm in length and 0.04 μm in width for Celgard® 2500. Porosity is 40% and pore density 9×10^9 pores/cm. However, the stretching temperature may not be critical. Indeed one patent calls for stretching at room temperature.[11] The objects become noticeably more opaque at this point and the apparent density decreases (Fig. 8.4). The extent of stretching controls both pore size and pore size distribution (Fig. 8.5). Films which are stretched only 100% have a bimodal distribution of pore sizes with many pores greater than 0.15 μm. These are more permeable than the films which are stretched 300% because the latter contain primarily pores below 0.1 μm. Stretching

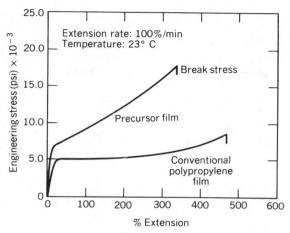

FIGURE 8.1. Stress–strain properties of precursor film prepared from isotactic polypropylene (from Bierenbaum et al.[8]; reprinted with permission from *Industrial Engineering Chemistry, Product Research Development*, © 1974 American Chemical Society).

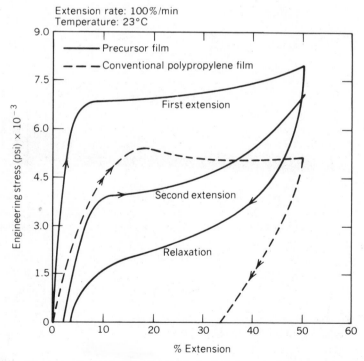

FIGURE 8.2. Recovery of precursor film from high elastic deformation (from Bierenbaum et al.[8]; reprinted with permission from *Industrial Engineering Chemistry, Product Research Development* © 1974 American Chemical Society).

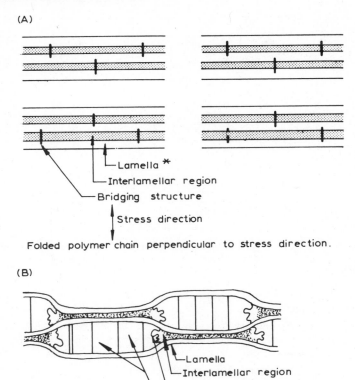

FIGURE 8.3. Schematic representation of semicrystalline morphology of (A) Celgard® precursor (extruded and annealed film), and (B) microporous Celgard® film after stretching) (from Bierenbaum et al.[8]; reprinted with permission from *Industrial Engineering Chemistry, Product Research Development* © 1974 American Chemical Society).

in excess of 300% results in a precipitous loss of porosity. Finally, because the newly stretched porous films are still elastic, they are set at a temperature just below the T_m while still under tension. This minimizes subsequent loss of porosity due to creep.

The surface structure of Celgard® 2500 shows rows of elongated pores separated by unstretched lamellae (Fig. 8.6). The stretched lamellar pores are aligned horizontally, that is, in the original machine direction. Fibrillar bridging structures separate the pores from each other and the rows of pores alternate with the unstretched lamellar crystallites. The cross-sectional view of the bulk structure indicates the presence of a 0.5-μm thick surface region whose density is greater than that of the substructure (Fig. 8.7).

The three-dimensional composite view of Celgard® 2500 (Fig. 8.8) clearly shows the pores defined by drawn fibrils to be slits with the major axes parallel to the

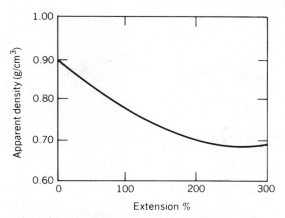

FIGURE 8.4. Apparent density of microporous polypropylene film as a function of extension (from Bierenbaum et al[8]; reprinted with permission from *Industrual Engineering Chemistry, Product Research Development* © 1974 American Chemical Society).

machine direction and the film surface. The longest dimension of the pore depends on the distance between the lamellar microcrystallites.

Although Celgard® is thin (0.025 cm thick) it can be laminated to itself to increase its stiffness and ease of handling. Its physical properties, reflecting the folding endurance characteristics of unmodified PP, are outstanding (Table 8.5). Its compatibility with various chemicals is what would be expected of unmodified PP films with a high surface area (Table 8.6).

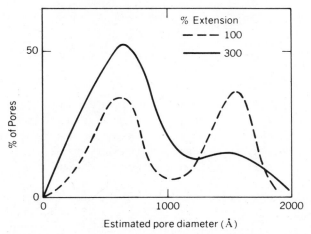

FIGURE 8.5. Pore-size distribution in microporous polypropylene films (from Bierenbaum et al.[8]; reprinted with permission from *Industrial Engineering Chemistry, Product Research Development* © 1974 American Chemical Society).

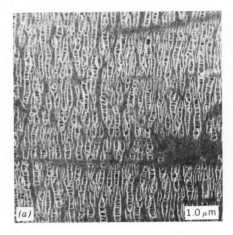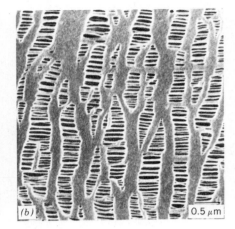

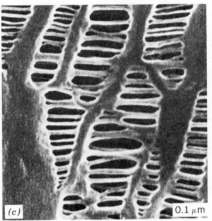

FIGURE 8.6. High-resolution secondary electron images of Celgard® 2500 surface (from Sarada et al.[10]).

Celgard® is available in both film and hollow-fiber form. Celgard® 2400 and Celgard® 2500 are hydrophobic films with effective pore size (pore-width dimensions) of 0.02 and 0.04 µm, respectively. Two-ply forms are also available as are various composite laminates to nonwoven polyproplyene fabrics. The corresponding hydrophilic (surfactant-containing) grades are Celgard® 3400 and Celgard® 3500.

The two hydrophobic microporous hollow-fiber grades, Celgard X-10 and X-20, differ in porosity, ~20 and 40%, respectively, but not in effective pore size (0.03 µm). They both have MW cutoffs of approximately 100,000 daltons. Celgard X-10 is available in 100, 200, and 240-µm ID 25-µm wall thickness. One particular area where these fibers are expected to dominate is in hollow-fiber blood oxygenators.

Gore-Tex® microporous poly(tetrafluoroethylene) (PTFE),[12, 13] is one of the most

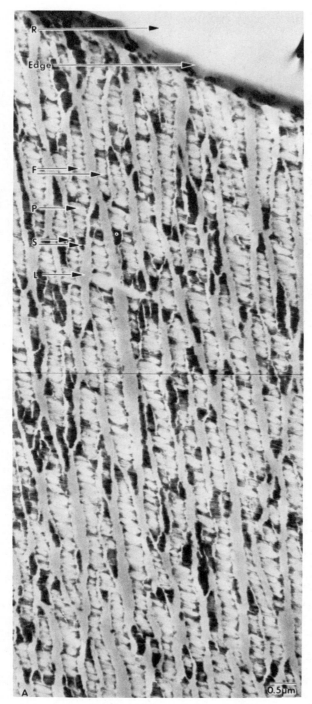

FIGURE 8.7. TEM micrograph of surfactant/OsO_4 treated cross section of Celgard® 2500 (from Sur-ada et al.[10]).

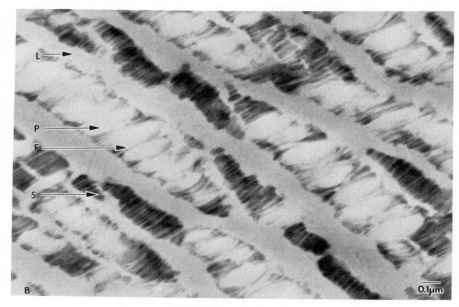

FIGURE 8.7. *(Continued)*

important of the porous membranes manufactured by a process other than phase inversion. Gore-Tex® also resembles Celgard® in that they both contain characteristic slitlike pores (Figs. 8.9 and 8.10). This is because both processes utilize stretching to introduce porosity. The fact that the slits in Gore-Tex® are not always parallel to one another is attributable to the fact that biaxial stretching is sometimes employed. Since PTFE cannot be melt extruded, a highly (~98.5%) crystalline dispersion polymer of 500,000 MW and fine (0.1 μm) fiberlike structures, Teflon® 6A, is mixed together with 15–25% of a lubricant such as naphtha or kerosene and

TABLE 8.5 TYPICAL PHYSICAL PROPERTIES OF CELGARD FILM[a]

Property	Value	Test Method
Tensile strength, MD[b]	20,000 psi	ASTM D882
TD[c]	2,000 psi	
Tensile modulus, MD	2×10^5 psi	ASTM D882
Elongation, MD	40%	ASTM D882
Tear initiation, MD	1 lb	ASTM D1004
MIT fold endurance	10^5	ASTM D643
Mullen burst	20 points	ASTM D774

[a]From Bierenbaum et al.[8] Reprinted with permission from *Industrial Engineering Chemistry Product Research Development*, © 1983 American Chemical Society.
[b]MD = machine direction.
[c]TD = transverse to machine direction.

TABLE 8.6 COMPATIBILITY OF CELGARD FILM WITH VARIOUS COMPOUNDS[a,b]

Acids		Halogenated Hydrocarbons	
H$_2$SO$_4$ (concd)	A	Carbon tetrachloride	C
Alcohols		Tetrachloroethylene	C
Ethyl alcohol	A	(perchloroethylene)	
Ethylene glycol	A	Hydrocarbons	
Isopropyl alcohol	A	Benzene	B
Ether alcohols		Hexane	B
Butyl Cellosolve	B	Toluene	B
(2-butoxyethanol)		Ketones	
Methyl Cellosolve	A	Acetone	A
(2-methoxyethanol)		Methyl ethyl ketone	A
Bases		Oils	
KOH (40%)	A	10W30 motor oil	B
Ethers		Miscellaneous	
1,4-dioxane	B	*NN*-dimethylacetamide	A
Fuels		*N,N*-dimethylformamide	B
Gasoline	B	Nitrobenzene	B
Kerosene	B	Tetrahydrofuran	B
		Freon TF	B

[a]From Bierenbaum et al.[8] Reprinted with permission from *Industrial Engineering Chemistry Product Research Development*, © 1983 American Chemical Society.
[b]The compatibility statements are based on 72 h of exposure at room temperature (25°). Key: A, good (no effect); B, slight swell; C, material swells, separation characteristics should be evaluated.

then ram extruded. The lubricant is then removed by heating, after which the sheet is reduced in thickness by passing between calender rolls at 80°C. Uniaxial or biaxial stretching is followed by sintering at 327°C. During the sintering process the amorphous content increases and serves to "lock in" and strengthen the pores in the stretched membrane. The Gore-Tex® process is versatile and capable of producing membranes with pore-size and porosity ranges which rival those of phase-inversion membranes (Table 8.7). Gore-Tex® membranes are also available in laminated composites with a variety of support substrates. They represent the most chemically inert and hydrophobic synthetic polymeric membranes and are unique in their ability to filter organic solutions and hot inorganic acids and bases which are vital to the electronics industry.

8.3   SINTERED-PARTICLE MEMBRANES

Sintering refers to any change in shape undergone by a small particle or a cluster of particles of uniform composition when held at an elevated temperature.[14] In producing membranes by the sintering process, finely divided particles (spherical or fibrous in shape) are heated to a temperature at or below the melting range of the material. As the exterior surface of the particles soften or melts, capillary pres-

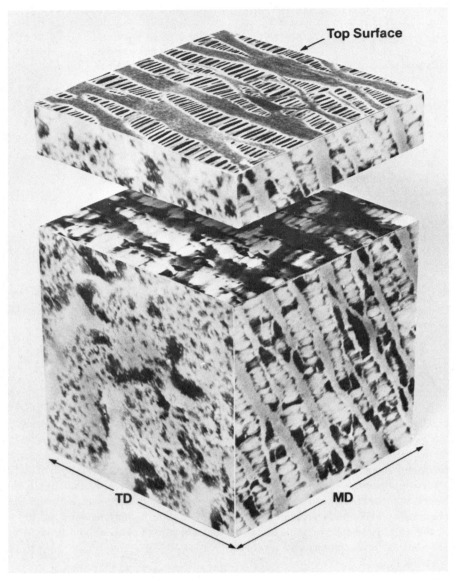

FIGURE 8.8. Three-dimensional representation of Celgard® 2500 microstructure (from Sarada et al.[10]).

sure tends to rearrange the solid particles in such a way that maximum packing and a minimum of surface area result. The decrease in surface area of the particles which occurs upon heating provides the driving force for densification.[15] Other factors are decreasing surface rugosity and the resultant slippage of particles past one another. Subsequently, bridges are formed, which collapse owing to the solu-

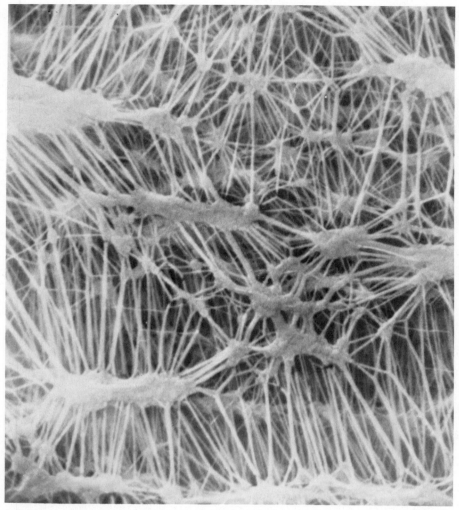

FIGURE 8.9. SEM photograph of surface of 0.2-μm Gore-Tex® membrane (W. L. Gore and Assoc.[12]).

tion of small amounts of material at the contact points. When this occurs, substantial rearrangement of neighboring particles can take place, with an ultimate disposition of solid particles consistent with high density. Although sintering is usually described as a single geometric process, such aspects of sintering as densification, rounding of particle surfaces, pore closure, and aggregation of particles into fewer though larger units are all to some degree independent of one another. Thus sintering is not described adequately in terms of any one of these changes alone.[16] The mechanisms of material transport involved in the sintering of polymeric materials include viscous or plastic flow, volume diffusion, surface migration, and

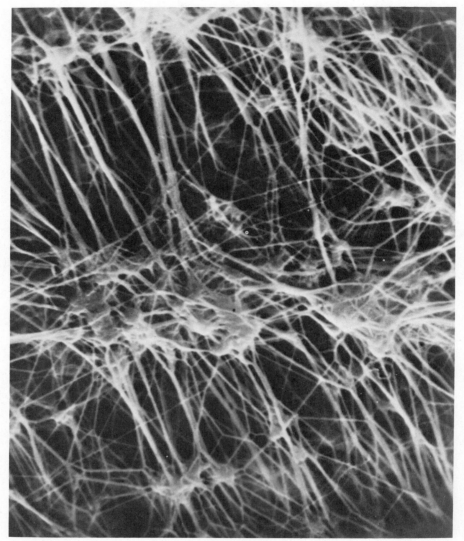

FIGURE 8.10. SEM photomicrography of surface of 1.0-μm Gore-Tex® membrane (W. L. Gore & Assoc.[12]).

creep. The evaporation and condensation mechanism sometimes involved in the sintering of low-formula-weight materials such as sodium chloride[17] is not probable in the case of macromolecules. Each of these various mechanisms has a distinct function to perform in the overall geometric change that constitutes sintering.

The sintering process for preparation of polymeric membranes is largely restricted to polymers with a flexible configuration. Excluded are rigid macromolecules or molecules having numerous bonds with an association energy which ex-

TABLE 8.7 GORE-TEX® MEMBRANE PROPERTIES[a]

Pore Size (μm)	Typical Thickness (in.)	Typical Porosity (%)	Typical Flow Rates		Minimum Water Entry Pressure (psi)	Minimum Bubble Point (psi)[d]
			Air[b]	Methanol[c]		
0.02	0.003	50	2.9	1.9	350	40
0.2	0.0025	78	75	50	40	13
0.45	0.003	84	170	110	20	7
1.0	0.003	91	530	350	10	3
3	0.001	95	1200	800	2	1
5	0.001	95	5700	3800	0.5	0.7
10–15	0.0005	98	14,600	9700	0.25	0.4

[a]From W.L. Gore and Assoc.[12]
[b]Air flow rate: Milliliters per minute per cm² of membrane area with a pressure drop of 4.88 in. H_2O at 21°C.
[c]Methanol Flow Rate: Milliliters per minute per cm² of membrane area with a pressure drop of 27.5 in. Hg at 21°C.
[d]Membranes wetted with anhydrous methanol at 21°C.

ceeds the temperature at which general pyrolytic breakdown occurs. For sintering of polymeric substances to occur, particle surfaces must soften sufficiently to permit interdiffusion of macromolecular segments into adjacent particles. This does not imply, however, that actual surface melting is a prerequisite of sintering, since sintering of poly(tetrafluoroethylene) at 327°C occurs without melting. Intercrystalline linkages may result from both folded-chain and extended-chain configurations. The sintering temperature of a given material depends on a number of factors, among which the nature and molecular weight of the polymer, crystallinity, the presence or absence of plasticizers and other additives, pressure and the nature of the atmosphere are important. Sintering temperature increases with increasing polymer molecular weight and polarity. It also increases with increasing crystalline size and crystalline content because molecular displacement is constrained by these parameters. Plasticizers tend to decrease, and antiplasticizers to increase, the sintering temperature. Inasmuch as pressure raises the freezing range of polymers, it will also increase the sintering temperature.

Many variations of the sintering process can influence the strength and permeability of the resultant membranes. Sintering loose granules results in weaker but more permeable membranes than the sintering of compressed powders. Where pressure is employed, a nonsinterable additive is often added which can be extracted from the membrane subsequent to sintering; for example, starch granules have been added to powdered polyethylene for subsequent leaching by water.[18] Densification can be minimized by the rapid application of thermal energy to heat-particle surfaces without affecting their interiors. Microwave sintering and electrical resistance have been employed to this end. In the latter instance the particles may be covered with a conductive coating such a graphite for subsequent removal.[19]

Permeability through sintered polymeric membranes increases with increasing particle size. Air permeability doubles as granule size increases from 0.18 to 1.0 mm.[20] Because the effective membrane area is low and pore-size distribution can vary with particle-size distribution and packing arrangement, the sintering process for membrane fabrication is generally restricted to substances for which alternate processes are not available. Until recently, the low solubility and/or polarity of such polymers as polyethylene and poly(chlorotrifluoroethylene) virtually limited production of membranes of these materials to melting or sintering processes. Although membranes produced by the sintering process are far from ideal insofar as their pore statistics are concerned, their relatively high strength and resistance to compaction, coupled with the chemical inertness of their polymer substrates, ensures significant utilization in certain specialized separations.

8.4 PARCHMENT PAPER AND CELLOPHANES

Cellulose membranes encompass barriers ranging from the papers, whose intact fibrillar structure is inversely proportional to the extent of beating, through glassine (finely beaten and compressed fibrils), to parchment paper and the cellophanes. In the last two, the fibers have been dissolved to varying extents before reprecipitation to form membranes. Parchment paper is produced from paper by passage through a bath of 67–78% sulfuric acid prior to immersion in water.[21] The acid transposes the surface fibers: into a gelatinous mass which forms a protective coating for the unaltered fibers beneath. After the acid has been removed by washing, the parchmentized web is treated with a solution of glucose or glycerol to reduce brittleness, and dried. Because most of the voids between the fibers are lost as a result of the increased surface contact and the resultant greater adhesion and compaction of the gelatinous mass, parchment paper is relatively impervious to air and water.

Cellophanes are prepared by the more or less complete dispersion of cellulose fibers, either as the xanthate (cellulose–CSSNa) in viscous solutions, or as cellulose–copper–ammonium complexes in cuprammonium hydroxide (cuoxam). Strictly speaking, only cellulose membranes from viscose solutions are referred to as cellophane; those from cuoxam solutions are called Cuprophane. Other solvents exist but have apparently not been utilized in commercial production. Structural differences between cellulose membranes which have been prepared from viscose and cuoxam solutions have been observed; the latter results in more homogeneous cross-sectional structures.[22] Similar differences have also been observed for viscose and cuoxam fibers.[24] The fact that the water uptake for cuoxam cellulose membranes is higher than that for viscose membranes suggests differences in their microcrystalline morphology. These can in turn be attributed to differences in solvent power and hence degree of dispersion within the sols prior to the precipitation.

The cellophanes are more permeable than parchment paper because of their greater degree of swelling (Table 8.8). Changes in the relative humidity have pronounced effects upon the permeability of water vapor through cellophane (Table 8.9).

TABLE 8.8   SWELLING BEHAVIOR OF PARCHMENT PAPER AND
CELLOPHANE MEMBRANES[a]

	Parchment Paper[b]	Cellophane[c]
Thickness increase, dry to wet (%)	45	100
Area increase, dry to wet (%)	18.5	23.6
Weight increase, dry to wet (%)	47.5	60.5[d]

[a]Adapted from Manegold and Viets.[24]
[b]Dialysis parchment G155:100, Schleicher and Schüll.
[c]Kalle.
[d]Modern cellophanes vary from 45 to 82 for viscose to 99 to 134 for cuoxam.[20]

TABLE 8.9   INFLUENCE OF RELATIVE HUMIDITY ON DIFFUSION OF
WATER VAPOR THROUGH CELLOPHANE[a]

Relative Humidity	Permeability (moles cm^2/s difference for 1-cm thickness) $\times 10^{-13}$
100	2700
50	300
0	1.5

[a]From Spurlin,[25] from data of Hauser et al.[26] and Doty et al.[27] (© 1954).

8.5   LEACHING-PROCESS MEMBRANES

The compounding of solid pore formers into a polymer melt or solution followed
by the extrusion and solidification of the resultant mass in thin-film form and,
finally, preferential leaching of the pore formers by a nonsolvent for the membrane
matrix are the essential steps of the leaching process for the preparation of porous
membranes. The purest form of the leaching process involves the milling of finely
divided fillers such as colloidal silica and salt granules into a melt. Although probably not true in every case, the filler is seen as inert in that its dispersion is primarily physical and (other than wetting) does not require a great deal of interaction
between filler particles and the polymer matrix. A further characteristic assigned
to the leaching-process membranes is that porosities tend to be low, typically less
than 40%.

However, when dealing with leached membranes with solution precursors, distinctions between the leaching and other processes which also begin with solutions
becomes somewhat arbitrary. Any given example might properly fit into any of
several membrane categories. Thus the formation of solutions and membranes from
physically compatible blends of two polymers and the subsequent evaporation of
the solution to yield an interpenetrating polymer network (IPN), has been categorized by this author as the PAPI variation of the phase-inversion process (Section

7.5). The basis for this assignment was the appearance of two interspersed liquid phases (Sol 2) prior to gelation. Likewise, the leaching of preformed polymers from membranes which have been formed as a result of polymerization of monomers has been treated separately (Section 8.6). In this case low membrane porosity and the desirability of emphasizing polymerization rather than phase inversion dictated the assigned category.

Perhaps the most promising of the leaching processes are those in which the poreformers are low-MW surfactants (preferably ionic types) which form high-MW, evently dispersed micelles in the fluid state and maintain this structure in the solid polymer matrix.[28] After the swollen solid matrix has been leached, the pores constitute the volumes previously occupied by the surfactant micelles. The surfactant must be added to the membrane precursor solution or suspension in micellar form which means that it must be added in amounts exceeding the critical micelle concentration (CMC). The preferred amounts of surfactants are from 10 to 200% (based on the weight of the membrane polymer). Porosity increases with increasing surfactant concentration (Table 8.10). The membrane control in Table 8.10 was clear, indicating minimal porosity. Turbidity increased, but not to opacity, as porosity increased in this series of ultragel membranes. High-porosity opaque microgel membranes resulted when 200% sodium dodecyl benzene sulfonate was added to the viscose solutions of the same concentration as shown in Table 8.10. The resultant microfilters were capable of retaining 10^9 *Pseudomonas diminuta*/cm² which suggests a pore size ~0.2 μm. The membrane polymers should be such that they will not cold flow at room temperature or the micelle extraction temperature. The preferred liquid carriers are water, the lower alcohols, and toluene. After the films are solidified they are swollen in a liquid which has the effect of breaking down the micelles to individual surfactant molecules to permit their ready extraction.

This surfactant leaching process has been applied to a variety of solutions including cellulose and methoxymethylated nylon 6/6 and to polyacrylic, poly(vinyl acetate), and polyethylene—paraffin lattices. In the last mentioned case, lauryl pyridinium chloride in toulene was utilized as the surfactant micelle and a microporous polyethylene membrane was the result.

TABLE 8.10 EFFECT OF SODIUM LAURYL SULFATE CONCENTRATION IN VISCOSE SOLUTION[a] UPON THICKNESS AND PERMEABILITY OF CELLULOSE ULTRAGEL MEMBRANES[a]

Membrane	[SLS] (% of polymer)	Thickness (in.) × 10^3	Permeability (gH₂O/in² at 20 in. Hg)
Control	—	10.3	0.369
A	10	10.2	0.585
B	20	10.9	0.712
C	50	13.9	0.864

[a]From Bridgeford.[28]
[b]7% cellulose + 5.8% NaOH.

8.6 POROUS MEMBRANES FORMED DURING POLYMERIZATION

The formation of addition-type thermoset polymers can be carried out under conditions which will lead to the formation of either dense or porous membranes. In the case of polymers such as styrene (S)–divinylbenzene (DVB), two techniques are available for the synthesis of *macroreticular*, or porous, membranes. Increasing the DVB (cross-linker) concentration beyond ~ 12% causes the separation of insoluble cross-linked gel particles from the solution, apparently in much the same manner as occurs in the formation of Sol 2 prior to gelation during phase inversion. The second approach is analogous to leaching (Section 8.5) and PAPI (Section 7.5) processes in that preformed polymers are added to the initial solution for leaching in a postpolymerization step. In the case of S–DVB, polystyrene is added to the monomer mix and removed by leaching with water usually after sulfonation of the S–DVB *and* the included polystyrene. As previously indicated (Section 4.2), the polymerization of hydroxyethylmethacrylate (HEMA) in solution can result in the formation of porous hydrogels of both ultragel and microgel types. The formation of insoluble gel particles from HEMA occurs at high monomer conversions and is probably attributable to cross-linking as a result of interaction between free radicals produced via chain-transfer reactions.

8.7 NUCLEATION-TRACK MEMBRANES

In 1959 Silk and Barnes[29] discovered that fission fragments from radioactive decay, which impinge upon and penetrate solids, can, under certain conditions, result in narrow trails of radiation-damaged material. Such tracks can often be enlarged by etching with suitable reagents. Fleischer et al.[30] subsequently exploited this phenomenon to develop a class of nucleation-track (Nuclepore®) membrane filters characterized by cylindrical pores with a narrow pore-size distribution.[31]

Tracks can be developed in nonconducting substances, whether they are inorganic (e.g., mica), or organic (e.g., polymeric) films. When a massive charged particle such as a heavy positive ion passes through inorganic material, it propels electrons out of atoms in the crystal lattice, creating an ion-explosion spike. The resulting wake of positive ions causes mutual repulsion, thereby disrupting the regular lattice. The relative sensitivities of different crystalline inorganic materials increase with decreasing values of $\epsilon E V^{4/3}$, where ϵ is the dielectric constant, E is Young's modulus, and V is the average volume per atom. There is an energy loss threshold for track production corresponding to approximately one ionization per atomic plane crossed by the charged particle. For most organic polymers, however, where less than one ionization per atom plane is produced, broken chemical bonds are a more probable explanation of track formation than the ion-explosion-spike model. The density of ionization damage caused by a nuclear particle is directly proportional to the square of the electric charge of the particle and approximately inversely proportional to the square of its velocity. The explanation of the latter phenomenon lies in the increasing probability of collision between the oribiting electrons and the ionic particles with increasing length of time spent in close proximity to one another.

For each material in which tracks can be produced, there is a characteristic critical value of energy-loss rate dE/dx (see Table 8.11). Because the maximum possible level of dE/dx increases with increasing atomic number, for every substance there is a lower limit for the masses of heavy ions which can produce tracks. Particles which give up energy less rapidly than this critical value fail to produce etchable tracks.

Because fission fragments disperse in random directions, they must align to produce parallel tracks in a polymeric film. This is achieved with a collimated beam of particles by separating a uranium or californium source from the film, evacuating the intervening space, and exposing the assembly to thermal neutrons. The larger the separation and the smaller the source area, the more complete the resulting hole alignment will be.

Damage tracks in polymeric films are composed of end-group species which are more highly reactive than the bulk polymer as a result of greater chemical activity and increased accessibility. The tracks consequently can be developed by immersion in acid or base. Not all tracks which intersect on etched surface will necessarily be revealed, however. If the rate of attack perpendicular to the surface is V_G and the velocity of attack along the track is V_T, the tracks making an angle less than $c = \sin^{-1}(V_G/V_T)$ with respect to the surface will not be seen since the surface will dissolve away more quickly than tracks can develop.[33] Tracks lengthen and widen in linear fashion with increased time of solution until some maximum length is reached. They have a conical shape, the cone angle θ_c, depending on the ratio R of the etching rate V_T along the track to the rate V_G on undamaged surfaces. For large values of R, θ_c ($= 2\csc^{-1}R$) is smaller, and the pits approach cylindrical channels in shape some $50\,\text{Å}$ in diameter. If the exposure is prolonged, the etching continues at a reduced rate corresponding to that of attack upon the undamaged material. Where R approached unity, the track will be a shallow pit.

The choice of etchants varies, of course, with the chemical nature of the polymeric film (Table 8.12), concentration, temperature, and the orientation of the attacked surface. The length of the fission track varies with the nature of the source;

TABLE 8.11 TRACK REGISTRATION IN VARIOUS POLYMERS[a]

Polymer	Etching Conditions	Critical Rate of EnergyLoss ($MeV/mg\ cm^2$)	Lightest Detectable Particle
Poly(ethylene terephthalate) (Mylar)	6 N NaOH, 10 min, 70°C	4	O
Bisphenol-A polycarbonate (Lexan, Merlon)	6 N NaOH, 8 min, 70°C	4	C
Cellulose acetate butyrate	6 N NaOH, 12 min, 70°C	2	He
Cellulose nitrate	6 N NaOH, 2–4 h, 23°C	2	H

[a]From Fleischer et al.[32]

TABLE 8.12   ETCHING SOLUTIONS FOR VARIOUS POLYMERIC FILMS[a]

Polymer	Etching Solution
Bisphenol-A polycarbonate	aq. NaOH, sp. gr. 1.3
Poly(ethylene terephthalate)	aq. NaOH, sp. gr. 1.3
Cellulose acetate	aq. KOH, sp. gr. 1.4
Cellulose nitrate	aq. KOH, sp. gr. 1.4
Poly(methyl methyacrylate)	Aqua regia and HF (6:1)

[a]From Fleischer and Price[33]; © 1963 by the AAAS.

tracks produced by fission fragments from ^{252}Cf penetrate polycarbonate film to a maximum depth of 20 μm, whereas those from ^{235}U penetrate only 10–12 μm.

Some polymers are very sensitive to environmental conditions. The quality of the track images in cellulose acetate butyrate, cellulose acetate, and nitrocellulose depends on how the material was prepared. Track registration requires the presence of oxygen (presumably to produce reactive carboxyl groups). Because of disruption of the microcrystalline structure by nucleation tracks, cellulose acetate must be annealed prior to etching if uniform channels are to result. Other peculiarities are often encountered. Nitrocellulose yields much better tracks when etched slowly at room temperature than when etched rapidly at 70°C, even though the tracks are quite stable at the higher temperature. The presence of trace impurities may also play a dominant role in determining the geometry of attack. The etching rate may also vary in a complicated way with concentration, and here too the geometry of attack may be altered. Polycarbonate films, on the other hand, are relatively insensitive to environmental conditions.

Minimum hole diameters are obtained by partial annealing prior to ecthing and by using weaker etchants. Maximum membrane thickness is limited by the distance traversed by the incident particles. This distance can be increased, however, by using heavy particle accelerators. Maximum pore density is limited by the fact that membranes become excessively brittle and radioactive at high doses. For polycarbonate films the maximum tolerance dose amounts to approximately 10^{11} fission fragments per square centimeter, corresponding to 0.5% of the total surface area in 25-Å holes. This dose is obtained from a 5000-Å pure ^{235}U source in about 10 min at a neutron flux level $\sim 10^{13}$ nV. If natural uranium were employed, a day's exposure would be required.

A small number of pores fail to penetrate the membrane completely. This effect arises from the fact that a portion of the energy of the fission fragments is expended in traversing the radioactive source before entering the evacuated space separating the membrane from the source. This difficulty can be minimized, but not eliminated, by decreasing the thickness of the radioactive source. One use for which nucleation-track membranes are unsurpassed is as collection filters for materials which are to undergo microscopic investigation. Here they provide a more neutral background than other porous membranes (Fig. 8.11).

Although the only membrane with straight cylindrical pores perpendicular to

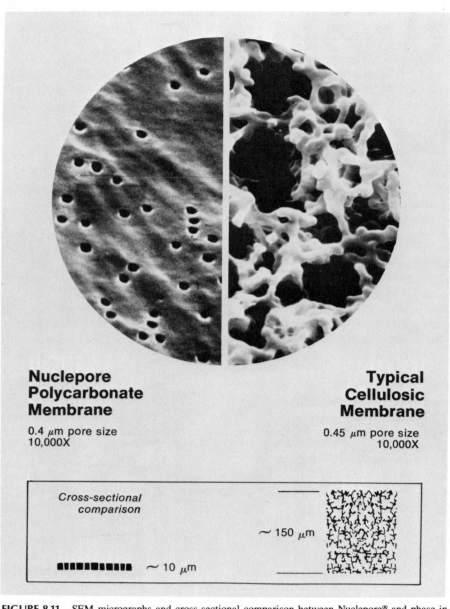

Nuclepore Polycarbonate Membrane

0.4 μm pore size
10,000X

Typical Cellulosic Membrane

0.45 μm pore size
10,000X

Cross-sectional comparison

~ 150 μm

~ 10 μm

FIGURE 8.11. SEM micrographs and cross-sectional comparison between Nuclepore® and phase-inversion membranes of equivalent nominal pore size (Nuclepore Corporation[31]).

both surfaces which is commercially available is the irradiation-track type, several other routes to comparable structures have been evaluated and found deficient in one or more respects. Among these are membranes from bundles of hollow fibers sliced perpendicular to the lumina[34] which proved impractical because of handling difficulties. The ionotropic gels of Thiele[35] after intensive development ultimately proved incapable of yielding pores smaller than about 2 μm in diameter. The utilization of lasers is deemed possible by some, but difficulties in the attainment of pores of sufficient fineness and density appear likely to arise.

REFERENCES

1. W. Brown, *Biochem. J.*, **9**, 591 (1915).

2. H. Gregor and K. Sollner, *J. Phys. Chem.*, **50**, 53 (1946).

3. J. McBain and S. Kistler, *J. Gen. Physiol.*, **12**, 187 (1928).

4. H. Spandau and R. Kurz, *Kolloid Z.*, **150**, 109 (1957).

5. A. Michaels, R. Baddour, H. Bixler, and C. Choo, *Ind. Eng. Chem. Process Des. Dev.*, **14**, (1962).

6. A. Keller, *J. Polym. Sci.*, **15**, 31 (1955).

7. B. Sprague, paper presented at the U.S.–Japan Joint Seminar on Polymer Solid State, Cleveland, Ohio, October 9–13, 1972.

8. H. Bierenbaum, R. Isaacson, M. Druin, and S. Ploven, *Ind. Eng. Chem. Prod. Res. Dev.*, **13**,(1), 2 (1974).

9. M. Druin, J. Loft, and S. Plovan, U.S. Patent 3,801,404 (1974).

10. T. Sarada, L. Sawyer, and M. Ostler, *J. Membrane Sci.*, **15**, 97 (1983).

11. K. Kamada, S. Minami, and K. Yoshida, U.S. Patent 4,055,696.

12. W. L. Gore and Assoc., Inc., Membrane Division, Elkton, MD.

13. R. Gore, U.S. Patents 3,953,566, 3,962,153 (1976).

14. C. Herring, *J. Appl. Phys.*, **21**, 301 (1950).

15. W. Kingery, *J. Appl. Phys.*, **30**, 301 (1959).

16. F. Rhines, Plansee Proceedings, Third Seminar, Reute/Tyrol, 1958, pp. 38–54; *Chem. Abstr.*, **54**, 94141 (1960).

17. V. Kargin, A. Gorina, and T. Koretskaya, *Vysokomolekul, Svedin*, 1, 1143 (1959).

18. *Chem. Eng. News*, **41**(17), 48 (1963).

19. A. Weber, German Patent 1,005,696 (April 4, 1957).

20. *Chem. Abstr.*, **60**, 10235a (1964).

21. E. Sutermeister, *Chemistry of Pulp and Paper Making*, Wiley, New York, 1920.

22. C. Jayne and K. Balser, *Ind. Chim. Belg.*, **32**, 365 (1967).

23. H. Mark, in *Cellulose and Cellulose Derivatives*, E. Ott, H. Spurlin, and M. Grafflin, Eds, Chap. 4, Wiley-Interscience, New York, 1954.

24. E. Manegold and K. Viets, *Kolloid Z.*, **56**, 7 (1967).

25. H. Spurlin, in *Cellulose and Cellulose Derivatives*, E. Ott, H. Spurlin, and M. Grafflin, Eds., Chap. 9, Wiley-Interscience, New York, 1954.

26. P. Hauser and A. McLaren, *Ind. Eng. Chem.*, **40**, 112 (1948).

27. P. Doty, W. Aiken, and H. Mark, *Ind. Eng. Chem. Ed.*, **16**, 686 (1944).

28. D. Bridgeford, U.S. Patent 3,852,224 (December 3, 1974).

29. E. Silk and R. Barnes, *Phil. Mag.*, **4**, 1970 (1955).

30. R. Fleischer, P. Price, and R. Walker, *Sci. Amer.*, **220**, 30 (1969).

31. Nuclepore® Corporation, Pleasanton, CA.

32. R. Fleischer, P. Price, and R. Walker, *Rev. Sci. Instrum.*, **34**, 510 (1963).

33. R. Fleischer and P. Price, *J. Geophys. Res.*, **69**, 331 (1964).

34. L. Akobjanoff, *Nature*, **178**, 104 (1956).

35. H. Thiele, German Patent, 1,011,853 (July 11, 1957).

9 LIQUID AND DYNAMICALLY FORMED MEMBRANES

Thin semipermeable barriers in the liquid state are known as *liquid membranes*. Although comparatively few liquid membranes are composed of synthetic polymers they are included in this volume because they represent a logical extrapolation to the most inherently permeable condensed state of matter. Liquid membranes are encountered in a variety of forms. Unsupported, or *emulsion-type* membranes, sometimes improperly called liquid surfactant membranes, consist of submillimeter-sized spherical droplets which are mixed with solutions of the material to be recovered. Supported liquid membranes are of two types. *Liquid surfactant* membranes are the semipermeable barriers consisting of a surfactant layer and any associated bound water molecules which spontaneously concentrate at the interface between solid polymeric membranes and feed solutions. *Immobilized* liquid membranes are those which are confined within a microporous solid. *Dynamically formed* membranes, although specifically distinct from liquid membranes, are also included because they initially exist in the liquid state. They are formed into a semipermeable barrier layer from suspensions of finely divided particles when the latter are allowed to impinge upon a porous surface.

While relatively little commercial use is made of these at present, the inherently high permeability of membranes in the liquid state coupled with—in certain instances—unsurpassed permselectivity, suggest that they may play a role in the future.

9.1 EMULSION-TYPE LIQUID MEMBRANES

Emulsion-type membranes are made by forming an emulsion of two immiscible phases and then dispersing the emulsion in a third or continuous phase. Although

311

the encapsulated and continuous phases are usually miscible with one another, the membrane phase must be immiscible with both of them. If the continuous phase is water, the emulsion is of the water-in-oil type (Fig. 9.1a). If the continuous phase is oil, the emulsion is of the oil-in-water type (Fig. 9.1b). To stabilize oil-in-water emulsions, saponin and glycerol are added to the membrane phase prior to emulsification. The saponin acts as a surfactant while the glycerol is a film-strengthening agent preventing film rupture.[1] Water-in-oil–type emulsions are also stabilized by the addition of surfactants. Since the essential barrier of liquid membranes consists of the organic liquid or water which is merely stabilized by the surfactant, the use of the term liquid surfactant membrane in this context appears somewhat inappropriate. The variables in emulsion-type liquid membranes are perhaps most easily understood by reference to a specific process such as the extraction of ammonia from municipal wastewater.[2]

To extract ammonia from wastewater, a water-in-oil–type emulsion is utilized. The liquid membrane consists of a blend of paraffinic solvents. Permeant diffusivity varies inversely with the viscosity of the paraffins which comprise the liquid membrane. Twenty weight percent sulfuric acid is the reagent phase. The paraffins and aqueous sulfuric acid are emulsified to yield oil droplets 0.1–0.5 mm in diameter containing micrometer-sized droplets of aqueous sulfuric acid. After the emulsion

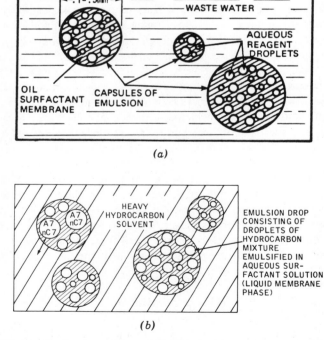

(a)

(b)

FIGURE 9.1. (a) water-in-oil-type emulsion and (b) oil-in-water type emulsion (from Downs and Li[2]).

is mixed with the wastewater, ammonia permeates the liquid paraffin membrane and enters the reagent-phase droplets where it reacts with sulfuric acid to form ammonium sulfate. Since the NH_4^+ ion is insoluble in paraffin, it is retained within the reagent-phase droplets. Once the sulfuric acid has been depleted, the reagent-phase droplets contain a solution of ammonium sulfate. The spent emulsion may be de-emulsified by a variety of methods, such as treatment with solvents,[3] centrifugation followed by high shear,[4] and electrostatic coalescence.[5]

The emulsion-type liquid membrane process is known to depend on a number of process variables: treat ratio, internal reagent utilization, globule size, internal-phase droplet size, leakage, swelling, feed pH, contact time, and staging.

With a feed containing 4000 ppm ammonia and reagent-phase droplets of 20% H_2SO_4, the stoichiometric-volume treat ratio of feed to reagent phase is 19 to 1. However, the rate of reaction increases if sulfuric acid is present in excess (Table 9.1).

Leakage refers to the loss of internal-reagent-phase droplets into the external feed phase. Leakage increases with decreasing globule diameter (Table 9.2). The

TABLE 9.1   EFFECT OF TREAT RATIO ON EXTRACTION EFFICIENCY AND INTERNAL REAGENT UTILIZATION[a]

Feed-to-Internal-Reagent Treat Ratio	% Extraction	% Reagent Utilization
36.6/1[b]	53.2	100.0
18.3/1	93.8	96.7
12.2/1[c]	98.3	76.3
9.2/1[c]	98.3	56.0

[a] From Downs and Li.[2] Reprinted from *Journal of Separation Process Technology*, p. 19, by courtesy of Marcel Dekker, Inc.
[b] Insufficient internal reagent for complete removal of ammonia.
[c] Insufficient ammonia for complete utilization of internal reagent.

TABLE 9.2   EFFECT OF MIXING INTENSITY ON GLOBULE SIZE, LEAKAGE, AND EXTRACTION[a]

Mixing Intensity (rpm)	Globule Diameter (μm)[b]	Leakage (% of total internal phase)	Extraction (% of total ammonia)
450	420	2.0	61.8
600	310	2.0	76.4
900	210	3.7	93.2
1100	180	7.0	94.9
1400	c	18.3	97.1

[a] From Downs and Li.[2] Reprinted from *Journal of Separation Process Technology*, p. 19, by Courtesy of Marcel Dekker, Inc.
[b] Globule size was measured photographically.
[c] Too small to measure accurately by photographic technique.

latter is controlled by mixing intensity. However, extraction efficiency also increases with decreasing globule diameter (increasing membrane surface area) so that compromises are necessary. Swelling, the opposite of leakage, refers to the amount of water which transfers from the feed phase to the emulsion phase. The number of stages and the flow directions of feed liquid membrane relative to one another influence extraction efficiency in a predictable manner. Efficiency increases with countercurrent flow and an increasing number of stages. It also increases if the emulsions are allowed to settle between stages, but then so does leakage.

The efficiency of liquid membranes is enhanced in one of two ways. The first involves the maintenance of the maximum concentration difference of the diffusing species between the feed solution and the internal reagent phase. This is effected by reaction of the diffusing species with the reagent in the microdroplets to yield a substance which is insoluble in the liquid membrane and hence will not diffuse back into the feed solution. For example, weak bases such as ammonia and weak acids such as phenol are converted by sulfuric acid and sodium hydroxide, respectively, to ammonium and phenolate ions. The second approach is to increase the solubility of the diffusing species in the liquid membrane by the incorporation of carrier materials in the membrane.[6-10] In the case of the extraction of chromium, for example, a tertiary amine can be included in the liquid membrane. The amine will react with dichromate ions at the interface between the emulsion droplets and the feed phase. The amine–dichromate ion complex will then be carried to the reagent-phase microdroplet interface where sodium hydroxide will strip the dichromate ions and release the amine to repeat the process.

The separation of aromatic hydrocarbons such as toluene from aliphatic hydrocarbons such as heptane is possible because of the higher permeation rates of the former through the aqueous phase. A water-in-oil–type emulsion is required. The toluene–heptane mixture is emulsified together with saponin and glycerol.[1] The emulsion is added to kerosene and the toluene preferentially leaves the microdroplets and enters the kerosene phase. The relative permeability of toluene to heptane, α, increases with the stirring rate: $\alpha = 6$ at 55 rpm and increases to ~ 20 at 370 rpm. The separation of olefins from paraffins can be enhanced by the utilization of carriers such as cuprous ammonium acetate which form olefin complexes whose solubility in the liquid membrane is greater than that of the olefin by itself.

9.2 SUPPORTED LIQUID MEMBRANES

Two possible modes for the support of liquid membranes exist: Placement of a dense or dense-skinned membrane in series with but *external* to the liquid membrane, between it and the product solution; *and internal* containment of the liquid membrane within a microporous support.

Certain surface-active feed additives are capable of forming membranes not by intrusion into a porous support but by concentrating themselves at the interface between a liquid solution and a dense membrane in the solid state which they cannot permeate. Because the concentrated interfacial layers remain in the liquid state, they may be considered as *liquid-surfactant membranes*. The bilayer consisting of dense and liquid surfactant membranes constitutes a composite structure whose members act in series to influence permeability and permselectivity.

Liquid-surfactant membranes form spontaneously as a result of the surfactant capacity of the dissolved molecules.[11] Surface activity is, of course, related to the hydrophilic/hydrophobic balance of these molecules; the hydrophilic portion tends to keep the surfactant in solution, while the hydrophobic portion tends to remove it. The result is a compromise in which the surfactant molecules align themselves at interfaces, the hydrophilic moieties facing towards the solutions and the hydrophobic moieties away form them. For this reason such membranes are invariably formed at every impermeable interface encountered by a surfactant solution. It has long been recognized that surface films of soluble surfactants influence material transport across phase interfaces in much the same manner as insoluble films.[12]

The first and most widely used liquid-surfactant membrane for hyperfiltration desalination was poly(vinyl methyl ether) (PVME) first utilized in this capacity by Martin.[13] A few parts per million of this additive in the saline feed solution result in a dramatic increase in permselectivity with but a small decrease in permeability. This phenomenon was attributed by Michaels et al.[14] to the physical blockage of large pores by feed additive molecules, but Markley et al.[15] have since accepted the present author's liquid-surfactant membrane rationalization.[11,16,17] A lower rate of migration of salt through the membrane results from the decrease in the amount of salt per unit of membrane water. The bound water is capable either of interacting with free water or moving from site to site in the direction of a pressure gradient. The net result is a greater transport of water than salt, thereby leading to salt rejection by the membrane.

The addition of increasing amounts of surfactant feed additives causes a progressive diminution in the permeation rates of both salt and water. This diminution continues until the critical micelle concentration (CMC) of the given feed additive in the given medium has been attained (Fig. 9.2 and Table 9.3). The reason for this

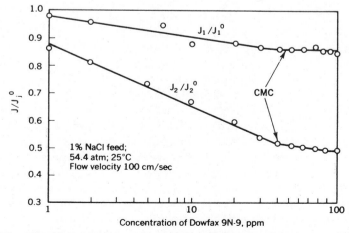

FIGURE 9.2. Effect of surfactant concentration on normalized transport rates of salt J_2/J_2^0 and water J_1/J_1^0 (from Kesting, et al.[11]).

TABLE 9.3 RELATIONSHIP BETWEEN CMC AND NORMALIZED WATER AND SALT TRANSPORT[a] RATES[b]

Surfactant	CMC in 1% aq. NaCl (ppm)	Minimum Surfactant Concentration at Which Fully Developed Liquid Membranes Exist, Deduced from		Average Values of Normalized Transport Rates at the CMC	
		J_1/J_1^0	J_2/J_2^0	J_1/J_1^0	J_2/J_2^0
9N-6	25	20	20	0.75	0.56
9N-9	43	40	40	0.85	0.52
9N-15	50	60	70	0.90	0.47
PVME	6	5	5–6	0.83	0.45

[a] 1% NaCl feed at 54.4 atm and 25°C, feed-flow velocity 100 cm/s.
[b] From Kesting et al.[11]

is the progressive coverage of the cellulose acetate membrane–saline solution interface by the building surfactant layer. This layer offers increasing resistance to the transport of both water and salt up to the CMC, at which concentration its coverage of the underlying cellulose acetate membranes is complete. The minor changes in material transport rates which occur at concentrations in excess of the CMC are presumably second-order effects primarily associated with increasing density within the already fully developed surfactant layer.

The considerable variation in the permeability and permselectivity of the various liquid-surfactant membranes is believed due to two factors: (1) the hydrophilic/hydrophobic balance; and (2) the packing arrangement within and between individual micelles. Surfactants with a high hydrophilic/hydrophobic ratio produce liquid membranes with a higher resistance to salt and a lower resistance to water than those with a low ratio (Fig. 9.3 and Table 9.3). All effective feed additives evidence a strong interaction between their hydrophilic moieties, such as the ether groups, and water molecules.[18] Note, the shift of 0.15 to 0.30 μm to longer wavelength of the C–O–C asymmetric stretching bands in the presence of water for poly(vinyl methyl ether) (Fig. 9.4) and a poly(oxyethylene)nonylphenol containing 9 moles of ethylene oxide per nonylphenol moiety (Fig. 9.5).

Surfactants for which no such interaction is indicated (Fig. 9.6) do not function as effective feed additives. Another factor insofar as anionic surfactants are concerned is the negative charge of the membrane itself, which, although of a low order, would nevertheless tend to repel similarly charged substances. However, a polyoxyethylene–polyoxypropylene block copolymer, which experienced a modest 0.1-μm shift in this band (Fig. 9.7), was completely ineffectual as an additive. For this reason the formation of hydrogen bonds between water and the surfactant may be considered as a necessary but not sufficient requisite to its effectiveness as a feed additive. It appears probable that structural regularity within the liquid-surfactant membranes is as necessary as it is in the case of the solid membranes which

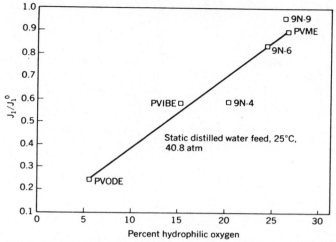

FIGURE 9.3. Effect of surfactant hydrophilicity on normalized rates of water transport (from Kesting, et al.[11]).

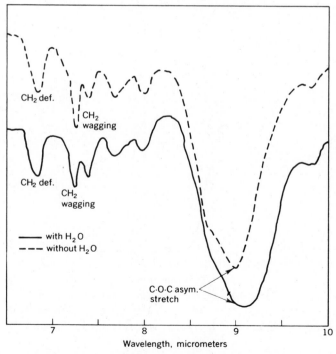

Figure 9.4. Infrared spectra of poly(vinyl methyl ether) in the presence and absence of water (from Kesting and Subcasky[18]); reprinted from *Journal Macromolecular Science*, p. 151, by courtesy Marcel Dekker, Inc.).

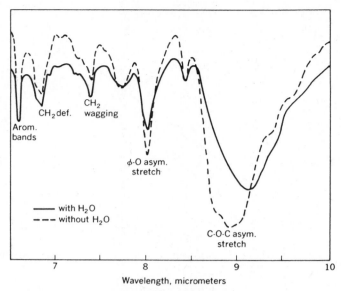

FIGURE 9.5. Infrared spectra of poly(oxyethylene)nonylphenol in the presence and absence of water (from Kesting and Subcasky[18]); reprinted from *Journal Macromolecular Science*, p. 151, by courtesy of Marcel Dekker, Inc.).

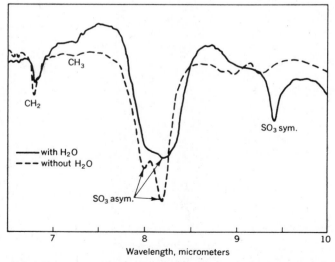

FIGURE 9.6. Infrared spectra of sodium dodecyl sulfate in the presence and absence of water (from Kesting and Subcasky[18]); reprinted from *Journal Macromolecular Science*, p. 151, by courtesy of Marcel Dekker, Inc.).

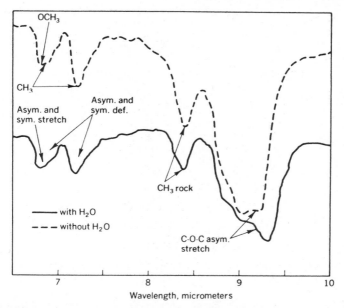

FIGURE 9.7.   Infrared spectra of a polyoxyethylene–polyoxypropylene block copolymer in the presence and absence of water (from Kesting and Subcasky[18]); reprinted from *Journal Macromolecular Science*, p. 151, by courtesy of Marcel Dekker, Inc.).

it covers. It is felt that uniformly distributed, closely spaced regions of strongly bonded water must be present if an effective barrier to salt transport is to result. If nonbonded water, as in the case of sodium dodecyl sulfate, or water bonded to blocks that are widely separated by blocks that do not bind water, as in the case of the polyoxyethylene–polyoxypropylene block copolymers, is present, continuity between bound-water domains will be broken and salt will be able to permeate through regions of ordinary water.

The formation of suitable surfactant membranes is a highly efficient means of increasing permselectivity since permeability is greater for a given permselectivity than it is when simple annealing is used to accomplish this end (Fig. 9.8). The converse statement, that permselectivity is higher for a given permeability than for simple annealing, also holds. The efectiveness of liquid surfactant membranes versus that of heat treatment can be related by an effectiveness factor e such that

$$e = \frac{\log S_2 - \log S_1}{J_{w2} - J_{w1}}$$

where S and J_w refer to the salt-reduction factor and product flux, respectively, and the subscripts 2 and 1 refer to the respective values of those parameters in the presence and absence of the additive. Additive treatments yielding values of e greater than 0.0436 are more effective in improving membrane performance than heat treatment (Fig. 9.8).

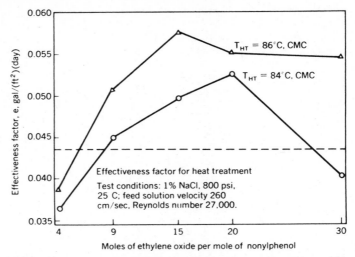

FIGURE 9.8. Effectiveness of poly(oxyethylene)nonylphenol additives for tubular membranes with 1% NaCl feed solutions (from Subcasky[19]).

Liquid-surfactant membranes can be precoated onto a dense supporting membrane if an alternative to spontaneous formation from the feed solution is desired. This can be accomplished, for example, by coating the membrane substrate with a 0.1% solution of poly(vinyl methyl ether) in carbon tetrachloride so that after evaporation of solvent an average of 42 μg of the polymer is deposited per square centimeter of membrane area. Values for the surfactant composite and untreated membranes (of comparable permselectivity) for brackish water (Table 9.4) and seawater (Table 9.5) indicate that 35–40% more product was obtained by utilizing liquid membranes than by simple annealing.

The fractional loss of the surfactant liquid membranes per permeation through its supporting membrane depends upon the size and shape of the molecules of the

TABLE 9.4 POLY (VINYL METHYL ETHER) LIQUID SURFACTANT MEMBRANE PRECOATED COMPOSITE VERSUS UNTREATED MEMBRANE FOR BRACKISH-WATER DESALINATION[a,b]

Permeability Data	Liquid-Membrane Precoated Composite		Untreated Membrane	
	At 1000 hr	At 10,000 hr	At 1000 hr	At 10,000 hr
Product flux (gal/ft^2 day)	23.16	20.90	16.16	15.02
Total product (gal/ft^2)	1010	9116	695	6464
Average flux (gal/ft^2 day)	24.24	21.88	16.69	15.15

[a] From Subcasky.[19]
[b] Test conditions: 10,610 ppm NaCl; 800 psig; 25°C; feed-solution velocity, 460 cm/s; Reynolds number, 44,500.

TABLE 9.5 POLY (VINYL METHYL ETHER) LIQUID-SURFACTANT MEMBRANE
PRECOATED COMPOSITE VERSUS UNTREATED MEMBRANE FOR
SEAWATER DESALINATION[a,b]

Permeability Data	Liquid-Membrane Precoated Composite		Untreated Membrane	
	At 1000 hr	At 10,000 hr	At 1000 hr	At 10,000 hr
Product flux (gal/ft^2 day)	9.85	8.20	6.88	6.35
Total product (gal/ft^2)	446	3712	297	2741
Average flux (gal/ft^2 day)	10.69	8.91	7.13	6.58

[a] From Subcasky.[19]

[b] Test conditions: 3.5% NaCl feed; 1500 psig; 25°C; feed-solution velocity, 460 cm/s; Reynolds number, 44,500.

former and the pore characteristics of the latter, as well as upon the environmental conditions (Table 9.6).

Despite the fact that the test conditions favor the retention of poly(vinyl methyl ether), the greater permeability of the poly(oxyethylene)nonylphenol is clearly apparent. The low permselectivity of the former is attributable to its high molecular weight and to the fact that it has multiple hydrophobic sites (methylene groups), each of which represents a potential point of contact with the underlying cellulose acetate membrane. The poly(oxyethylene)nonylphenol, on the other hand, is a linear molecule with a single hydrophobic moiety for contact with the solid membrane substrate.

As with dynamically formed membranes, surfactant liquid membranes need not be continuously present in the feed solution. After initial formation, surfactant liquid-membrane integrity can be maintained by intermittent addition (Fig. 9.9).

Liquid barriers may be internally supported or immobilized in a number of ways:

TABLE 9.6 LIQUID-MEMBRANE ADDITIVE PERMEABILITY[a]

Operating Conditions and Performance Characteristics	Additive Poly (vinyl methyl ether)	Poly (oxyethylene nonylphenol 9 moles ethylene oxide)
Additive concentration in feed solution (ppm)	20	50
Concentration of NaCl in feed solution (%)	3.5	1.0
Operating pressure (psig)	1500	800
Membrane annealing temperature (°C)	92	86
Additive concentration in product (ppm)	<0.01	8.5
Additive rejection (%)	100	~83

[a] From Subcasky.[19]

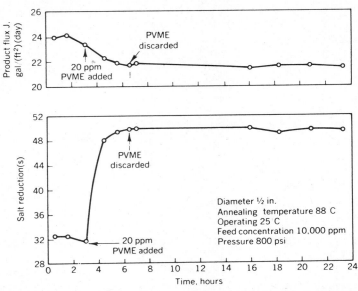

FIGURE 9.9. Effect of intermittency on membrane performance with a feed additive (from Baerg[20]).

Conversion of a liquid layer such as poly(ethylene glycol) to a rigid gel by the addition of a gelling agent such as Cabosil® or Cellosize®[21] deposition of a film of a polymer such as poly(vinyl methyl ether) which is potentially soluble in an aqueous feed solution and controlling swelling by the introduction of covalent cross-links; and confinement of the liquid membrane within a microporous support. Of these three types of *immobilized liquid membranes*, the third has thus far received the most attention.

The advantages of immobilized liquid membranes relative to solid membranes are higher diffusivities and solubilities and, particularly when *facilitated* or *coupled* transport is employed, high selectivity. Facilitated transport is a process in which the transport of a solute across a membrane is augmented as a result of a reversible reaction with a carrier which shuttles between opposite sides of the membrane. This principle is also operative in ion-selective membrane electrodes (Chapter 3). *Coupled transport* involves the movement of metal ions from one aqueous solution against a concentration gradient to a second solution by coupling the transport of the metal ion in one direction to that of a hydrogen ion in the same or opposite directions. The principal difficulties with this type of liquid membrane are loss of the membrane from the microporous support and the fact that the support must usually be thin (1–2 μm) for the various processes to be economically feasible.

The permeation of CO_2 and H_2 will serve as examples of facilitated transport of gases and the permeation of dichromate and uranyl sulfate as examples of coupled transport of metal ions, across immobilized liquid membranes.

A concentrated (6.4 M $CsHCO_3$) aqueous HCO_3^-/CO_3^{2-} solution can be utilized as a liquid membrane to facilitate the recovery of CO_2 from gases which are inert in such solutions.[22] At the side of the membrane which is adjacent to the higher partial pressure of CO_2, CO_2 reacts to form HCO_3^-. The HCO_3^- ion moves across the membrane where it decomposes to yield carbon dioxide. The partial pressure of CO_2 on the high-pressure side was 4 cm Hg, whereas on the low-pressure side it varied between 0.4 and 0.2 cm Hg. At 25°C, the average CO_2 permeability was 75×10^{-9} and the average O_2 permeability, 0.05×10^{-9}, a CO_2/O_2 separation factor of 1500. This separation factor was many times higher than any achieved through solid polymeric membranes. One of the reasons for this high separation factor is that concentrated HCO_3^-/CO_3^{2-} solution decreased the O_2 permeability by a factor of 40 compared to water. The use of catalysts such as sodium arsenite can increase the uptake of CO_2 by the liquid membrane with the result that the separation factor can be further increased by a factor of 3. The permeability of H_2S as HS^- is also facilitated by the HCO_3^-/CO_3^{2-} liquid membrane. Near the low-pressure side of the membrane, HS^- combines with a proton so that it emerges from the low-pressure side as H_2S.

The removal of toxic materials such as the dichromate ion from electroplating rinsewaters is of increasing importance for pollution control. Reuse of water and the toxic materials are important secondary considerations. Smith et al.[23] have evaluated coupled transport through internally supported liquid membranes as a possible solution to this problem (Fig. 9.10).

In the coupled transport of the dichromate ion, trioctylamine (R_3N), diluted with approximately two parts of an aromatic diluent, is used as the complexing agent. The amine, which is soluble in the organic phase, forms an organosoluble complex $(R_3NH)_2Cr_2O_7$ with $Cr_2O_7^{2-}$ and $2H^+$ which it extracts from the acidic feed solution. The complex traverses across the membrane to the basic product solution

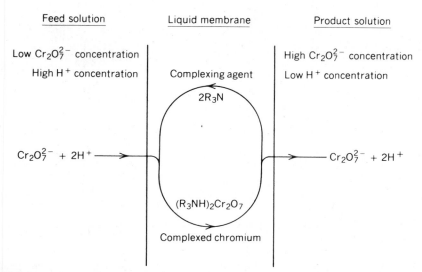

FIGURE 9.10. Coupled transport of dichromate ion across a liquid membrane containing a tertiary amine (from Smith et al.[23]).

where the complex gives up its 2H's in the form of $2H^+$, which then combine with OH^- ions. This simultaneously releases the $Cr_2O_7^{2-}$ to the product solution and frees the amine to repeat the shuttle process. The driving force for the reaction is the transport of H^+ ions across the membrane to neutralize OH^- in the product solution. The desired $Cr_2O_7^-$ transport is therefore obtained by coupling its transport to that of H^+ ions. The membrane which has been most studied for this application is Celgard 2400® (Section 8.2). Another likely polypropylene membrane— because its structure should mitigate against the loss of the liquid membrane for which it provides internal support—is Accurel® (Section 7.4).

The flux of chromium through the membrane is strongly dependent upon the concentration of amine (Fig. 9.11). Two competing factors are involved: The chromate–amine complex gradient (and hence the flux of chromium) increases with increasing amine concentration; viscosity, however, increases (and hence chromium flux decreases) with increasing amine concentration. The net result is that the maximum chromium flux is obtained at an amine concentration of approximately 30 vol %. The ability of coupled transport to move a solute against its own concentration gradient is remarkable. Beginning with a feed concentration of 5 g/L chromium and a product solution concentration of 150 g/L, transport was carried out until the feed solution contained only 16 ppm Cr^-, a ratio of about $10^5 : 1$. The flux of chromium likewise decreased only slightly with increasing chromium concentration in the product.

The transport of uranyl sulfate, $UO_2(SO_4)_2^{2-}$, across a liquid membrane by cou-

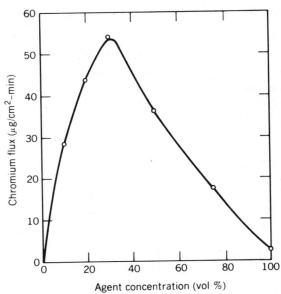

FIGURE 9.11. Chromium flux as a function of agent concentration in the membrane (from Smith et al.[23]). [2]Agent: Alamine 336 in Aromatic 150; feed solution: 2.0 g/L Cr (as CrO_3); product solution: 40 g/L NaOH.

pled transport takes place by the same mechanism.[24] The concurrent transport of H^+ ions as part of the amine–uranyl sulfate complex is the driving force. Uranyl sulfate may be separated from a solution containing vanadium, iron, and molybdenum. The first two are present as cations which, therefore, do not form complexes with the amine. Since molybdenum is present as an anion, it does permeate the membrane. It may subsequently be separated from uranium by selective oxidation of uranyl sulfate to insoluble UO_4, uranium yellow cake.

9.3 DYNAMICALLY FORMED MEMBRANES

Precoating filters for separation of colloidal particles was the earliest example of *in situ* formation of membranes upon porous supports. Because of the increased interest in hyperfiltration separations, attempts have been made to extend this concept to barriers capable of separating small molecules and ions. Structures of this type are called *dynamically formed membranes*.[25-29] Just as filter aids can be added directly to a colloidal slurry to form a precoat, so the components of dynamically formed membranes can be added to a feed solution to form a membrane at the interface between the porous support and the feed solution. The concentrations of substances required in the feed solutions to form such membranes are low, typically on the order of 50 ppm or less. They need not be added continuously; after formation the intermittent addition of as little as 1 ppm is often sufficient to maintain their integrity.

The advantages of dynamic formation are obvious: ease of initial formation, ease of maintenance, high product flux, and low cost. Its disadvantages are attributable in part to the inadequacies of the porous supports upon which they are formed and in part to the low degree of order in the membranes themselves, which is responsible for their low permselectivity (relative to separately formed membranes). Carbon black has been used as a porous support, but because a substantial fraction of its porosity is present as large pores, membrane homogeneity and permselectivity have not been good.[29] The utilization of ultrafiltration membranes as porous supports for dynamically formed hyperfiltration membranes is not an acceptable solution to the problem of suitable support substrates because of its effect upon the economics of the process. In the current state of the art, supports with pores of 1 μm or less give the most satisfactory results, although attempts are also being made to plug 5-μm pores with filter aids prior to membrane deposition. A particularly promising approach is the use of asymmetric alumina ceramic tubes containing small (~ 1 μm) pores at the membrane–feed-solution interface and large pores (~ 30 μm) in the substructure of the supports.

Most, but not all, of the polymeric substances which have been employed as dynamically formed membranes are of the ion-exchange variety; that is, they contain positively or negatively charged groups or both. Examples are humic acid (produced from decaying oak leaves), leonhardite (an oxidized form of lignite), starch sulfate, carboxymethylcellulose, homogenized Sephadex (cross-linked dextran), guar gum, sulfonated polystyrene, polyacrylic acid), and quaternized polyethyl-

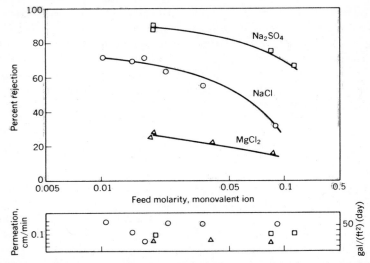

FIGURE 9.12.   Desalination characteristics of dynamically formed humic acid membranes (from Kraus et al.[29]).

eneimines. Neutral substances such as poly(vinyl pyrrolidone), hydroxyethylcellulose, poly(ethylene oxide), and block copolymers containing poly(ethylene oxide) and either poly(ethylene terephthalate) or poly(propylene oxide) have proved less encouraging.

Humic acid has been utilized by itself (Fig. 9.12) and with bentonite (Fig. 9.13). In addition to its ion-exchange properties, the bentonite probably acts as a filter

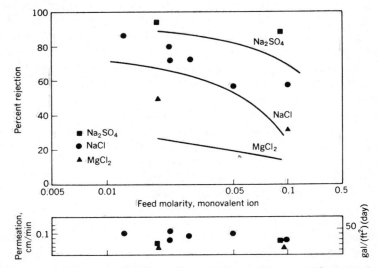

FIGURE 9.13.   Desalination characteristics of dynamically formed humic acid–bentonite membranes (from Kraus et al.[29]).

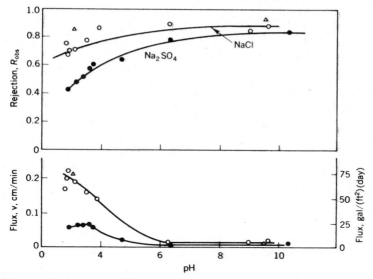

FIGURE 9.14. Desalination characteristics of dynamically formed poly(4-vinyl pyridine) membranes (from Kraus et al.[29]).

aid to circumvent the problem of abnormally large voids at the surface of the porous support.

Desalination by a poly(4-vinylpyridine) membrane dynamically formed upon a porous poly(vinyl chloride) support illustrates the pH dependence of both permeability and permselectivity of positively charged membranes (Fig. 9.14). Negatively charged membranes, such as highly anionic guar derivatives, exhibit the opposite effect—namely rejection increases with increasing pH.[28] Neutral membranes are less dependent upon the hydrophilic/hydrophobic balance. A predominantly hydrophilic (Pluronic F 108) block copolymer of poly(ethylene oxide) and poly(propylene oxide) (80:20) exhibits a flux of 100 gal/ft² day and a rejection of 20% for a 0.05 M NaCl feed solution at 800 psi.[29] A predominantly hydrophobic material (Pluronic L 101) containing only 10% of poly(ethylene oxide) exhibited the same flux but only 12% rejection under identical conditions. The fact that the flux did not vary with the hydrophilicity of these two materials is an indication that substantial physical inhomogeneities and even discontinuities exist in the membranes. Permeability decreases and permselectivity increases with increasing molecular weight of the polymeric additive, a phenomenon which is undoubtedly related to increasing pore-plugging efficiency with increasing particle size.

Membranes can also be dynamically formed and then rigidized. Eyrand et al.,[30] for example, formed membranes with 0.06-μm pores, by filtering titanium oxide into a sintered stainless-steel support having channels of 30 μm and coating the composite structure in vacuum with copper. Of current active interest are hydrated zirconium oxide (ZOSS) and zirconium oxide polyacrylic acid (ZOPA) coated onto sintered stainless steel.[31]

REFERENCES

1. N. Li, *Ind. Eng. Chem. Process Des. Dev.*, **10**, 215 (1971).
2. H. Downs and N. Li, *J. Separ. Proc. Technol.*, **2**(4), 19 (1982) (Marcel Dekker, Inc., N.Y.).
3. N. Li, T. Hucal, and R. Cahn, U.S. Patent 4,001,109 (January 4, 1977).
4. N. Li, U.S. Patent 4,125,461 (November 14, 1978).
5. E. Hsu, N. Li, and T. Hucal, U.S. Patent pending.
6. E. Matulevicius and N. Li, *Sep. Purif. Methods*, **4**, 73 (1973) (Marcel Dekker, Inc., N.Y.).
7. N. Li, R. Cahn, and A. Shrier, U.S. Patent 3,779,907 (1973).
8. T. Kitagawa, Y. Nishikawa, J. Frankenfeld, and N. Li, *Environ. Sci. Technol.*, **11**(6), 602 (1971).
9. F. Frankenfeld and N. Li, in *Recent Developments in Separation, Science*, Vol. 3, p. 285, N. Li, Ed., CRC Press, Boca Raton, FL, 1977.
10. D. Schiffer, H. Hochhauser, D. Evans, and E. Cussler, *Nature*, **250**, 484 (1974).
11. R. Kesting, W. Subcasky, and J. Paton, *J. Colloid Interface Sci.*, **28**, 156 (1968).
12. A. Schwartz, J. Perry, and J. Berch, *Surface Active Agents and Detergents*, Vol. 2, Interscience, New York, 1958, p. 418.
13. F. Martin, personal communication (March 1963).
14. A. Michaels, H. Bixler, and R. Hodges, *J. Colloid Sci.*, **20**, 1034 (1965).
15. R. Markley, R. Cross, and H. Bixler, Office of Saline Water (U.S. Dept. Interior) R. & D. Report 281, (December 1967).
16. R. Kesting, A. Vincent, and J. Eberlin, Office of Saline Water (U.S. Dept. Interior) R. & D. Report 117, (August 1964).
17. R. Kesting, "Reverse Osmosis Process Using Surfactant Feed Addition," OSW patent application SAL-830 (November 1965).
18. R. Kesting and W. Subcasky, *J. Macromol. Sci.*, **A3**(1), 151 (1969) (Marcel Dekker, Inc., N.Y.).
19. W. Subcasky, Final Report to Office of Saline Water (U.S. Dept. Interior) on Contract 14-01-0001-1317 (January 1969).
20. W. Baerg, Third Quarterly Report to Office of Saline Water (U.S. Dept. Interior on Contract 14-01-0001-1438 (March 30, 1968).
21. W. Robb and D. Reinhard, U.S. Patent 3,335,545 (1967).
22. W. Ward, in *Recent Developments in Separation Science*, N. Li, Ed., Vol. 1, p. 153, CRC Press, Boca Raton, FL, 1972.
23. K. Smith, W. Babcock, R. Baker, and M. Conrad, in *Chemistry in Water Reuse*, W. Cooper, Ed., Chap. 14, Ann Arbor Science Pub., Ann Arbor, MI, 1981.
24. W. Babcock, R. Baker, E. Lachapelle, and K. Smith, *J. Membrane Sci.*, **7**, 71 (1980); **7**, 89 (1980).
25. A. Schor, K. Kraus, J. Johnson, and W. Smith, *Ind. Eng. Chem. Fundam.*, **7**, 44 (1968).
26. K. Kraus, H. Phillips, A. Marcinkowsky, and J. Johnson, *Desalination*, **1**, 225 (1966).
27. A. Schor, K. Kraus, W. Smith, and J. Johnson, *J. Phys. chem.*, **72**, 2200 (1968).
28. J. Johnson et al., Quarterly Report to Office of Saline Water (U.S. Dept. Interior) from ORNL (June 15, 1969).
29. K. Kraus, A. Schor, and J. Johnson, *Desalination*, **2**, 243 (1967).
30. C. Eyrand et al., U.S. Patent 2,022,187 (February 20, 1962).
31. G. Spencer, Paper presented at National ACS Meeting, St. Louis, April 1984.

10 BIOLOGICAL MEMBRANES

Just as the study of synthetic and biological polymers took place independently of one another with relatively little cross-fertilization, so also have synthetic and biological membranes traditionally followed their own independent courses. However, the time for mutual exclusivity is past. Indeed the application of what may be termed *synthetic biological membranes* to functions which have hitherto been the exclusive concerns of synthetic polymeric membranes is already underway. Synthetic biological membranes can be said to *mimic* the functions of natural biological membranes in the synthetic membrane field. They form the basis for a new research area which has been entitled *membrane mimetic chemistry* by Fendler.[1] In the present chapter the salient structural and functional features of natural biological membranes will be shown to provide a rationale for the development of certain synthetic biological membranes, the practical application of which to biomedical, controlled-release, and energy-related problems lies at the present frontier of membranology.

10.1 NATURAL BIOMEMBRANES

Biological membranes perform a number of tasks which are essential to the very existence of life:

1. They provide boundaries to the cells which constitute the fundamental unit of life in both simple and complex organisms.
2. They constitute semipermeable barriers which by a variety of transport mechanisms control the composition of matter within the cell.
3. They surround subcellular organelles which contain within them additional

329

enzyme-containing membranes that catalyze reactions or serve to transmit or store energy.

All biological membranes are composed of lipids and proteins.[2] Simple membranes contain proteins whose molecules associate with one another only randomly; complex membranes contain proteins with a definite geometric relationship to one another. In common with the more advanced synthetic membranes, biomembranes exhibit an inhomogenity in depth. However, biomembranes also exhibit two additional degrees of asymmetry: inhomogeneity within a plane parallel to the surface and *sidedness,* that is, difference in composition between the exo (cell exterior surface) and endo (cell interior surface) cytoplasmic portions of the bilayer. The composition of both lipids and proteins is different in each layer. The lipids vary in composition but generally consist of amphiphilic liquids containing long (16–24 C atoms) nonpolar aliphatic tails and polar heads. Because of their molecular structure, lipids are surface active. They tend, in the presence of water, to align themselves in continuous bilayers with their heads in both surfaces contiguous to the aqueous phase and their tails normal to the plane of both surfaces and extending to the center of the membrane (Fig. 10.1). The proteins which comprise between 50 and 70% of the dry weight of membranes tend to be globular in shape

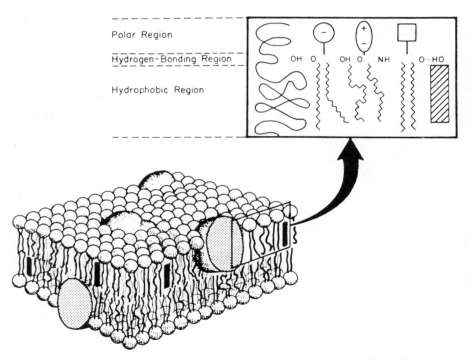

FIGURE 10.1. Fluid-mosaic model of a cell membrane (from Fendler[3]; reprinted with permission from *Account of Chemical Research,* © 1980 American Chemical Society).

with molecular weights of from 12,000 to 15,000 and diameters between 30 and 50 Å.[1-4] Because the thickness of the lipid bilayer is only 40 ± 5 Å, whereas the average membrane thickness is 70 ± 10 Å, it is apparent that a portion of each globular protein protrudes from the lipid bilayer into the surrounding aqueous medium. Whatever passive permeability occurs is through the lipid bilayer, whereas active transport is predominantly through the globular proteins. The latter also serve as structure-reinforcing elements, analogous in this respect to the microcrystallites which are sometimes encountered in synthetic polymeric membranes. Another analogy between biological and synthetic membranes is the existence of the former in crystalline and liquid crystalline phases, and of the latter in glassy and rubbery states, below and above a characteristic transition temperature, respectively. The rigidity of crystalline biological membranes is due to more permanent associations between the protein molecules both in the same layer and in the adjacent layer. The association of protein molecules with one another in the more fluid liquid crystalline membranes is less permanent. In the latter, continua may exist in either protein or lipid phases as a result of intermittent formation and breaking of interprotein bonds. The proteins in the outer layer are usually different from those in the inner layer. This leads to a form of asymmetry which is of significance to material transport phenomena.

Considerable variation is found in membrane lipids. Fatty acids, sterols, phospholipids, glycolipids, and many other types are common. Distinctions can be made between bacterial, animal, and plant varieties. However, in spite of their many differences all biomembrane lipids have a number of important features in common. Since they are all amphiphatic, polar-group orientation into, and nonpolar group orientation away from, the aqueous environment is a universal characteristic. Although some interaction between polar headgroups on neighboring lipids is possible, the shielding effects of water and soluble electrolytes tend to keep this to a minimum. Likewise, the interactions between the nonpolar tails owing to dispersion forces are so weak that considerable freedom of movement is permitted. The liquid nature of biomembrane lipids naturally results in low resistance and thereby accounts for the fact that most passive permeation occurs through lipid, rather than through protein, domains. The polymethylene chains of lipid tails can be visualized as hydrophobic cylinders whose diameters vary somewhat depending upon the average conformations of neighboring methylene groups. Cylinder diameters are minimized, and interaction between neighboring atoms maximized, if the chains assume a stretched all-trans conformation (Fig. 10.2). The cross-sectional area of closely packed n-hydrocarbon chains is ~ 18 Å^2, which corresponds to a diameter of ~ 4.1 Å. However, the area of lipid molecules increases with the presence of charged groups (20.6 Å^2 for stearic acid), and double bonds (32 Å^2 for oleic acid). Area also increases with increasing temperature and characteristic transition temperatures are observed as cooperative melting occurs as a result of the formation of gauche configurations in the polymethylene chains. Because of their amphiphatic nature, all biolipids exhibit surface activity which allows them in the presence of an excess of water ($>$ 15 water molecules/lipid molecule) to spontaneously form bilayers which may take various forms.

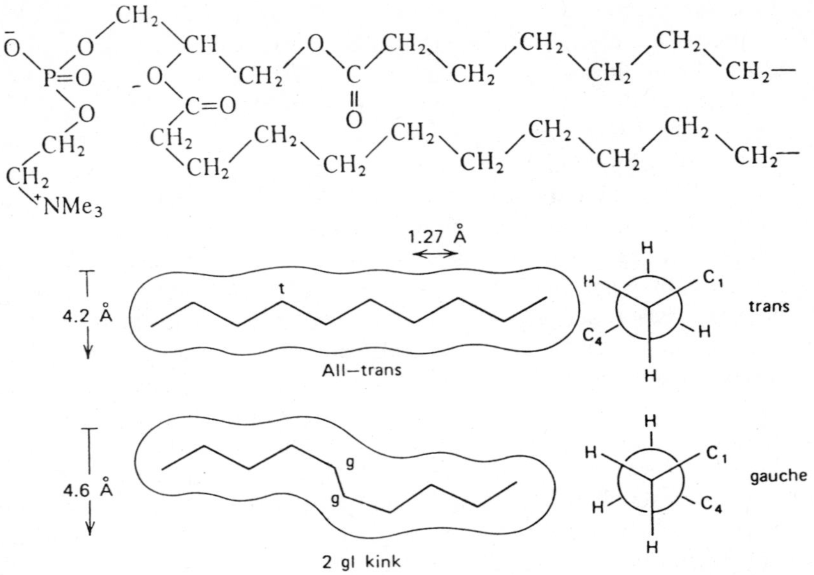

FIGURE 10.2. Structure of a phospholipid molecule emphasizing the probable relative orientation and conformation of the polar groups and the acyl chains. (from Jain and Wagner[2]; © 1980).

Proteins constitute up to 70% by weight of biomembranes. Although proteins are not covalently bonded to the lipids which constitute the continuous phase of the bilayer, the two do tend to influence each other's structure and function. Proteins may be separated from biomembranes by three separate techniques:

1. Replacement of the phospholipids with other surfactants.
2. Solvent extraction of the lipids.
3. Disruption of water structure by salts and other solutes (Chapter 4).

Although the polypeptide backbone of proteins is relatively polar (see the closely related polyamides in Chapter 4), nevertheless, the polarity of proteins varies considerably both within a given protein and from one protein molecule to another. This is due both to differences in chain folding (conformation) and to differences in the hydrophobicity of amino acid groups (Table 10.1). As a result, membrane proteins tend to be less polar than water-soluble globular proteins.

There is no doubt that both proteins and lipids are asymmetrically distributed and/or oriented in biomembranes and that this fact strongly influences material transport. Both proteins and lipids retain their *sidedness,* that is, they do not flip-flop across the bilayer. However, proteins are able to engage in lateral movement within their assigned layers. This facile lateral diffusion is probably related to the hydrophobic nature of membrane proteins (relative to water-soluble proteins) which in turn makes for relatively weak interactions. Lateral diffusion is also related to defect structures which become particularly apparent in the vicinity of the phase

TABLE 10.1 HYDROPHOBICITY OF AMINO ACID SIDE CHAINS[a]

Side Chain	$\delta \, \Delta F^b$ (cal/mole)
Tryptophan	−3400
Phenylalanine	−2500
Tryosine	−2300
Leucine	−1800
Valine	−1500
Methionine	−1300
Alanine	−500
(−C), NH−)	+4100
(−C=O · · ·H−N)	−1400

[a]From Tanford.[5]; © 1972.
[b]These numbers represent the additional free energy of transfer from water to ethanol or dioxane at 25°C.

transition temperature. Protein asymmetry appears to arise during biosynthesis. Those that are found on the exterior surface of the cells, *ectoproteins,* generally contain carbohydrates; those that are found on the interior (cytoplasmic) surface of the cell membranes, *endoproteins,* do not. Carbohydrates appear to stabilize or "lock-in" exoproteins and may also play a role in cell surface recognition. There is more protein at the interior surface of the bilayer than at the exterior surface.

Proteins can gather into reversible aggregates or patches depending on such conditions as changes in pH or temperature, the addition of anesthetics, and glycerolization. Certain proteins appear to be confined to specific regions of the cell.

Many enzymes (catalytic proteins) require the presence of lipids to function.[6] This is shown by deactivation and activation with phospholipid removal and replacement. Protein–lipid interactions cover the range from covalent bonding to ionic and hydrophobic interactions. It is likely that solubility considerations influence the stability of certain conformations which in turn enhance enzymatic activity.

10.2 MATERIAL TRANSPORT THROUGH BIOLOGICAL MEMBRANES

Material transport through biological membranes can occur by a variety of mechanisms: passive permeability, facilitated transport, active transport, and bulk transport.

Passive permeability through biomembranes occurs as a result of concentration differences on either side of the membrane and is governed by a solution–diffusion mechanism (Chapter 2). In diffusional studies involving biomembranes, permeability coefficients are determined by a variety of techniques such as the exchange of a radiolabeled solute in the absence of a concentration gradient and the coupling of a reaction of the permeant with another reaction. The transport of ions is followed by conductivity or electrode potential measurements.

The activation energy E_a for the self-diffusion of water is 4–5 kcal/mole, whereas the E_a for the diffusion of water through the human adult RBC (red blood cell) is 6 and through the liposomes of egg PC (phosphatidyl choline) E_a is 8–9.[7] Thus the rate-limiting step for water permeation through biomembranes is not simply self-diffusion. Passive permeability of small nonelectrolyte solutes occurs in successive steps: transfer across the interface, dehydration of the solute, and diffusion through hydrocarbon chains. The acyl chains form kinked conformations and develop free volume which can accommodate solutes. Polarity decreases and fluidity increases from the membrane surface towards the center of the membrane. The permeability coefficients for ions and hydrophilic solutes through biomembranes are much lower than for water or small nonelectrolytes. This is due to the large ΔF (~ 40 kcal/mole) required to bring an ion from an aqueous solution of dielectric constant 80 into a lipid phase with a dielectric constant of ~ 2.[8] The same situation is encountered in ion exclusion by hyperfiltration membranes (Chapter 2). Indeed the values of ion permeability through neat gel or liquid crystalline phases are so low that it is apparent that permeation through biomembranes occurs primarily by other than a passive mechanism.

Facilitated transport refers to permeability which occurs at specific sites in the biomembrane. Facilitated transport can occur in a variety of systems (Fig. 10.3). Because there is a finite number of discrete sites, they can become saturated at high concentrations and hence solute transport will not exceed a certain maximum rate. A distinction can be made between *simple facilitated transport* in which permeability occurs *with* the concentration gradient and *active facilitated transport* in which permeability occurs *against* the concentration gradient. Facilitated transport systems lower the energy barriers for solute transport and can occur by carrier or gated pore channel mechanisms. A carrier is a substance which binds to a permeant solute at one surface and then transports it to the opposite surface. Certain ionophores, although not themselves the carriers normally encountered in biomembranes, nevertheless well exemplify certain aspects of carrier-facilitated transport. For example, both valinomycin and the crown ethers (Chapter 3) can encom-

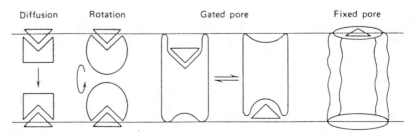

Diffusion   Rotation   Gated pore   Fixed pore

FIGURE 10.3. Site-mediated facilitated transport systems (from Jain and Wagner[2]; © 1980).

pass specific ions within their structures. By shielding the ion from the surrounding environment, an ionophoric carrier effectively increases ion solubility and thereby facilitates ion transport across lipid bilayers. In other words, ionophores sequesterions into lipid soluble complexes. To enter into an ionophore, a cation must exchange its waters of hydration for equivalent coulombic neutralizing groups such as ester or amido acyl oxygen groups. For flexible ionophores, ion selectivity is minimal. However, for inflexible ionophores, such as valinomycin, the precise spacing of the ester carbonyl groups which line the lumen of the cylinder is critical, and ion selectivity can be very high indeed. As examples, the selectivity of valinomycin for K^+ ($r = 1.33$ Å) over Na^+ ($r = 0.95$ Å) is $>10^4$, whereas the selectivity of the flexible ionophore nonactin is only 15. In free valinomycin the carbonyls are believed to be oriented away from the center of the cylinder, thereby granting the ions free access to the cavity. Once the ions are inside, the carbonyl groups orient themselves toward the center thereby stabilizing the ionophore:ion complex. In media of low dielectric constant the conformation with the ester groups pointing inward is stable.[9] The reverse is true in polar media. The rate at which the complex transverses the membrane can vary between 100 and 3000/s. The nature of the lipids in the biomembrane also affects carrier-mediated ion transport. The surface charges of the polar lipid groups can act to shield ions in the vicinity and hence to inhibit uptake of the ions by the carriers. These charges are influenced by such environmental factors as pH, ionic strength, temperature, and so on. Lipid structure also strongly influences solubility and viscosity. Recall the inhibiting effect of high viscosity on ion mobility in internally supported liquid membranes (Chapter 9).

Most transport between adjacent cells occurs through permanent pores or channels situated at gap junctions. The channels take the form of short intercellular pipes or *connexons*. The latter in turn are composed of two properly aligned hollow cylindrical proteins known as *intramembranous particles* or IMPs. A connexon forms a continuous channel which extends from the interior of one cell, through two bilipid membranes, into the interior of another cell. The cross section of connexons are hexagonal and appear to be made up of six protein subunits which can tilt and slide relative to one another so as to exert a *gating effect* similar to the operation of an iris diaphragm behind a camera lens.[10,11] Single IMPs penetrate cell membranes and constitute the ordinary gated channel for facilitated transport between cell cytoplasm and cell exterior.

Active transport is the general term used to describe a variety of mechanisms by which solute is transported across a biomembrane *against* its concentration gradient. One common mechanism is *co-* or *coupled transport* in which the uphill transport of one solute is coupled to a downhill movement of another solute (Chapter 9). Important examples of transport are the sodium ion-coupled system for sugars and amino acids,[12] and the proton-coupled permease system for lactose in bacteria. The sodium–potassium pump introduces two K^+ ions into, and removes three Na^+ ions from, a cell as the result of the hydrolysis of adenosine triphosphate ATPase which transports two calcium ions for each molecule of ATP hydrolyzed. Proton pumps by which protons are transferred as a result of various redox reactions appear to be the primary mode of energy transfer across bacterial, mitochondrial, and chloroplast membranes.

The final mode of material transport across biomembranes is *bulk transport.* Bulk transport of extracellular material into a cell occurs by *endocytosis.* Invagination of the material into the plasma membrane is followed by fusion of the latter and encapsulation of material within the confines of the cytoplasm. If the encapsulated material is fluid, the process is known as *pinocytosis* and if solid, it is known as *phagocytosis.* Exocytosis is the reverse process whereby material which is on the inside of the plasma membrane is removed to the outside.

10.3 SYNTHETIC BIOMEMBRANES

In order of increasing concentration, synthetic surfactants can associate in a number of different aggregative forms: monolayers, spherical micelles, rodlike or cylindrical micelles, and liquid crystals (Fig. 10.4). The surface-active phospholipids of biomembranes, however, tend instead to form lipid bilayers in which approximately 11 molecules of water are bound to each phospholipid molecule. The bilayer arrangement is thermodynamically favorable because the cross-sectional area of

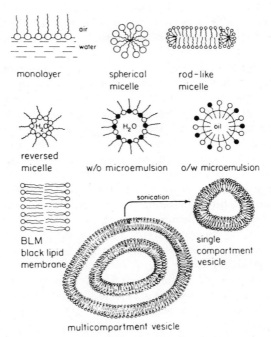

FIGURE 10.4. An oversimplified representation of organized structures of surfactants in different media (from Fendler[3]; reprinted with permission from *Account of Chemical Research,* © 1980 American Chemical Society).

the polar group is about the same as that of two acyl chains. Configurations other than bilayers require excessive crowding of the hydrophobic tails. Lipid bilayers can be prepared in various configurations: *bimolecular lipid* (also known as *black lipid*) *membranes* (BLMs) and liposomes. These will now be considered in turn.

Black lipid membranes are of importance, not because they exist in nature as such, but because they represent a reproducible prototype of a standard lipid bilayer. They thus represent a convenient model for the study of physical and passive transport properties of the idealized lipid bilayer portion of biomembranes. Black lipid membranes are prepared by coating a hydrophobic septum (containing a pinhole and separating two aqueous solutions) with a solution of lipid in an organic solvent (Fig. 10.5). The initial coating is thick and appears gray. Thinning to ~ 40 Å occurs within a few minutes resulting in interference colors which finally turn black. It is now believed that BLMs are dynamic rather than static in nature and hence tend to fluctuate between swollen and shrunken structures (Fig. 10.6). Although both BLMs and the liposomes possess bimolecular lamellar structures, the former have the advantage that the compartments on both sides of the septum are sufficiently large to permit the introduction of electrodes and the removal of samples for analysis.

The expected value for the thickness of a BLM composed of two 18-carbon chains (46 Å) and two polar groups (14 Å) is 60 Å. Electron micrographs reveal a dual *railroad track* structure which closely resembles that found in biological membranes. The interfacial tensions of both BLMs and biological membranes vary between 0.2 and 3 dyn/cm, which is comparable to the interfacial tension between the bulk lipid and water. This means that the density of lipids in the bilayer is roughly the same as their bulk density. The electrical capacities of both BLMs and biomembranes lie between 0.4 and 0.8 μ $\mathcal{F}/cm^2$. Capacity measurements permit the determination of the thickness of a bilayer dielectric with which has come the realization that this thickness is less than the total membrane thickness. This has been rationalized by postulating the penetration of water molecules into the bilayer to a depth of three or four methylene groups from the polar end group. However, a tilt of $20°–30°$ of the alkyl chain from the perpendicular would also account for this difference.

Many electrical measurements have been made with BLM membranes. Capacitances, conductivities, dielectric breakdown voltages, and membrane potentials have been determined. Conductivities are utilized primarily to determine the effect of additives. Both ionic and nonionic additives can alter the transport behavior of BLMs. Water permeabilities vary between 5×10^{-4} and 100×10^{-4} cm/s and are determined by observing the volume flux of water once a concentration gradient has been established across the membrane.[2] The composition difference between BLMs account for differences in permeability. The permeabilities of polar nonelectrolytes across BLMs are much lower than that of water itself. Passive ion permeabilities through BLMs are much lower than through actual biomembranes. This has been ascribed to the higher dielectric constants of the latter and is related to the phenomenon of ion exclusion from hyperfiltration membranes (Chapters 2, 4).

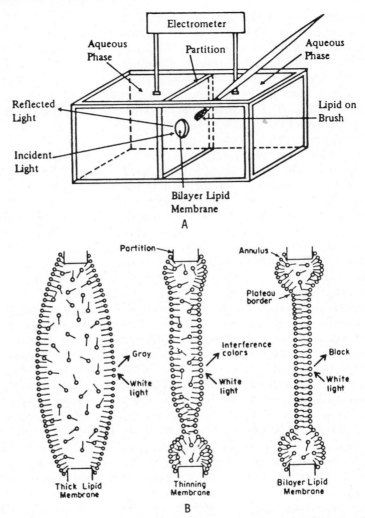

FIGURE 10.5. Preparation of BLMS. (A) Surfactant (lipid) solution is painted across a pinhole separating two aqueous compartments. Electrodes in the aqueous phases permit the measurements of bilayer conductance and other electrical properties. (B) Time-dependent thinnning of a BLM in aqueous medium indicating the patterns of reflected light. (from Fendler[1]; © 1982).

Although BLMs represent perhaps the most elementary form of lipid bilayer model, they are by no means the only one (Fig. 10.4). *Vesicle* is the general term used to describe spherical or ellipsoidal single- or multicompartment closed bilayer structures of any chemical composition. *Liposomes* are vesicles which are composed of naturally occurring or synthetic phospholipids. The term liposome originally referred to liquid crystalline particles which resulted from the dispersion of phospholipids in water, but is now recognized that, depending largely upon the

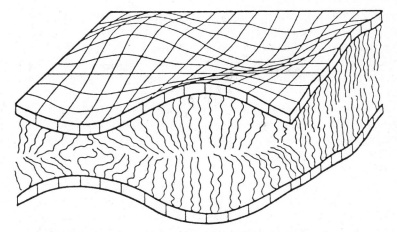

FIGURE 10.6.   Breathing model of a BLM (from Bach and Miller[13]).

methods of dispersal, a variety of structures may be formed. The distinguishable types include multilamellar vesicles (MLV), unilamellar vesicles (SUV), and single-walled large vesicles (LUV). The last mentioned are also known as cellules or spherules. The various liposomes differ from one another in compartment shape and size and phospholipid packing density and in the number of lipid bilayers. Multilamellar vesicles result when phospholipids are mixed above the liquid crystalline transition temperature with greater than 30 wt % water. They contain 8–14 lamellae and are between 5 and 50 μm in size. Although MLV preparations tend to be heterogeneous, they can be made more homogeneous by passage below the phase transition temperature through irradiation track membranes (Chapter 8). The preparation of MLVs by a phase-inversion process has also been reported.[14] In this procedure an aqueous buffer is introduced into an ether solution of the phospholipid and the ether is allowed to evaporate. The MLVs produced by this technique have larger compartments than those which are formed by the aqueous swelling of anhydrous phospholipids.[14] Sonication of MLVs causes their breakdown into small unilamellar vesicles which are 180–400 Å spherical sacs bounded by a single lipid bilayer. Small unilamellar vesicles are unstable at low temperatures and slowly coagulate and transform into MLVs. Dispersion of phospholipids in a large excess of aqueous medium yields (0.5–10 μm in diameter) LUVs. The latter are much less stable than biological cells of comparable size and tend to revert to MLVs. This instability is believed to be due to the absence of proteins. Large single-walled vesicles have been utilized for entrapping macromolecules such as enzymes,[15,16] and to reconstitute proteins in their functional states. Rhodopsin, cytochrome c oxidase, and acetylcholine receptors have been incorporated into LUVs which have displayed biological activity.[17]

Vesicles of polymerizable surfactants have been prepared and utilized as biomembrane analogs.[18] Such surfactants contain unsaturated moieties such as alkenes or styrenes in either head or tail groups. After formation into vesicles by either phase inversion or sonication, they can be stabilized by cross-linking their double bonds under the influence of UV light or free-radical initiators such as azobisiso-butyronitrile or potassium persulfate. Polymerized vesicles appear to resist morphological changes for periods of several months. They apparently are perfectly sealed and retain encapsulated ^{14}C-labeled glucose more efficiently than do unstabilized vesicles.

There are many potential applications of synthetic biomembranes. In some the properties of the protein and/or lipid fractions strongly support the use of biomembrane analogs. In others it is possible to choose between synthetic polymeric or synthetic biomembranes. This section includes a number of examples in each category.

Direct conversion of solar energy into electricity can be accomplished by solid-state photovoltaics based on doped silicon crystals. However, the cost of these modules is too high to be commercially feasible. Several alternatives involving synthetic biomembranes have been proposed. One of these utilizes bacteriorhodopsin from the "purple membrane" of the halophilic bacterium *Halobacterium halobrium*.[19] When bacteriorhodospin absorbs visible light it undergoes a cyclic reaction by which protons are taken up at the cytoplasmic side of the bacterial membrane and released at the exterior surface. It therefore functions as a light-driven proton pump to generate transmembrane electrical potential and pH differences. If bacteriorhodopsin molecules are enclosed in polymerizable vesicles and properly oriented by electric field or surface forces they constitute a synthetic biomembrane which functions as a wet photovoltaic cell. Bacteriorhodopsin has been oriented between transparent polyacrylic acid and polyacrylamide films. The resultant cell produced photocurrents measured at short circuit as high as 20 μA/cm^2, with open-circuit voltages as high as 50 mV.[19]

The surfactant vesicle type of synthetic biomembrane has a number of advantages over other membrane types.[18] Because they are highly charged, appreciable surface areas result. Appropriate distances are maintained between proton donor and acceptor sites. Precise spacing is essential for efficient energy transfer without self-quenching. Positively charged lysopyrene (proton donor) DODAC SUVs readily accept a high concentration of negatively charged pyranine (proton acceptor) molecules (Fig. 10.7). Energy transfer efficiencies of up to 43% were achieved largely because of the high pyranine concentration on the vesicle surface.

Because of the excellent biocompatibility of properly designed synthetic biomembranes, liposomes and surfactant vesicles have been widely investigated as drug encapsulants.[21] Both synthetic biomembranes and synthetic polymeric membranes are being exploited in *artificial cells*.[22] The latter are encapsulated systems which can be incorporated into the body to effectively mimic its natural functions. Among the materials which have been included within artificial cells are enzyme systems, cell extracts, biological cells, adsorbents, and so forth. Red blood cell (RBC) substitutes have received a considerable amount of attention. Silicone-encapsulated microspheres were found to be rapidly removed from circulation.[23] However, liposomes incorporating fluorocarbon fluids are currently undergoing clinical trials. Encapsulation of enzymes, ion-exchange resins, and activated charcoal in cellulose

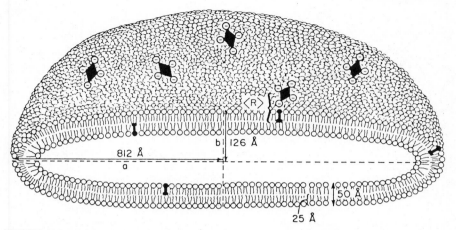

FIGURE 10.7. Schematic representation of a well-sonicated cationic DODAC surfactant vesicle. Proposed positions of lypopyrene ($\bullet$—$\bullet$) and pyranine ($\ominus \blacklozenge \ominus$) and the average distance between ($\langle R \rangle$) are also indicated (from Nomura et al.[20]; reprinted with permission from *Journal of the American Chemical Society*, © 1980 American Chemical Society).

nitrate followed by coating with albumin resulted in artificial cells which effectively removed organic uremic waste metabolites, but not water, electrolytes, or urea. The utilization of these cells in connection with standard hemodialysis reduces the time required for hemodialysis and produces fewer side effects.

Encapsulation of ion-exchange resins by albumin and of activated charcoal with cellulose nitrate and albumin results in artificial cells which are of use in detoxification.[24] Tyrosinase and other enzymes have been encapsulated in artificial cells to function as an artificial liver.[25] Finally, rat-islet cells have been encapsulated and implanted intraperitoneally. These cells produce sufficient insulin to maintain normal glucose levels in the diabetic animals while remaining impervious to rejection by the body.

REFERENCES

1. J. Fendler, *Membrane Mimetic Chemistry,* Wiley, New York, 1982.
2. M. Jain and R. Wagner, *Introduction to Biological Membranes,* Wiley, New York, 1980.
3. J. Fendler, *Acc. Chem. Res.,* **13,** 7 (1980).
4. D. Green and G. Vanderkooi, "Structure and Function of Biological Membranes," in *Colloidal and Morphological Behavior of Block and Graft Copolymers,* G. Molau, Ed., Plenum, New York, 1971.
5. C. Tanford, *The Hydrophobic Effect,* Wiley-Interscience, New York, 1972.
6. B. Fourcans and M. Jain, *Adv. Lipid Res.,* **12,** 147 (1974).
7. See Reference 2, p. 170.

8. A. Parsegian, *Nature,* **221,** 844 (1969).

9. D. Haynes, T. Wiens, and B. Pressman, *J. Membrane Biol.,* **18,** 23–38 (1974).

10. N. Univin and R. Henderson. *Sci. Am.,* **250**(2), 78 (1984).

11. R. Schultz and S. Asunmaa, *Recent Progr. Surface Sci.,* **3,** 291 (1970).

12. S. Schultz and P. Curran, *Physiol. Rev.,* **80,** 637 (1970).

13. D. Bach and I. Miller, *Biophys. J.,* **29,** 183 (1980).

14. F. Szoka, Jr. and D. Papahadjopoulos, *Proc. Natl. Acad. Sci. USA,* **75,** 4194 (1978).

15. J. Reeves and R. Dowben, *J. Cell Physiol.,* **73,** 49 (1969).

16. J. Reeves and R. Dowben, *J. Membrane Biol.,* **3,** 123 (1970).

17. A. Darszon, C. Vandenberg, M. Schoenfeld, M. Ellisman, N. Spitzer and M. Montal, *Proc. Natl. Acad. Sci. USA,* **77,** 239 (1980).

18. J. Fendler, "Polymerized Surfactant Aggregates as Biomembrane Analogs—Characterization and Utilization," Paper presented at Membrane Conference, Bend, Oregon, November 1983.

19. S. Caplan, "Solar Energy Conversion Using Membrane-Bound Bacteriorhodopsin or Photosynthetic Complexes," Paper presented at Membrane Conference, Bend, Oregon, November 1983.

20. T. Nomura, J. Escabi-Perez, J. Sunamoto, and J. Fendler, *J. Am. Chem. Soc.,* **102,** 1484 (1980).

21. See Reference 1, p. 506.

22. T. Chang, "Membrane Technology in Artificial Cells," Paper presented at Membrane Conference, Bend, Oregon,, November 1983.

23. T. Chang, *Trans. Am. Soc. Artif. Internal Organs,* **12,** 13 (1966): *26,* 354 (1980).

24. T. Chang, E. Chirito, P. Barre, C. Cole, and M. Hewish, *Trans. Am. Soc. Artif. Internal Organs,* **21,** 502 (1975).

25. T. Chang, *Lancet* II, 1371 (1972).

INDEX